To Renew Books
PHONE (925) 969-3100

Dual Attraction

DUAL ATTRACTION

Understanding Bisexuality

Martin S. Weinberg
Colin J. Williams
Douglas W. Pryor

New York   Oxford
OXFORD UNIVERSITY PRESS
1994

Oxford University Presss

Oxford   New York   Toronto
Delhi   Bombay   Calcutta   Madras   Karachi
Kuala Lumpur   Singapore   Hong Kong   Tokyo
Nairobi   Dar es Salaam   Cape Town
Melbourne   Auckland   Madrid

and associated companies in
Berlin   Ibadan

Published by Oxford University Press, Inc.,
200 Madison Avenue, New York, New York 10016

Library of Congress Cataloging-in-Publication Data
Weinberg, Martin S.
Dual attraction : understanding bisexuality /
Martin S. Weinberg, Colin J. Williams, Douglas W. Pryor.
p.   cm.   Includes index.
ISBN 0-19-508482-9
1. Bisexuality—United States—Longitudinal studies.
2. Sexual behavior surveys—United States.
I. Williams, Colin J.   II. Pryor, Douglas W.
III. Title.
HQ74.W44   1994   306.76'5'0973—dc20   93-5620

2 4 6 8 9 7 5 3 1

Printed in the United States of America
on acid-free paper

Dedicated with Much Love
to

Barbara, Ellana, and Marion
Huw and Sally
Amy and Hannah
and to the memory of
David Lourea

Preface

Bisexuality is a topic long ignored by sex research. Despite the early investigations of Alfred Kinsey, who discovered a continuum of sexualities, people who are not exclusively heterosexual in their sexual feelings and behaviors have usually been considered "homosexual." As a result, a great deal of attention has been paid to explaining why some people are homosexual or to the differences between heterosexuals and homosexuals. This trend has reinforced the belief that sexual preference exists as a dichotomy, that there are two natural classes of people whose sexual desire is fixed. Bisexuals, who do not fit in either category, have generally been seen as confused, dishonest, or in transition to becoming homosexual. We believe that sexual preference is much more complicated than this and will never be fully understood until the phenomenon of bisexuality is rigorously examined in its own right.

Our study illustrates how the social world shapes our sexuality. We show that sexual preference is a more complex, fluid, and emergent outcome than today's biological determinism dictates. We introduce the reader to a group of people who define themselves as "bisexual" and who organize their sexuality around that label. We write about them dealing with their dual attraction, struggling to put together satisfying relationships, and confronting a rapidly changing sexual environment. This environment became dominated by AIDS during the course of this study. How they adapted their sexuality to the emerging health crisis provides the final theme to the book.

We are grateful to the many people who helped us with this project. First of all, Maggie Rubenstein for her persistence in getting us to undertake the study, the late David Lourea who as President of the Bisexual Center helped us in every way he could, and The Edward L. Anderson Jr. Foundation, Inc. who funded the research. In addition, we would like to express our appreciation to twelve members of the Bisexual Center for their assistance as volunteers in conducting interviews, and for endless other tasks they performed in the data collection phase of the research. We cannot name them for reasons of confidentiality, but we will always

remember them and hope that some day soon the need for such concealment will be unnecessary.

We would also like to thank a number of students at Indiana University who provided invaluable assistance to the research: Cassandra Cal-
han and Elizabeth Bennett who assisted us in the field; and Tammy
Taylor, Lynn Gibson, and Jon Young who helped coordinate the data
analysis. Special thanks to Jon for the tremendous amount of work he did
on the project. The following students also worked on the project at
various times: Neal Carnes, Kim Gamson, Kari Hallett, Amy Kopel, Lilly
Komanov, Liisa McCallister, Lee Mitchell, Heather Morgan, Christine
Ostapiuk, Kathleen Petroff, Diane Phillips, Laurie Ragland, Denise Riggs,
Michelle Rubinstein, Phelissa Scott, Lisa Turner, and Lisa Wilder.

Throughout the data analysis and write-up we often relied on our
colleague Brian Powell for advice; his counsel was always helpful. Our
thanks also to Susan Duke and Chris Martindale who gave us so much of
their time in typing numerous drafts as well as the final manuscript.

Finally we would like to acknowledge Indiana University for providing a safe haven within which to do sex research. We feel heirs to the
similar support they gave Alfred Kinsey over fifty years ago.

Contents

Part III   After AIDS

Dual Attraction

1

The Riddle of Bisexuality

In 1991 the *New York Times* and other media gave extensive coverage to a research report that suggested an answer to the question: How do people become homosexual?[1] The researcher, the neuroscientist Simon LeVay, claimed to have found anatomical differences between the brains of homosexual and heterosexual men and concluded that sexual preference is predetermined. Later that year, similar journalistic attention was directed at two researchers who claimed, through the study of twins, that homosexuality was genetically based.[2] In an article titled "Are Some People Born Gay?" it was flatly stated, "Science is rapidly converging on the conclusion that sexual orientation is innate."[3] Then, in 1993, it was suggested that a "homosexuality" gene had been discovered on the X chromosome.[4] These reports received quick and spirited rejoinders, not just about their methodological limitations, but also their social and political implications. Thus came questions such as: If homosexuality is inborn, does this mean that antigay prejudice will decline as homosexuals are not responsible for their same-sex attractions or behaviors? Or, if homosexuals can be readily identified by brain scans, will this mean that there will be efforts to discover the prehomosexual child with the goal of corrective intervention?

An intriguing question is why such studies attract the attention they do. Clearly, *any* research on sexuality will interest the American public, living as we do in a society at once fascinated by and anxious about the erotic. But homosexuality especially fascinates the public. Not only does it stand apart from "normal" sexual relations, but many people believe they can actually recognize "homosexuals," unlike other sexual nonconformists (e.g., mate swappers). Indeed, whether Americans want to see homosexuals or not, they are confronted with militant gays and lesbians demanding social acceptance and civil rights: probably few New Yorkers remained unaware of the demand of Irish gays and lesbians to march in the St. Patrick's Day Parade against the wishes of parade organizers in 1992 and 1993. And of course the AIDS crisis has made all Americans more aware of homosexuality.

Many scientists share the public's interest in homosexuality, and a good number of them also see it as an anomaly to be corrected. They have invested great efforts in trying to explain it. But in so doing, they have usually assumed that there is a special or unique category of people who in fact *are* "homosexual" and that this group can readily be compared with another equally pure group of people who *are* "heterosexual." Researchers then look for differences between the two groups, and when they find them, they claim that these differences explain a homosexual "sexual orientation." Sexual attraction is reduced to an either/or question. People are either heterosexual or homosexual according to the criteria dictated by the researcher.

But life is not so simple. People do not always have an *exclusive* preference for their own *or* the opposite sex. Sometimes they are attracted to their own and the opposite sex during the same period in time. Or they may live many years attracted to the opposite sex, but then change their interests to the same sex. Or they may dart around between opposite and same-sex partners with no discernible pattern. These "bisexual" patterns of sexuality are not uncommon. Alfred Kinsey's pioneering studies of sexual behavior found that nearly *half* of all men and about a *quarter* of all women are *not* exclusively heterosexual or homosexual in their sexual feelings or behaviors.[5] More important, Kinsey reported that the direction of a person's sexual attraction or behavior does not necessarily remain stable; a considerable number of people change the heterosexual and homosexual mix in their sexuality during their lives. Although Kinsey's conclusions have not gone unchallenged, surprisingly, they have been generally ignored by most studies of sexual preference.

The existence of bisexuality, and its many patterns, thus remains a major riddle. If, as LeVay argues, there are differences in the size of the third interstitial nucleus of the anterior hypothalamus in "heterosexuals" and "homosexuals" (with it being smaller for "homosexuals"), what would one expect to find in the case of people attracted to both sexes in varying degrees? If homosexuality and heterosexuality are genetically predetermined, does that mean we need to look for a bisexual gene as well? Is there a biological composition in people that determines each of the different patterns of bisexuality and that allows for radical changes in sexual preference over a lifetime? We do not mean to deny a neurological, hormonal, or genetic component to sexuality, but it is difficult to explain people in between the heterosexual or homosexual dichotomy in this way.

When we began our investigation ten years ago, almost nothing was known about bisexuality. Even today, we are struck by the virtual absence of research on people attracted erotically to both sexes, especially in light of research that suggests that bisexuality is fairly widespread. In this book, we look at people who have such a dual sexual preference and who are trying to adapt themselves to a society that supports the belief that people must be attracted to one sex *or* the other. We use the term "sexual preference" to emphasize that people take an active part in constructing their sexuality, as opposed to "sexual orientation," which suggests that sexual preference is established at birth and fixed thereafter. By this we do not mean that people have complete freedom of choice or that they suddenly decide to be one way or another. But sexual attraction is far more complex than biology allows, and it can be more fully understood by focusing on what people themselves think and do—how they recognize and act on their desires. We are especially interested in how society shapes the choices people make about sex and the penalties it exacts when those choices run contrary to its dictates. Our approach follows the lead of Alfred Kinsey, who said:

> The problem [of sexual preference] is, after all, part of the broader problem of choices in general: the choice of the road one takes, of the clothes that one wears, of the food that one eats, of the place in which one sleeps, and of the endless other things one is constantly choosing. A choice of a partner in a sexual relation becomes more significant *only because society demands that there be a particular choice in this matter,* and does not so often dictate one's choice of food or of clothing.[6]

Simply put, the approach that we take, in contrast to the biological one, emphasizes the standpoint of the people we are examining and tries to capture how they construct their sexual lives. In this book we will focus on how a community of people experience and organize their sexuality with both male and female partners. What does it mean to them when they call themselves "bisexual," and how do they pursue their sexuality? What reactions do they encounter, and to what extent are obstacles such as AIDS a factor in their behavior? To us, these questions are every bit as interesting as the search for elusive biological causes.

This book is based on a comprehensive and long-term study of people who identified themselves as "bisexual." The setting was San Francisco during the 1980s. This was an ideal place for our investigation. We had been doing sex research in San Francisco for over fifteen years and were familiar with the city and its sexual underground. Visiting the city almost every year, we had built up associations with people interested in research on human sexuality professionally or personally. Through these contacts we learned of an organization called the Bisexual Center, which we were told was the largest and most active bisexual organization in the country. The leaders of the center were eager to support a study of bisexuality because they knew little attention had been paid to their sexual preference, and they felt that scientific knowledge might help those struggling with dual attractions.

San Francisco is a natural laboratory for the sexual. More than any other city in the Western world, it is a recognized originator of diverse sexual behaviors, values, ideologies, and organizations. It caters to a wide range of interests, providing an encouraging environment and many opportunities for those who want to experiment sexually. Dozens of organizations like the Bisexual Center provide social support and meeting places for people with varying sexual desires. Moreover, there is a multitude of groups, newsletters, magazines, and activities in which people explore the meaning of sexuality. Because of the very atypicality of the setting, sexual variation is maximized, and most aspects of a sexual phenomenon can be studied in one place.

Many members of the Bisexual Center were involved in a variety of intimate and sexual relationships with both men and women, providing a range of interesting cases for our research. The organization consisted of people who had joined together to help solve the problems they faced in coming to grips with their unconventional sexuality. We should note that unlike the people at the Bisexual Center, most persons in the United States who behave bisexually do not adopt a bisexual identity, i.e., do not call themselves "bisexual."[7] Some people who behave bisexually are confused and think that they may be in the process of becoming homosexual. Others simply deny their same-sex feelings and behaviors in order to preserve their self-image as heterosexuals. Since the people in our study are members of the minority of bisexually behaving people who actually come to see themselves as "bisexual," care should be taken in generalizing from our results to *all* bisexuals.[8] On the other hand, if everyone has the potential to be bisexual, the people we met at the Bisexual Center provide a glimpse of the satisfactions and frustrations that are possible if such a sexual preference is adopted.

In 1983, we conducted a fieldwork study and an interview study with members of this group, exploring in general what it was like to be "bisexual." Then, to examine how their social, psychological, and sexual profiles compared with those of people with different sexual preferences, we studied members of three organizations: one for people who defined themselves as homosexual (the Pacific Center) and two others from which we used people who considered themselves heterosexual (the San Francisco Sex Information service and the Institute for Advanced Study of Human Sexuality). There were ties between the groups and their members. The same person was involved in the creation of two of them; they had the same basic philosophy (a very permissive attitude toward sexuality); and each shared the same basic function (providing social support for persons who wanted to explore their sexuality). We conducted a mail questionnaire study with people from all three of these organizations in 1984 and 1985 in order to compare self-identified "bisexuals" with self-identified "homosexuals" and "heterosexuals."

What we did not foresee was a development that had tremendous implications for research on sexual preference: the emergence of AIDS.

Not only was this a personal tragedy for many people; it also radically altered the social context of their sexual behavior, especially when bisexuals began to be blamed for bringing AIDS from the homosexual to the heterosexual community. As the effects of the disease spread, we realized that it was crucial to explore how AIDS had affected the bisexuals we had originally studied. On a personal level, we were concerned with how they were faring. But we also wanted to examine how large-scale changes in society—disease, death, and additional stigma—could affect sexual preference. So in 1988 we returned to San Francisco and once again interviewed many of the same people we had studied before the emergence of the disease.

Our initial forays into the world of bisexuality revealed a diversity of unusual sexual activities, mixes and numbers of partners, and changes and vacillations in the direction of sexual desire and in self-labeling. We also found many very ordinary things—love, warm friendships, intimate relationships, jealousies, marriages, children, social conflicts, sexual politics, and personal regrets—all this in addition to concerns about the looming problem of AIDS. When we left this world in the late 1980s we did so with the conviction that sexual preference is far more complex than most people believe, and that social factors affect people's sexuality in unrecognized ways.

In this book we show, for example, that the majority of bisexuals established heterosexuality first in their lives. Homosexuality was something they later "added on." They became bisexual over time, depending on a range of sexual and cultural experiences. And many did not define themselves as bisexual until years after their first dual attractions.

Men and women experienced their bisexuality in different ways, especially in the mix of their sexual behaviors and romantic feelings. For men it was easier to have sex with other men than to fall in love with them. For women it was easier to fall in love with other women than to have sex with them. There were many other behavioral, physical, and emotional differences between men and women as well. Both bisexual men and women, however, seemed to share the same traditional ideas about gender; for example, sex with men was described as more physical, with women as more intimate. Each gender was said to have something different or unique to offer, which formed the basis of bisexuals' dual attractions.

Rather than one type of bisexual, we found five types. Few were like the stereotype—equally attracted to both sexes. And a surprising number of people who defined themselves as "heterosexual" or "homosexual" reported that they were not exclusive in their sexual preference.

Many of the bisexuals changed in their sexual feelings, sexual behaviors, and/or romantic feelings over periods as short as three years. They changed more than heterosexuals or homosexuals did, though a surprising number of the latter were not completely "anchored" in their sexual preference either. Indeed, once heterosexuals and homosexuals had lifted the anchor of an exclusive sexual preference and experienced sex with both men and women, subsequent changes seemed easier for them. Multiple relationships and nonmonogamy were widely accepted practices among the bisexuals. Sexual freedom was the dominant ideology, but this did not stop a large number of the bisexuals from being married. Overall we found marriage and bisexuality not necessarily to be incompatible, and many unmarried bisexuals said that they hoped to marry in the future for common reasons: love, sex, and children.

Most bisexuals reported pervasive feelings of confusion and uncertainty about their sexuality at some point in their lives. Before defining themselves as bisexual, many were confused because they didn't know of the label "bisexual," or they found it difficult to apply if they had not yet engaged in sexual behavior with both men and women. Even after adopting the identity, they remained confused because of negative social reactions, sexual experiences weighted more to one sex, and stereotypical images about bisexuals. Yet despite feelings of confusion about their sexual preference and the hostility that many experienced, few said they regretted being bisexual, before or after AIDS. Instead, many bisexuals were proud of the journey they had traveled and saw themselves as better people because of it.

When it came to their social image, bisexuals were more selective about disclosing their sexual identity to others than were homosexuals— but they were more open about being "out" than heterosexuals who had unconventional sex lives. Some saw it as necessary to hide behind a heterosexual front, but doing this often left them feeling angry and frustrated. Even in San Francisco, a bastion of sexual freedom, bisexuals routinely experienced discrimination and rejection. Surprisingly, homosexuals were said to be just as negative as heterosexuals, but often in more subtle ways. Some homosexuals saw bisexuals as people who were really homosexual but afraid to admit it, or as unloyal lovers who might switch to the other sex on a whim. The AIDS crisis, however, has led to improved relations between bisexuals and homosexuals, given their shared vulnerability to death.

These and other of our findings have important implications not only for our understanding of bisexuality but for our understanding of heterosexuality and homosexuality as well. This study will show that learning plays a significant part in helping people traverse sexual boundaries, past and present intimate relationships influence a change in sexual preference, and that bisexual activity is inseparable from a social environment that provides varied sexual opportunities. Thus a mosaic of factors far more complex than usually considered is involved in explaining sexual preference. Our research finds that sexual preference is much less fixed and much more complicated and fascinating than most current thinking holds.

Findings such as these will increase in importance as society begins to recognize the wider spectrum of human sexuality and how its expression is closely intertwined with the social environment. We believe that a close examination of bisexuality represents the route to a better understanding of sexuality in general, preparing the way for the calmer and more positive approach to sex that must surely follow the end of the AIDS crisis.

I

BISEXUAL LIVES

2

Bisexuals in San Francisco

The Bisexual Center was established in the mid-1970s by a group of friends in San Francisco. The founders wanted to provide a safe haven for people who were interested in being sexual with both men and women and who wanted to get together and talk about their sexual preference. These first members met at various locations throughout the San Francisco Bay Area for weekly discussion groups. As knowledge of the groups and their purpose spread, the number of participants grew quickly. Advertisements were placed in local underground newspapers, and information was passed by word of mouth. Eventually, yearly memberships were offered and a regular mailing list was assembled. In the seven years following its inception, the Bisexual Center grew into a tax-exempt nonprofit organization situated at a permanent location with a yearly membership of about three hundred persons. At the time of our study, the center was located on the middle floor of a trilevel flat in a residential neighborhood near Haight-Ashbury.

The Bisexual Center offered a full slate of programs and services. On Tuesday and Wednesday nights, there were open rap (discussion) groups. On Monday evenings there were special closed support groups, which were limited in size and had preset discussion topics. A special group for women only also gathered on that night. Other more unusual assemblages included the Bi-Bye Booze Group for bisexual alcoholics and even a Sadomasochistic Bisexual Alcoholics Anonymous Support Group. Weekend workshops were conducted every couple of months, focusing on special issues like "coming out" or "bisexual intimacy." The Bisexual

Center also sponsored its own parties on many holidays and a members-only raft trip during the summer. Donations were collected at all of the events (usually a nominal donation of one to three dollars), and a news-letter called the *Bi-Monthly* was mailed to subscribers. The Bisexual Center was managed by a staff of approximately fifteen volunteers. Six peo-ple worked as facilitators for the various support groups, and overall policy was set by a board of directors, eight members who met more or less once a month.

The Bisexual Center articulated its philosophy and objectives in the form of a one-page flyer, which said:

- We are people seeking to love and share intimately with both women and men. Self-defined as bisexuals (although such labels are limiting), we are working to create for ourselves and others a strong sense of community.
- The Bisexual Center is united in struggling for the rights of all women and men to develop as whole, androgynous beings.
- We support relationships between persons of the same and other sex. These relationships may include relating spiritually, socially, emotionally, sensually, sexually, and intellectually. We also sup-port persons choosing a celibate lifestyle.
- We support people who have been oppressed because of sexual preference, gender, age, or ethnic group.
- We encourage and support people struggling . . . for the right to engage in free expression of consenting sexual activity. We support the open expression of affection and touch among people, without such expression necessarily having sexual implications.

We decided to do this research after repeated requests by one of the founding members of the Bisexual Center. She felt that bisexuals and bisexuality had been glossed over in many earlier studies of sexuality. Other members of the center, however, did not accept us as readily. Not only were we challenged because we were ostensibly not bisexual our-selves, but many center members also had ideas of what we should and should not be studying and how we ought to go about doing things.

Two things helped us become accepted and gain the trust of the members. First, because housing is expensive and scarce in San Fran-cisco, the board of directors let us use one of their offices at the center as a temporary living space. During our six-month stay there, we assisted in the day-to-day operations of the center, answering the door and the phone, handing out literature, answering questions about events, helping with the newsletter, preparing food for parties, and so on. We came to be seen as people not only interested in our own research but genuinely concerned about the members and the organization.

Second, we got members involved in the research process itself. We constructed a short questionnaire that asked members to tell us what they felt a study of bisexuality should include. We mailed out this brief survey as an insert in the *Bi-Monthly* newsletter and handed it out to people who came to the Bisexual Center. Approximately 150 people completed the survey.

As we were getting to know people and establishing trust, we were also carefully observing everything we could at the center. The best place to gather information was in the rap groups, in which members gave detailed personal accounts of their lives as bisexuals. At first, we were only permitted access to two weekly "open" mixed-sex rap groups. Fortunately, these were the most popular groups that convened, with anywhere from two to thirty people present on a given night. Some of the participants had long defined themselves as bisexual, while others had just experienced their first same-sex encounter and felt confused. Occasionally the heterosexual spouse of a newly recognized bisexual attended to learn about bisexuality. Sometimes curious gays or lesbians would drop in and listen to a discussion.

The "open" rap groups were almost always run by one male and one female facilitator who worked as a team. The facilitators would introduce themselves and the Bisexual Center (and us as researchers), then lay down guidelines for discussion. The subjects of the discussions varied, including such topics as when or if to tell others about your bisexuality; dealing with sexually transmitted diseases; dealing with people's fears about bisexuality; learning to love both sexes; and handling jealousy.

People typically attended these sessions in cycles. They would enter the group, attend for five, seven, maybe ten weeks, and then stop coming. Usually they came to deal with a particular bisexual issue in their lives and to talk about that problem for a number of weeks. Once the issue had been resolved, they would stop attending sessions. About half of those who came to the rap sessions were relative newcomers who had never before been to the Bisexual Center. The remaining half were primarily previous participants who had new problems they needed to work out. Occasionally people would stop by for a one-time visit, but this was the exception rather than the rule.

After about a month of attending open rap groups, we began to expand our range of observations. We were patient and took things slowly, letting those around us get used to our presence before saying we wanted to do more. As trust between us and the leadership of the Bisexual Center developed, we found that simply suggesting that we wanted to look at new things was more or less well received. We began to attend some of the special all-day weekend workshops that the Bisexual Center sponsored. We also made a point of going to all the social events at the organization: potluck dinners; holiday parties with dancing, drinking, and snacks; and "games night," when people would bring their favorite games such as Scrabble or Monopoly. Eventually we were allowed to observe "closed groups." Every month we attended the board of directors meetings. After about eight weeks, we had managed to secure detailed observations of almost every function the Bisexual Center sponsored.

Finally, because of our around-the-clock presence at the Bisexual Center, we were often privy to the informal give-and-take between people that occurred during off hours. Sometimes friends would stop by to chat together, and they often let us sit with them and listen. At other times, people would stop in to pick up information about an upcoming organizational event. If no one was around, we greeted and talked with them. If a volunteer was present, we often asked what the visit was about after they left. Finally, one of us was always present when the newsletter committee met to assemble the *Bi-Monthly,* and we would hear a great deal of talk about the things the committee members had done during the week, and so on.

Although it was not always easy, we found that sustaining a balance between our status as outsiders and that of participants in the events at the Bisexual Center helped us obtain information. Being outsiders from Indiana worked to our advantage, because many members of the Bisexual Center felt they could tell us private information that they normally kept secret. For example, we were surprised to learn that some people who did volunteer work at the Bisexual Center and who were some of the community's most active and outspoken members had never had a same-sex experience, and some of them even felt that bisexuality was an illness or sickness. As outsiders, we were safe people with whom to talk.

The Bisexual Center was not the only place we made observations. We soon found out that the bisexuals who came to the center often had very active and diverse sex lives that involved a network of institutions we will refer to as the "sexual underground." Our fieldwork extended into this larger social arena, which included, for example, gay bars, adult movie houses, sex clubs, sexual support centers (including the Bi Center), volunteer groups that provided services related to sexuality, sex therapy training programs, etc. Practically everyone in the study knew these places, and some frequented them from time to time. People could find partners, experiment with new ways of being sexual, and learn new values through participation in the sexual underground. Over time it became clear that this subculture was important in sustaining the overall aura of sexual liberalism that gives San Francisco its distinctive character. The following observations made at a swing club and an SM club may help to illustrate how some of our thoughts about bisexuality were shaped.

The swing club was inconspicuously located in a two-story house along a neighborhood street in Oakland. While we saw many scenes worth reporting, we will describe one to convey the atmosphere of the setting and the relevance for our research. Three people were involved; a man was engaging in rear-entry vaginal intercourse with a woman who was performing oral sex on another woman. At the same time other women were engaging in oral sex with women who had engaged in sex with men earlier. There were no men engaging in same-sex activity (which is generally not allowed in swing clubs). We subsequently found that a number of the women in our study had first engaged in bisexuality at a swing house. This clued us in to the importance of different opportunities in the development of bisexuality. We also had an interest in the extent to which bisexuals engaged in swinging or group sex. To what degree did participating in such activities produce or neutralize jealousy? And how did bisexuals deal with jealousy if it appeared?

The SM club was located in a large warehouse south of Market Street. The front door opened into a conventional bar, which was connected to a large back room that contained, among other props, racks, suspensions, ropes, chains, shackles, clamps, handcuffs, paddles, whips, leather and rubber paraphernalia, a prison, and a shiny new motorcycle. This club sponsored a "bisexual night" (every third Friday). Again, we will describe one scene to convey a sense of what we saw and its relevance to our research. A woman, dressed in leather and with breasts bared, was playing a dominant role. She was whipping one man on the buttocks with a horse whip, fondling his genitals during stop periods, and demanding that he perform fellatio on another man. This other man was handcuffed to a bar suspended a foot above his head. Both men were wearing rubber panties to generate heat and retain bodily excretions. This setting raised questions for us about the range of sexual behaviors in which bisexuals might be involved. It also led us to wonder in what other ways bisexuals were able to step out of traditional gender roles in their exploration of sexuality.

Besides attending sex clubs, many members also participated in educational and support organizations that were part of the sexual underground. The Institute for Advanced Study of Human Sexuality was one such place. This was a small private educational organization, located near downtown San Francisco, that offered a Ph.D. program in the study of sex. We spent about a week at the institute, which was staffed by a small faculty, a few of whom had international reputations. We found that many of the people we met at the Bisexual Center had taken classes in sexuality offered by the institute.

The San Francisco Sex Information service was another popular organization among those who attended the Bisexual Center. This was a telephone hotline that answered anonymous callers' questions about sex. We visited the one-bedroom apartment that housed the switchboard a number of times, and we learned that many of the volunteers were familiar with or had been to the Bisexual Center.[1] Since everyone who worked the switchboard had received training, one of us went through the organization's "sexual attitude reassessment" (SAR) training program directed at examining and liberalizing sexual attitudes. It was fifty hours long and consisted of numerous presentations and small-group discussions. The people who spoke included gays, lesbians, bisexuals, heterosexuals, transsexuals, a transvestite, swingers, sex surrogates, and even a sexware salesperson. To give an example, one bisexual-identified male, an active member of the Bisexual Center, gave a speech about how people in society were anal phobic and how this limited the horizons of their sexuality.

We also visited the Pacific Center, a support organization that provided much the same programs and functions as the Bisexual Center, but for gays and lesbians. It was located near downtown Berkeley in a two-story house. Like the Bisexual Center, the Pacific Center sponsored rap groups, offered counseling, published a newsletter, and helped plan gay and lesbian events. Some of the people at the Bisexual Center had attended rap groups at the Pacific Center at one time or another. We were told, however, that because it mainly served the gay and lesbian community, the Pacific Center offered little support when it came to the issue of bisexuality. This organization had a much larger membership and was more visible in the sexual politics of San Francisco than the Bisexual Center.

Our observations in these service-based organizations gave us many new ideas and provided critical information about the bisexual experience. We were surprised, for example, to observe firsthand the dislike many gays and lesbians felt for bisexuals and vice versa. The communities were divided by conflict. Many gays felt that bisexuals were really homosexuals who were afraid to admit it. On the other hand, bisexuals felt that gays and lesbians ignored the fact that bisexuals also faced discrimination for engaging in same-sex behaviors. We also began to wonder about the similarities and differences in sexual behavior and relationships among gay men, lesbians, bisexuals, and heterosexuals. It occurred to us, for example, to examine some of the accusations leveled against bisexuals: Were they "in transition" to homosexuality? Were they sexually "promiscuous"? Could they sustain stable relationships?

We spent time in still more sex-related settings to investigate the extent to which people at the Bisexual Center frequented other parts of the sexual underground. These included, for instance, gay bathhouses, lesbian bars, gay bars, female impersonator shows, a bar that was said to cater to bisexuals, bookstores that specialized in sexual materials, peep rooms, and local parks with popular tearooms (men's restrooms for sex). We also visited a urine club (for those interested in urine play), an enema club, a "scat" club (for feces play), and a fisting club (where one partner inserts his or her fist in the other partner's anus and/or vagina). From these observations, still more questions emerged: How is sex among bisexual males similar to or different from sex among homosexual males? Does each involve unique kinds of experience (i.e., what is often labeled "kinky sex")? How often is bisexuality associated with very unusual sexual practices such as urine play and fisting?

Thus, in San Francisco the boundaries of acceptable sexual behavior extended well beyond what most people expect and do. It was this commonality, coupled with a "sex-positive" ideology, that demarcated the boundary of the sexual underground. This social milieu was not occupied solely by those who considered themselves bisexual, but by people of all sexual persuasions. The sexual underground seemed critical to shaping sexual expression in general. Our particular concern was its role in facilitating the development of sexual preferences.

This description of the initial phase of our study has of necessity been rather general given the large number of places we observed and the wide range of people we spoke to. If we could convey the value of this field-work for our research in one particular experience, it would have to be a conference organized by the leaders of the Bisexual Center that was designed to facilitate communication between the gay and lesbian com-munities of San Francisco. We cannot reproduce the respectful hush that surrounded some of the speakers, the ripples of anger that swirled around others, or the give-and-take between speaker and audience or between audience members. What we can do, however, is to present what was said. In this way the reader can see some of the questions that are exam-ined in the rest of this book: Who and what am I? Why do I feel the way I do? Why is society afraid of me? Whom should I turn to for support? Who can I trust? All these issues centered on the basic fact that, for bisexuals, their sexual preference is at odds with what society feels is natural and normal.

(In the quotations that follow and throughout this book, all names given are pseudonyms.)

All we want to have as bisexual, lesbian, and gay people is the right to lead lives that work for us, to have good relationships, good work, and a sense of who we are. Some of the things we've heard about each other you may have run into in the past: that gay and lesbian people don't like bisexuals ever; that bisexual people can't be trusted ever; that we will run back to our heterosexual privilege; and so on. All the myths that ruin our lives are pretty much just that— myths.   (Mary: A Bisexual-identified Woman)

I'm frankly appalled and disgusted at the prejudice against bisex-uality that I'm aware of within the gay male community. I want to start off with trying to go into what I think are some of the biases. One of the ones I hear often [from gays] is "I was pressured into dating girls by my family and friends when I was in high school," and therefore that's what bisexuality is. It's a phony period of being pressured into conforming to society's standards, and it's a giving in to this pressure, therefore it's a lie; it's immaturity. So many gay men I know feel that; it's their experience, that's what happened to them; they hate it and are angry about it and therefore anyone who's still into that trip should come out of it and stop being phony. That's such a strong feeling that I have had some of myself, and I know a lot of my friends have it still. It's that generalization from our experience to everybody else which is illogical, but is still there. It's important to allow people to be in touch with their experience, but not allow them to impose that chauvinism on everybody else. Another belief is that straights run the world and oppress gays. Gays are finally making progress. Progress is fragile, so you bisexuals shut up and let us gays have our time now. We have a chance to really make some progress in understanding homosexuality so don't raise all of this sophisticated side issue crap; give us our chance. Even big shots in the academic world think that, and they're angry that somebody else is crowding in on the opportunity to understand homosexuality. I think that's bullshit, too. It's not legitimate to make progress in one area at the expense of another group.

Also, there is the belief that homosexuality, not heterosexuality, is what people are really discriminating against, so bisexuality is a nonissue: "Homosexuals really deal with oppression!" That [attitude] denies oppression of bisexuals and oppression of bisexuals by gays. For example, I can't compete against a real woman if I'm a gay male. So I hate bi men because they might give me up for a woman. That raises issues of low self-esteem among gay men.

It's important not to allow these fallacies to continue. I believe that bi's and gays have the same enemies, the same issues are used against both, therefore we're together. People practicing homosexuality are a minority, that hooks us into sexual minorities, and all minorities are oppressed. The only way to handle oppression is to seek common links, stand together and fight. There's too much overlap between what constitutes a bisexual or gay person to legitimately divide us. Also, we can use the concept of "biphobia" just as much as the concept of homophobia. There is a phobia about bisexuality. There are myths and stereotypes for that just as there are for homophobia.   (Hal: A Gay-identified Man)

I have not always been proud of the gay part of myself. There was a long period of time when I was terrified that I was a latent homosexual. I was incredibly homophobic and avoided any contact with men to the point of not taking gym classes in high school to avoid being around naked men. I focused my attention on heterosexual feelings and relationships. It took me a long time to feel comfortable with my homosexual feelings. That's true for a lot of gay and bi men and lesbian and bi women. It takes a long time for the process of feeling comfortable and ridding oneself of homophobia. People easily associate my bisexuality with my homophobia without recognizing that there are a lot of gay and lesbian brothers and sisters who are also homophobic for a period of time and no one discounts their process because they are gay men and lesbian women.

As a bisexual married man who is deeply committed, involved and eroticized to women as well as men, some things that I experience in the gay mecca of the world is that when I go out in public, I've had beer bottles thrown at me, had men run up to me in the street screaming, "Fucking faggot." I put my life in danger all of the time. I have also had the experience of getting invalidated by the gay community. I resent the concept that I take heterosexual privilege. I would like to tell you about heterosexual privilege as I experience it. I used to be a warm, touching person who reached out to others. As a bisexual man I feel that I can only be half a person in the straight community. And there's a subtle message in the gay community not to talk about my heterosexual relationships. (Dan: A Bisexual-identified Man)

This conference is an affirmation of the importance of self-definition. The right to choose or not choose a label. The right to choose whom we love. The lesbian and gay movement was not about the right to choose same-sex partners, the right to have same-sex feelings. It was about the right for all of us to have emotional liberation, to love whom we wish to love. Bi's have been a part of the lesbian and gay rights struggle throughout history—invisibly so, but there. In the early 1900s in Europe; they were there from Stonewall to now.

We all know what homophobia is—the fear of same-sex feelings/relationships, the fear of people who feel that way and do those things. What about biphobia? Biphobia, to me, *is* homophobia. Oppression is oppression is oppression. And bis are a minority within a minority. We all know how minorities are treated. (Eleanor: A Bisexual-identified Woman)

Some of my bi, lesbian, and gay friends who came from other parts of the country say that it must be easy to be gay, bi, or lesbian growing up in San Francisco. It wasn't easy for me. I went through nurses' training, undergraduate and graduate work in college, and worked at a mental health agency for several years, all the time lying about who I was, not even coming out to myself. Not until my early thirties did I begin to honor and love the same-sex feelings I had in me, that are certainly equal to if not more than, at times, my opposite-sex feelings. I was a little girl who never bought into the female script. We were told a big lie when we were born. Little boys were here and girls were there, and we stood as adversaries. But my intrapsychic maleness and femaleness had a war going on. I really felt different—as if I'd been born on another planet and put on earth fifty years too soon. I went to a lot of therapists who verified that I was sick. I had to make a choice—straight or gay. Everything else was role confusion, schizoid, denial, projection, etc. Yet I knew I had loving feelings for some women and some men. Well for thirty years I guess I've been coming out in this trendy, chic fad because I still have feelings for men and women, long-term committed relationships with several people who are very important to me. Some I see once a year, some each week. I also love the sexual, sensual, and affectionate energy I get from people I share love with. I'm not monogamous, yet I feel a commitment and a caring for the people in my life. To me it doesn't fit to have just one person in my life that I love. I have a network of intimate friendships—sexual, sensual, affectionate friends who support me.   (Mary: A Self-identified Bisexual Woman [second comment])

I have felt gay all along, along with feeling bi. My own direct experience was that when I was with men and close with men in a bar, bath, community, march, or on the beach, there was nothing that divided me from my gay brothers. And I wasn't going to let anything divide us. I felt that we were all part of the same thing. Feeling bisexual and being in a relationship with a woman did not stop me from feeling that connection with the gay community. The experience of fighting the [antihomosexual] ideas represented by Anita Bryant and John Briggs was an eye-opener in a lot of ways. We always spoke in terms of lesbian, gay, bisexual, and alternative lifestyle heterosexual people. We were a campaign of emotional liberation. Part of the flack that I've felt in our campaigns is nonrecognition of the diversity of the community we represented—particularly of the bisexual element. What I've seen in San Francisco is generally a nonrecognition, subtle discrimination, a sense of invisibility.

In terms of identifying, I feel like a citizen of the lesbian, gay, and bi community. Sexually, and in my heart, I identify as gay. Emotionally and personality-wise, I want to emphasize and have support for my relationships with men and with women, to identify with the community and as bi. People don't always fit into discreet categories. I insist on identifying with both. I don't want to close off any future options. It is politically correct for me to identify as bisexual because of the oppression. There are lots of lesbian/gay organizations. Are bi's part of those? Are they to be included? If so, where's the visibility? If not, where are the independent and separatist bi organizations? One negative consequence of the invisibility of the bisexual community is that we have not had enough bi input into all of the movements we're involved in. And bi's have not always felt a part of organizations.   (Alan: A Bisexual-identified Man)

The majority of lesbians, heterosexuals, and gay men still believe that bisexuality is a phase or bi's are really gay or don't exist. Those myths must be done away with. Over the years I've talked to a lot of lesbians who were devastated because the woman they loved returned to the "safe space" of the man they were with before. These lesbians transferred their sense of betrayal and bitterness to other women. This has a lot to do with the myth that bi women are not to be trusted. Many lesbians believe that bi's use that label to avoid stigma. Lesbians feel that although a bi woman lives with a woman she fantasizes about men and keeps open the option to return to men if her woman relationship dissolves. There's resentment that bi's can appear to be heterosexual when its convenient. They are seen only to want a sexual, not an emotional, commitment.

There is a certain concern with sexually transmitted diseases since bi women are seen to be having sex with men. And with the AIDS epidemic there is the fear of coming in contact with bi women who have contact with bi men. Since most lesbians have had heterosexual experience they believe that a bi phase is just that because they went through that phase and it's simply a process of coming out. But politically and emotionally speaking, bi's and lesbians share the same types of oppression—as women and regarding sexual preference. Many politically active lesbians are now defending women's rights to choose bisexuality as a lifestyle. But there is a danger in assuming that bisexuality is the "ideal state" for humans. There is a tendency on the part of some bi's to push that concept to the extreme and put down others. That only complicates matters. The work of organizations like the Bi and Pacific Centers is beginning to raise consciousness. My hope is that all of us who are, by certain standards, seen as heterosexuals who've gone wrong, can get together to help others understand the wide spectrum of human sexuality and the inhumanity of sexual oppression of lesbians, gays, and bi's. (Pam: A Self-identified Lesbian)

The bisexual world that we encountered in San Francisco was very complex. It intersected not only with the world of conventional (heterosexual) sex but with the worlds of unconventional sex (both heterosexual and homosexual). And for the most part we were only able to gain access to the more open bisexuals. Nonetheless, our efforts did provide us with a wide range of interesting questions to pursue.

After about four months of intensive observation, we began a more structured phase in our research, an in-depth interview study based on our observations and inquiries from people at the Bisexual Center. Many who came to the Bisexual Center for the first time were struggling with questions about their sexual identity: Am I bisexual? How do I know if I am bisexual? Others had returned to the Bisexual Center after a somewhat lengthy absence because they experienced self-doubts and continued confusion about their sexuality. Long-term members were concerned with questions of social stigma and acceptance. And of course there was always talk about relationships, sexual behaviors, sexually transmitted diseases, and the like. Depth interviews that included extensive probing questions were the most obvious way to pull together a profile of the lives of the people we had met.

The questions contained in the interview were wide-ranging. We asked about factors that were influential in leading people to define themselves as bisexual and about events that led them to call their bisexual identity into question. Other questions focused on the ways bisexuals met their sexual partners, what they found attractive in men versus women, their views about the ideal sexual-affectional relationship, the structure and organization of their involved relationships, and how they dealt with jealousy. In one portion of the interview we asked a comprehensive set of questions about sexual behavior. Last were items about relationships with homosexuals and heterosexuals, the experience of coming out as bisexual, and general feelings about being bisexual.

During the early months, from January through April of 1983, we collected the names and phone numbers of people who were willing to participate in an interview. Later on, each person was called, screened over the phone, and a time and place for the interview was scheduled. We used three criteria to screen the people willing to participate: (1) they had attended the Bisexual Center at least once; (2) they defined themselves as bisexual; and (3) they currently expressed more than incidental homosexual and heterosexual feelings and behaviors. This ensured a set of respondents who were arguably bisexual and who had to deal with sexual feelings for and/or behaviors with both sexes. Between April and June of 1983, we interviewed one hundred people: forty-nine men, forty-four women, and seven male-to-female transsexuals. Four additional transsexual bisexuals were obtained in 1988 to make up our transsexual sample.

We recruited and trained five male and three female volunteers from the Bisexual Center and the Institute for Advanced Study of Human Sexuality as interviewers.[2] All told, the volunteers and the research team each completed fifty interviews. To minimize the effects of interviewer bias, we took special care to match the bisexuals in the study with interviewers around whom they felt comfortable. In the end, the majority were interviewed by someone of the same sex, though a few specifically requested an interviewer of the opposite sex. For those who said that they would be ill at ease talking with a volunteer, we completed the interview personally. A few people requested that we do their interview because they felt it would be more confidential. Others indicated that they wanted a friend who was interviewing to do theirs. This resulted in some of the most detailed interviews of all.

The danger of having organizational members conduct the interviews lay in the possible reluctance of people being interviewed to disclose something negative about being bisexual, the Bisexual Center itself, a mutual acquaintance, and the like. Similarly, there was always the chance that persons being interviewed might exaggerate their sexual history to impress a bisexual interviewer or provide acceptable responses to fit with perceived interviewer expectations. To control for these effects, we compared interviews conducted by volunteers with those we did ourselves, and we found no major differences.

The interviews, most of which were conducted at the Bisexual Center,[3] yielded extremely comprehensive information on sexual behavior, relationships, and social histories. Altogether we asked each person 415 closed and 126 open questions,[4] and the interviews ran from 2 to 9 hours in length. Interviewers recorded the answers by hand on preprinted interview schedules. (Tape-recording was not practical both because of the time and cost that would have been involved in transcribing the answers and because many people felt uncomfortable with the idea.) The interviewers wrote what was said as close to verbatim as possible and always in the first person, using pacing techniques to manage the rate of talk by the interviewees.

Shortly after the depth interviews were completed, we left the field to analyze the results. We looked for general patterns in the answers to the open questions, and we analyzed the responses to the closed questions by computer, running tests of statistical significance to determine whether the differences we observed between men and women might be just the result of chance. (When we talk about "significant" differences throughout this book, we mean differences that are statistically significant.) When there was no significant difference between the women's and the men's responses on a question, we present a combined percent. Further details regarding the data analysis are presented in the appendices with the tables. When we present quotes from respondents the notation (M) refers to male and (F) to female.

The bisexuals in the interview study had a somewhat homogeneous social profile (Table 2.1). They were largely twenty-five to forty-four years old, college educated, with moderate incomes (between $10,000 and $30,000), almost all white, employed at the time of the interviews, more Protestant, but with similar lesser numbers of Jewish, Catholic, and other religions represented. The vast majority described themselves as not at all religious, as liberal to radical in their politics, and as moderate to strong in a general question about degree of acceptance of feminist ideology. Everyone in the study resided in the San Francisco Bay Area or within a short drive away.

We present our findings about bisexuality in the chapters that follow, beginning with a description of the process by which the people we interviewed became bisexual.

3

Becoming Bisexual

Becoming bisexual involves the rejection of not one but two recognized categories of sexual identity: heterosexual and homosexual. Most people settle into the status of heterosexual without any struggle over the identity. There is not much concern with explaining how this occurs; that people are heterosexual is simply taken for granted. For those who find heterosexuality unfulfilling, however, developing a sexual identity is more difficult.

How is it then that some people come to identify themselves as "bisexuals"? As a point of departure we take the process through which people come to identify themselves as "homosexual." A number of models have been formulated that chart the development of a homosexual identity through a series of stages.[1] While each model involves a different number of stages, the models all share three elements. The process begins with the person in a state of identity confusion—feeling different from others, struggling with the acknowledgment of same-sex attractions. Then there is a period of thinking about possibly being homosexual—involving associating with self-identified homosexuals, sexual experimentation, forays into the homosexual subculture. Last is the attempt to integrate one's self-concept and social identity as homosexual—acceptance of the label, disclosure about being homosexual, acculturation to a homosexual way of life, and the development of love relationships. Not every person follows through each stage. Some remain locked in at a certain point. Others move back and forth between stages.

To our knowledge, no previous model of *bisexual* identity formation exists. In this chapter we present such a model based on the following questions: To what extent is there overlap with the process involved in becoming homosexual? How far is the label "bisexual" clearly recognized, understood, and available to people as an identity? Does the absence of a bisexual subculture in most locales affect the information and support needed for sustaining a commitment to the identity? For our subjects, then, what are the problems in finding the "bisexual" label, understanding what the label means, dealing with social disapproval from two directions, and continuing to use the label once it is adopted? From our fieldwork and interviews, we found that four stages captured our respondents' most common experiences when dealing with questions of identity: initial confusion, finding and applying the label, settling into the identity, and continued uncertainty.

The Stages

Initial Confusion

Many of the people interviewed said that they had experienced a period of considerable confusion, doubt, and struggle regarding their sexual identity before defining themselves as bisexual. This was ordinarily the first step in the process of becoming bisexual.

They described a number of major sources of early confusion about their sexual identity. For some, it was the experience of having strong sexual feelings for both sexes that was unsettling, disorienting, and sometimes frightening. Often these were sexual feelings that they said they did not know how to easily handle or resolve.

> In the past, I couldn't reconcile different desires I had. I didn't understand them. I didn't know what I was. And I ended up feeling really mixed up, unsure, and kind of frightened.   (F)

> I thought I was gay, and yet I was having these intense fantasies and feelings about fucking women. I went through a long period of confusion.   (M)

Others were confused because they thought strong sexual feelings for, or sexual behavior with, the same sex meant an end to their long-standing heterosexuality.

> I was afraid of my sexual feelings for men and . . . that if I acted on them, that would negate my sexual feelings for women. I knew absolutely no one else who had . . . sexual feelings for both men and women, and didn't realize that was an option.   (M)

> When I first had sexual feelings for females, I had the sense I should give up my feelings for men. I think it would have been easier to give up men.   (F)

A third source of confusion in this initial stage stemmed from attempts by respondents trying to categorize their feelings for, and/or behaviors with, both sexes, yet not being able to do so. Unaware of the term "bisexual," some tried to organize their sexuality by using the readily available labels of "heterosexual" or "homosexual"—but these did not seem to fit. No sense of sexual identity jelled; an aspect of themselves remained unclassifiable.

> When I was young, I didn't know what I was. I knew there were people like Mom and Dad—heterosexual and married—and that there were "queens." I knew I wasn't like either one.   (M)

> I thought I had to be either gay or straight. That was the big lie. It was confusing. . . . That all began to change in the late 60s. It was a long and slow process. . . .   (F)

Finally, others suggested they experienced a great deal of confusion because of their "homophobia"—their difficulty in facing up to the same-sex component of their sexuality. The consequence was often long-term denial. This was more common among the men than the women, but not exclusively so.

> At age seventeen, I became close to a woman who was gay. She had sexual feelings for me. I had some . . . for her but I didn't respond. Between the ages of seventeen and twenty-six I met another gay woman. She also had sexual feelings towards me. I had the same for her but I didn't act on . . . or acknowledge them. . . . I was scared. . . . I was also attracted to men at the same time. . . . I denied that I was sexually attracted to women. I was afraid that if they knew the feelings were mutual they would act on them . . . and put pressure on me.   (F)

> I though I might be able to get rid of my homosexual tendencies through religious means—prayer, belief, counseling—before I came to accept it as part of me.   (M)

The intensity of the confusion and the extent to which it existed in the lives of the people we met at the Bisexual Center, whatever its particular source, was summed up by two men who spoke with us informally. As paraphrased in our field notes:

> The identity issue for him was a very confusing one. At one point, he almost had a nervous breakdown, and when he finally entered college, he sought psychiatric help.

Bill said he thinks this sort of thing happens a lot at the Bi Center. People come in "very confused" and experience some really painful stress.

Finding and Applying the Label

Following this initial period of confusion, which often spanned years, was the experience of finding and applying the label. We asked the people we interviewed for specific factors or events in their lives that led them to define themselves as bisexual. There were a number of common experiences.

For many who were unfamiliar with the term bisexual, the discovery that the category in fact existed was a turning point. This happened by simply hearing the word, reading about it somewhere, or learning of a place called the Bisexual Center. The discovery provided a means of making sense of long-standing feelings for both sexes.

Early on I thought I was just gay, because I was not aware there was another category, bisexual. I always knew I was interested in men and women. But I did not realize there was a name for these feelings and behaviors until I took Psychology 101 and read about it, heard about it there. That was in college. (F)

The first time I heard the word, which was not until I was twenty-six, I realized that was what fit for me. What it fit was that I had sexual feelings for both men and women. Up until that point, the only way that I could define my sexual feelings was that I was either a latent homosexual or a confused heterosexual. (M)

Going to a party at someone's house, and finding out there that the party was to benefit the Bisexual Center. I guess at that point I began to define myself as bisexual. I never knew there was such a word. If I had heard the word earlier on, for example as a kid, I might have been bisexual then. My feelings had always been bisexual. I just did not know how to define them. (F)

Reading *The Bisexual Option* . . . I realized then that bisexuality really existed and that's what I was. (M)

In the case of others the turning point was their first homosexual or heterosexual experience coupled with the recognition that sex was pleasurable with both sexes. These were people who already seemed to have knowledge of the label "bisexual," yet without experiences with both men and women, could not label themselves accordingly.

The first time I had actual intercourse, an orgasm with a woman, it led me to realize I was bisexual, because I enjoyed it as much as I did with a man, although the former occurred much later on in my sexual experiences. . . . I didn't have an orgasm with a woman until twenty-two, while with males, that had been going on since the age of thirteen.   (M)

Having homosexual fantasies and acting those out. . . . I would not identify as bi if I only had fantasies and they were mild. But since my fantasies were intensely erotic, and I acted them out, these two things led me to believe I was really bisexual. . . .   (M)

After my first involved sexual affair with a woman, I also had feelings for a man, and I knew I did not fit the category dyke. I was also dating gay-identified males. So I began looking at gay/lesbian and heterosexual labels as not fitting my situation.   (F)

Still others reported not so much a specific experience as a turning point, but emphasized the recognition that their sexual feelings for both sexes were simply too strong to deny. They eventually came to the conclusion that it was unnecessary to choose between them.

I found myself with men but couldn't completely ignore my feelings for women. When involved with a man I always had a close female relationship. When one or the other didn't exist at any given time, I felt I was really lacking something. I seem to like both.   (F)

The last factor that was instrumental in leading people to initially adopt the label bisexual was the encouragement and support of others. Encouragement sometimes came from a partner who already defined himself or herself as bisexual.

Encouragement from a man I was in a relationship with. We had been together two or three years at the time—he began to define as bisexual. . . . [He] encouraged me to do so as well. He engineered a couple of threesomes with another woman. Seeing one other person who had bisexuality as an identity that fit them seemed to be a real encouragement.   (F)

Encouragement from a partner seemed to matter more for women. Occasionally the "encouragement" bordered on coercion as the men in their lives wanted to engage in a *ménage à trois* or group sex.

I had a male lover for a year and a half who was familiar with bisexuality and pushed me towards it. My relationship with him brought it up in me. He wanted me to be bisexual because he wanted to be in a threesome. He was also insanely jealous of my attractions to men, and did everything in his power to suppress my opposite-sex attractions. He showed me a lot of pictures of naked women and played on my reactions. He could tell that I was aroused by pictures of women and would talk about my attractions while we were having sex. . . . He was twenty years older than me. He was very manipulative in a way. My feelings for females were there and [he was] almost forcing me to act on my attractions. . . .   (F)

Encouragement also came from sex-positive organizations, primarily the Bisexual Center, but also places like San Francisco Sex Information (SFSI), the Pacific Center, and the Institute for Advanced Study of Human Sexuality, all of which were described earlier.[2]

At the gay pride parade I had seen the brochures for the Bisexual Center. Two years later I went to a Tuesday night meeting. I immediately felt that I belonged and that if I had to define myself that this was what I would use.   (M)

Through SFSI and the Bi Center, I found a community of people . . . [who] were more comfortable for me than were the exclusive gay or heterosexual communities. . . . [It was] beneficial for myself to be . . . in a sex-positive community. I got more strokes and came to understand myself better. . . . I felt it was necessary to express my feelings for males and females without having to censor them, which is what the gay and straight communities pressured me to do.   (F)

Thus our respondents became familiar with and came to the point of adopting the label bisexual in a variety of ways: through reading about it on their own, being in therapy, talking to friends, having experiences with sex partners, learning about the Bi Center, visiting SFSI or the Pacific Center, and coming to accept their sexual feelings.

Settling into the Identity

Usually it took years from the time of first sexual attractions to, or behaviors with, both sexes before people came to think of themselves as bisexual. The next stage then was one of settling into the identity, which was characterized by a more complete transition in self-labeling.

Most reported that this settling-in stage was the consequence of becoming more self-accepting. They became less concerned with the negative attitudes of others about their sexual preference.

I realized that the problem of bisexuality isn't mine. It's society's. They are having problems dealing with my bisexuality. So I was then thinking if they had a problem dealing with it, so should I. But I don't.   (F)

I learned to accept the fact that there are a lot of people out there who aren't accepting. They can be intolerant, selfish, shortsighted and so on. Finally, in growing up, I learned to say "So what, I don't care what others think." (M)

I just decided I was bi. I trusted my own sense of self. I stopped listening to others tell me what I could or couldn't be. (F)

The increase in self-acceptance was often attributed to the continuing support from friends, counselors, and the Bi Center, through reading, and just being in San Francisco.

Fred Klein's *The Bisexual Option* book and meeting more and more bisexual people . . . helped me feel more normal. . . . There were other human beings who felt like I did on a consistent basis. (M)

I think going to the Bi Center really helped a lot. I think going to the gay baths and realizing there were a lot of men who sought the same outlet I did really helped. Talking about it with friends has been helpful and being validated by female lovers that approve of my bisexuality. Also the reaction of people who I've told, many of whom weren't even surprised. (M)

The most important thing was counseling. Having the support of a bisexual counselor. Someone who acted as somewhat of a mentor. [He] validated my frustration. . . . , helped me do problem solving, and guide[d] me to other supportive experiences like SFSI. Just engaging myself in a supportive social community. (M)

The majority of the people we came to know through the interviews seemed settled in their sexual identity. We tapped this through a variety of questions (Table 3.1). Ninety percent said that they did not think they were currently in transition from being homosexual to being heterosexual or from being heterosexual to being homosexual. However, when we probed further by asking this group "Is it possible, though, that someday you could define yourself as either lesbian/gay or heterosexual?" about 40 percent answered yes. About two-thirds of these indicated that the change could be in either direction, though almost 70 percent said that such a change was not probable.

We asked those who thought a change was possible what it might take to bring it about. The most common response referred to becoming involved in a meaningful relationship that was monogamous or very intense. Often the sex of the hypothetical partner was not specified, underscoring that the overall quality of the relationship was what really mattered.

Love. I think if I feel insanely in love with some person, it could possibly happen. (M)

If I should meet a woman and want to get married, and if she was not open to my relating to men, I might become heterosexual again. (M)

Getting involved in a long-term relationship like marriage where I wouldn't need a sexual involvement with anyone else. The sex of the . . . partner wouldn't matter. It would have to be someone who I could commit my whole life to exclusively, a lifelong relationship. (F)

A few mentioned the breaking up of a relationship and how this would incline them to look toward the other sex.

Steve is one of the few men I feel completely comfortable with. If anything happened to him, I don't know if I'd want to try and build up a similar relationship with another man. I'd be more inclined to look towards women for support. (F)

Changes in sexual behavior seemed more likely for the people we interviewed (Table 3.2) than changes in how they defined themselves. We asked "Is it possible that someday you could behave either exclusively homosexual or exclusively heterosexual?" Over 80 percent answered yes. This is over twice as many as those who saw a possible change in how they defined themselves, again showing that a wide range of behaviors can be subsumed under the same label. Of this particular group, the majority (almost 60 percent) felt that there was nothing inevitable about how they might change, indicating that it could be in either a homosexual or a heterosexual direction. Around a quarter, though, said the change would be to exclusive heterosexual behavior and 15 percent to exclusive homosexual behavior. (Twice as many women noted the homosexual direction, while many more men than women said the heterosexual direction.) Just over 40 percent responded that a change to exclusive heterosexuality or homosexuality was not very probable, about a third somewhat probable, and about a quarter very probable.

Again, we asked what it would take to bring about such a change in behavior. Once more the answers centered on achieving a long-term monogamous and involved relationship, often with no reference to a specific sex.

For me to behave exclusively heterosexual or homosexual would require that I find a lifetime commitment from another person with a damn good argument of why I should not go to bed with somebody else. (F)

I am a romantic. If I fell in love with a man, and our relationship was developing that way, I might become strictly homosexual. The same possibility exists with a woman.   (M)

Thus "settling into the identity" must be seen in relative terms. Some of the people we interviewed did seem to accept the identity completely. When we compared our subjects' experiences with those characteristic of homosexuals, however, we were struck by the absence of closure that characterized our bisexual respondents—even those who appeared most committed to the identity. This led us to posit a final stage in the formation of sexual identity, one that seems unique to bisexuals.

Continued Uncertainty

The belief that bisexuals are confused about their sexual identity is quite common. This conception has been promoted especially by those lesbians and gays who see bisexuality as being in and of itself a pathological state. From their point of view, "confusion" is literally a built-in feature of "being" bisexual. As expressed in one study:

> While appearing to encompass a wider choice of love objects. . . [the bisexual] actually becomes a product of abject confusion; his self-image is that of an overgrown young adolescent whose ability to differentiate one form of sexuality from another has never developed. He lacks above all a sense of identity. . . . [He] cannot answer the question: What am I?[3]

One evening a facilitator at a Bisexual Center rap group put this belief in a slightly different and more contemporary form:

> One of the myths about bisexuality is that you can't be bisexual without somehow being "schizoid." The lesbian and gay communities do not see being bisexual as a crystallized or complete sexual identity. The homosexual community believes there is no such thing as bisexuality. They think that bisexuals are people who are in transition [to becoming homosexual] or that they are people afraid of being stigmatized [as homosexual] by the heterosexual majority.

We addressed the issue directly in the interviews with two questions: "Do you *presently* feel confused about your bisexuality?" and "Have you *ever* felt confused . . . ?" (Table 3.3). For the men, a quarter and 84 percent answered "yes," respectively. For the women, it was about a quarter and 56 percent.

When asked to provide details about this uncertainty, the primary response was that *even after having discovered and applied the label "bisexual" to themselves, and having come to the point of apparent self-acceptance, they still experienced continued intermittent periods of doubt and uncertainty regarding their sexual identity*. One reason was the lack of social validation and support that came with being a self-identified bisexual. The social reaction people received made it difficult to sustain the identity over the long haul.

While the heterosexual world was said to be completely intolerant of any degree of homosexuality, the reaction of the homosexual world mattered more. Many bisexuals referred to the persistent pressures they experienced to relabel themselves as "gay" or "lesbian" and to engage in sexual activity exclusively with the same sex. It was asserted that no one was *really* bisexual, and that calling oneself "bisexual" was a politically incorrect and inauthentic identity. Given that our respondents were living in San Francisco (which has such a large homosexual population) and that they frequently moved in and out of the homosexual world (to whom they often looked for support) this could be particularly distressing.

> Sometimes the repeated denial the gay community directs at us. Their negation of the concept and the term bisexual has sometimes made me wonder whether I was just imagining the whole thing.   (M)

> My involvement with the gay community. There was extreme political pressure. The lesbians said bisexuals didn't exist. To them, I had to make up my mind and identify as lesbian. . . . I was really questioning my identity, that is, about defining myself as bisexual. . . .   (F)

For the women, the invalidation carried over to their feminist identity (which most had). They sometimes felt that being with men meant they were selling out the world of women.

> I was involved with a woman for several years. She was straight when I met her but became a lesbian. She tried to "win me back" to lesbianism. She tried to tell me that if I really loved her, I would leave Bill. I did love her, but I could not deny how I felt about him either. So she left me and that hurt. I wondered if I was selling out my woman identity and if it [being bisexual] was worth it.   (F)

A few wondered whether they were lying to themselves about their heterosexual side. One woman questioned whether her heterosexual desires were a result of "acculturation" rather than being her own choice. Another woman suggested a similar social dimension to her homosexual component:

> There was one period when I was trying to be gay because of the political thing of being totally woman-identified rather than being with men. The Women's Culture Center in college had a women's studies minor, so I was totally immersed in women's culture. . . .   (F)

Lack of support also came from the absence of bisexual role models, no real bisexual community aside from the Bisexual Center, and nothing in the way of public recognition of bisexuality, which bred uncertainty and confusion.

> I went through a period of dissociation, of being very alone and isolated. That was due to my bisexuality. People would ask, well, what was I? I wasn't gay and I wasn't straight. So I didn't fit.   (F)

> I don't feel like I belong in a lot of situations because society is so polarized as heterosexual or homosexual. There are not enough bi organizations or public places to go to like bars, restaurants, clubs. . . .   (F)

For some, continuing uncertainty about their sexual identity was related to their inability to translate their sexual feelings into sexual behaviors. (Some of the women had *never* engaged in homosexual sex.)

> Should I try to have a sexual relationship with a woman? . . . Should I just back off and keep my distance, just try to maintain a friendship? I question whether I am really bisexual because I don't know if I will ever act on my physical attractions for females.   (F)

> I know I have strong sexual feelings towards men, but then I don't know how to get close to or be sexual with a man. I guess that what happens is I start wondering how genuine my feelings are. . . .   (M)

For the men, confusion stemmed more from the practical concerns of implementing and managing multiple partners or from questions about how to find an involved homosexual relationship and what that might mean on a social and personal level.

> I felt very confused about how I was going to manage my life in terms of developing relationships with both men and women. I still see it as a difficult lifestyle to create for myself because it involves a lot of hard work and understanding on my part and that of the men and women I'm involved with.   (M)

> I've thought about trying to have an actual relationship with a man. Some of my confusion revolves around how to find a satisfactory sexual relationship. I do not particularly like gay bars. I have stopped having anonymous sex. . . .   (M)

Many men and women felt doubts about their bisexual identity because of being in an exclusive sexual relationship. After being exclusively involved with an opposite-sex partner for a period of time, some of the respondents questioned the homosexual side of their sexuality. Conversely, after being exclusively involved with a partner of the same sex, other respondents called into question the heterosexual component of their sexuality.

> When I'm with a man or a woman sexually for a period of time, then I begin to wonder how attracted I really am to the other sex.   (M)

> In the last relationship I had with a woman, my heterosexual feelings were very diminished. Being involved in a lesbian lifestyle put stress on my self-identification as a bisexual. It seems confusing to me because I am monogamous for the most part, monogamy determines my lifestyle to the extremes of being heterosexual or homosexual.   (F)

Others made reference to a lack of sexual activity with weaker sexual feelings and affections for one sex. Such learning did not fit with the perception that bisexuals should have balanced desires and behaviors. The consequence was doubt about "really" being bisexual.

> On the level of sexual arousal and deep romantic feelings, I feel them much more strongly for women than for men. I've gone so far as questioning myself when this is involved.   (M)

> I definitely am attracted to and it is much easier to deal with males. Also, guilt for my attraction to females has led me to wonder if I am just really toying with the idea. Is the sexual attraction I have for females something I constructed to pass time or what?   (F)

Just as "settling into the identity" is a relative phenomenon, so too is "continued uncertainty," which can involve a lack of closure as part and parcel of what it means to be bisexual.

We do not wish to claim too much for our model of bisexual identity formation. There are limits to its general application. The people we interviewed were unique in that not only did *all* the respondents define themselves as bisexual (a consequence of our selection criteria), but they were also all members of a bisexual social organization in a city that perhaps more than any other in the United States could be said to provide a bisexual subculture of some sort. Bisexuals in places other than San Francisco surely must move through the early phases of the identity process with a great deal more difficulty. Many probably never reach the later stages.

Finally, the phases of the model we present are very broad and somewhat simplified. While the particular problems we detail within different phases may be restricted to the type of bisexuals in this study, the broader phases can form the basis for the development of more sophisticated models of bisexual identity formation.

Still, not all bisexuals will follow these patterns. Indeed, given the relative weakness of the bisexual subculture compared with the social pressures toward conformity exhibited in the gay subculture, there may be more varied ways of acquiring a bisexual identity. Also, the involvement of bisexuals in the heterosexual world means that various changes in heterosexual lifestyles (e.g., a decrease in open marriages or swinging) will be a continuing, and as yet unexplored, influence on bisexual identity. Finally, wider societal changes, notably the existence of AIDS, may make for changes in the overall identity process. Being used to choice and being open to both sexes can give bisexuals a range of adaptations in their sexual life that are not available to others.

4

Bisexual Types

In the last chapter, in examining the development of sexual preference, we found that there is a considerable period of time between a person's initial sexual attraction to and behavior with members of both sexes and his or her self-labeling as bisexual. Yet the label bisexual can mean very different things to different people. For example, as noted above, many men and women have sexual feelings for both sexes without ever identifying themselves as bisexual. Similarly, someone who identifies himself or herself as bisexual and who has sexual experiences with both sexes may feel that a long-term, committed relationship is possible only with one sex. Many combinations of behavior, feelings, and self-identification are possible.

Here we look at what we believe makes up a bisexual preference: sexual feelings, sexual behaviors, and romantic feelings, each directed to some degree toward partners of both sexes. How similar or different are self-defined bisexuals from one another? Are there different "bisexualities" that emerge from the possible combinations of each dimension of sexual preference? Is there a tendency for a greater heterosexual or homosexual leaning among bisexuals? Finally, do gender differences mean that bisexuality is different for men and women?

Kinsey and Beyond

Much contemporary research on sexual preference stems from the work of Alfred Kinsey and the scale he used to develop a heterosexual-homosexual continuum.[1] Kinsey and his associates regarded sexual pref-

The Kinsey Scale

	Sexual Feelings	Sexual Activities	Romantic Feelings
Exclusively heterosexual	0	0	0
Mainly heterosexual with a small degree of homosexuality	1	1	1
Mainly heterosexual with a significant degree of homosexuality	2	2	2
Equally heterosexual and homosexual	3	3	3
Mainly homosexual with a significant degree of heterosexuality	4	4	4
Mainly homosexual with a small degree of heterosexuality	5	5	5
Exclusively homosexual	6	6	6

erence as a more complex phenomenon than had previous researchers. Rather than classifying people as either "heterosexual" or "homosexual," they saw a person's sexuality as often a mix of homosexuality and heterosexuality. This led them to construct their now famous ratings scale, a seven-point continuum from exclusive heterosexuality (0) to exclusive homosexuality (6). People were rated on this scale in terms of their sexual behavior and subjective reactions (feelings and fantasies about desired partners) so that a score reflected the balance of homosexuality and heterosexuality. Since Kinsey's research, however, much attention has been paid to both the theoretical and the methodological shortcomings of his scale.[2] We use the scale simply for descriptive purposes and as a starting point from which to provide a fuller picture of the phenomenon of bisexuality. Some of the difficulties in using the Kinsey scale must first be addressed.

Who Defines Sexual Preference?

While Kinsey and his colleagues provided the idea of a "continuum" of sexual preference, later researchers tended to treat people as discrete categories. Instead of the two traditional homosexual and heterosexual categories, now there were seven, enabling investigators to refer to people, for example, as "Kinsey 3s."[3] The complexity of a person's sexuality, which Kinsey had begun to unveil, actually was obscured by researchers as complex sexual reality became reduced to numerical categories. This led to the further presumption that all those who occupied the same category on the Kinsey scale were of the same sexual type.

The rating system has also raised the question of what numerical criteria researchers should use in classifying people as bisexuals. Given that all the points on the scale between 1 and 5 could be considered to encompass bisexuality, should the scale categories be seen as the same size and the distance between them equally large? For example, is a change from a 0 to a 1 equal to the degree of change from a 5 to a 4? And does use of the Kinsey scale lead to criticisms that individuals have been characterized as "homosexual" when they are *really* "bisexual?"[4]

The emergence of the concept of sexual identity allows for a different way of looking at sexual preference. People themselves, unlike sex researchers, clearly *do* use unqualified terms such as "homosexual" and "heterosexual" in referring to their own sexuality. These terms are meaningful to them—ordinarily more so than a numerical Kinsey rating.[5] We no longer see the need to insist on fixed numbers on the Kinsey scale to connote "bisexuality" or to make cumbersome assumptions about equal-size sectors or intervals on the Kinsey scale. Instead, our subjects' interpretations of what *they* think they are or how much change *they* will accept on the Kinsey scale before questioning their categorical identities, becomes the important issue.

Thus we have adopted an approach that investigates how people who do identify themselves as "bisexual" rate themselves on the Kinsey scale.[6] This helps us describe the range of "bisexualities" in our study. To be included in our study, individuals had to define themselves as "bisexual." However, we did not include anyone who did not also have significant sexual feelings for, and/or behaviors with, partners of both sexes because we wanted to understand how our subjects integrated their bisexuality into the rest of their lives.

Multidimensionality—The Elements of Sexual Preference

Kinsey used two elements, behavior and subjective response, to classify people sexually, but he combined them in *one* scale. Researches since Kinsey have recognized that sexual preference is too elusive a phenomenon to be captured in this way. Thus, Bell and Weinberg's study of homosexuals assumed that sexual behavior and sexual feelings may not always be the same for an individual and employed two Kinsey scales— one to measure each dimension.[7] Subsequent research included interpersonal affection—the love and affection involved in sexual preference.[8]

We employ three dimensions in our Kinsey scale ratings. the first is *sexual feelings,* the extent to which a person is sexually attracted to each sex. This would also be exemplified in fantasies, daydreams, unfulfilled desires, and anything else indicative of sexual feelings. Second, there is *sexual behavior*—the extent to which persons actually engage in sexual activity with people of each sex. Finally, we include what we call *romantic feelings,* the extent to which a person falls in love with people of each sex.

Though other dimensions may be relevant, these three seemed the most important to us.[9]

Putting It Together: Constructing Sexual Preference

A multidimensional view of sexual preference raises two important questions: what is the congruence between ratings on these dimensions, and how do they relate to the sexual identity a person chooses?

Researchers disagree about which dimensions of sexual preference are most important. One study found that sexual activity was much more important than affections or erotic fantasies as the dimension underlying the identity label chosen.[10] Another study found that self-identification was the best predictor of a group of seven dimensions of sexuality though such labeling was by no means always congruent with each of the individual scores on the seven dimensions.[11] Other studies show "no relationship between a particular pattern of sexual behavior and the taking on of a particular sexual identity."[12] Therefore, it becomes an empirical matter to investigate the combinations of feelings, attractions, and behavior that underlie sexual identities. In our case, this allows us to show what *types* of bisexualities exist, and whether there is overlap with heterosexual and homosexual identities.

In addition to its previously stated shortcomings, the Kinsey format can also be limiting in its inability to handle other important dimensions of sexual preference. Notably, by itself, it may fail to capture how bisexuality may take different forms. For example, there is *simultaneous* bisexuality (having separate relations with at least one man and one woman during the same period of time), and *serial* or *sequential* bisexuality (having sex with just men or just women over a period of time, and just the other sex over another period of time). This shows the danger of relying on relatively simple scales to capture the complexity of people's sociosexual relations and suggests why we wanted to study the features of our subjects' actual relationships.

Despite the disadvantages of the Kinsey scale, we decided to use the same format, but with *three* such scales, combined with our subjects' sexual identity, to measure sexual preference. We made this choice because our fieldwork suggested that our subjects did make such comparative judgments about different aspects of their sexual preference (sexual feelings, sexual behaviors, and romantic feelings) and that these distinctions were meaningful to them. This decision did not restrict us in taking into account some of the criticisms raised about the Kinsey scale, because in our interviews and fieldwork we did ask respondents the *basis* for their choice of, experience with, and combinations over time of male and female partners, so that the meaning of their self-placement on the scales could be clarified. This allowed us to understand some of the complexity of sexual preference, in that factors other than a simple choice of one sex or the other were found, and the situations in which these factors appeared were established.

Our use of these scales falls in the tradition of empirically assessing how the relevant dimensions combine and how this produces the variety of sexual profiles that people have who use the same label to describe their sexuality.[13] Our later comparisons with self-identified heterosexuals and homosexuals were conducted in the same vein—that is, we look at the issue of the congruence of dimensions among persons who self-identify differently. We also examine to what extent the notion is true that "bisexuals" are somehow "in them middle" and "heterosexuals" and "homosexuals" are at the end-poles in a variety of ways.

Developing Sexual Preference

As the previous chapter shows, sexual preference can be seen as developing through stages. One important way of looking at these stages is by pinpointing the age at which various aspects of sexual preference appear (Table 4.1). This provides a context for examining a person's location on the Kinsey scales in his or her adult life. We began by asking our bisexual respondents how old they were at the time of their first sexual attraction to, and sexual activity with, each sex. We also asked the age at which they first labeled themselves bisexual. Among both the men and the women, the average age of first sexual attraction to the same and opposite sex was a number of years earlier than the associated sexual behavior. Likewise, our subjects' first sexual activity with the same and opposite sex occurred long before they identified themselves as bisexual. Thus the general pattern, regardless of a person's sex, was attraction, behavior, then definition.

However, there were significant differences between the men and the women in relation to the heterosexual versus homosexual dimensions. On the average, the women's first heterosexual attraction and experience (at 11.6 and 14.7 years) occurred before their first homosexual attraction and experience (at 16.9 and 21.4 years). The men also generally experienced heterosexual attractions first and at a similar age to the women (11.7 years), but they completed their first homosexual attraction and experience (13.5 and 16.3 years) *before* they had their first heterosexual experience (17.3 years). Also, on the average, the men had their first homosexual attraction and experience *earlier* than the women (13.5 and 16.3 vs. 16.9 and 21.4 years). The men, on the other hand, had their first heterosexual experience later than the women (17.3 vs. 14.7 years). On the average, the men and women first labeled themselves as "bisexual" at about the same age (27.2 vs. 26.8 years). The span of time between the age of first homosexual experience and first use of the label bisexual, was shorter for the women than the men.

Overall, the data suggest to us that, for most of the bisexuals, bisexuality appears to be an addition to an already developed heterosexual interest. For some of the men, even though the development of their heterosexual attractions preceded their homosexual attractions, their homosexual activities preceded their heterosexual activities. As we will discuss later, we believe that the reason is simply that during their youth they had fewer opportunities for heterosexual sex than for homosexual sex.

Bisexuals and the Kinsey Scales

All of those in our study defined themselves as "bisexual," but just because they all labeled themselves as such does not allow us to assume that they would be identical in their sexual profiles. What does it mean when people label themselves "bisexual"? We begin to answer this question by using our three Kinsey scales.

The common-sense assumption is that a self-identified bisexual would score in the middle of the scales—a 3 (equally heterosexual and homosexual) on each dimension, but this is not usually the case. We now look at how our respondents scored themselves—first the men and then the women.

Bisexual Men

Sexual Feelings

None of the men scored their sexual feelings as exclusively heterosexual (0) or exclusively homosexual (6). Only 10 percent were in the categories of 1 and 5 (Table 4.2). Thus practically all the men indicated being between 2 and 4 on this dimension. The most common score was a 2. Some 45 percent identified themselves at this point on the Kinsey scale. Many of the men thus reported significant sexual feelings for both sexes, but a larger proportion reported being somewhat more heterosexual in their sexual feelings.

Sexual Behavior

Six percent of the men scored themselves as 0's, exclusively heterosexual, in their sexual behaviors, almost a half as 1's or 2's, 15 percent as 3's, and a third as 4's or 5's. None scored themselves as 6's, exclusively homosexual. The pattern that is indicated is a wide dispersion when it comes to sexual behavior—even more than for sexual feelings. The data indicate the presence of a strong homosexual group (those who scored 4 and 5) as well as a substantial heterosexual group (those who scored 1 and 2) in their sexual behaviors.

Romantic Feelings

The pattern for romantic feelings among the men shows a more apparent heterosexual leaning than for sexual feelings. Some 40 percent of the men scored 0 or 1 in their romantic feelings compared with 4 percent in their sexual feelings. In all, 38 percent scored themselves as 2 or 3, and about 25 percent as 4's or 5's in their romantic feelings. No subjects ranked themselves as exclusively homosexual on this dimension.

Bisexual Women

Sexual Feelings

Like the men, about 90 percent of the women fell between the categories of 2 and 4 on the dimension of sexual feelings, with the remaining 10 percent scoring themselves as 1 or 5 (Table 4.3). Unlike the men, however, almost twice as many women as men scored themselves as 3's— i.e., having an equal balance of heterosexual and homosexual feelings. (More men scored themselves as 2's.) Thus, overall, the women scored somewhat less heterosexual than the men in their sexual feelings. The male-female difference, however, fails to reach statistical significance (Table 4.4).

Sexual Behavior

With regard to sexual behavior, 14 percent of the women scored themselves 0—exclusively heterosexual (more than twice the number of men). Over 40 percent scored themselves as 1's or 2's (a similar percentage to that of the men). Another 17 percent scored themselves 4's or 5's (compared with a third of the men), and 2 percent as 6's (compared with none of the men). Thus, about 20 percent of the women and a third of the men are on the homosexual side of the Kinsey scale (4–6). The greater weighing of homosexual behavior for the men, however, is not significantly different from the weighting of homosexual behavior for the women.

Romantic Feelings

No woman scored her romantic feelings as exclusively heterosexual (compared with 15 percent of the men). About 20 percent of the women scored themselves as 1's or 2's (compared with more than a third of the men). An even larger percentage, almost 45 percent, scored themselves as 3's (compared with a quarter of the men). Over a third of the women scored themselves as 4's to 6's (compared with about a quarter of the men). Thus, the women are significantly more homosexual in their romantic feelings than are the men.

The mean scores for the men averaged a 2.8 on sexual feelings, a 2.6 on sexual behaviors, and a 2.3 on romantic feelings. Thus, they show the greatest heterosexual leaning on romantic feelings. For the women the mean scores on the different dimensions are 3.1 for sexual feelings, a 2.3 for sexual behaviors, and a 3.2 for romantic feelings. Thus they show the greatest heterosexual leaning on sexual behavior. In terms of statistical significance, though, the women are only significantly different from the men in the more homosexual direction of their romantic feelings.

Bisexual Types

While these aggregate scores suggest some general leanings among bi-sexuals, they do not fully capture the varieties of bisexuality. Another critical question is how the three dimensions of sexual preference blend together to form a more complete sexual profile among people who define themselves as bisexual. Do the three dimensions cluster to-gether? If someone is strongly heterosexual in sexual behaviors, does the same hold true for sexual feelings and romantic feelings? Are there common profiles among our respondents suggesting a typology of bi-sexuals?

To examine these types of questions, we calculated composite three-digit Kinsey scale scores for each person in our study using the three dimensions of (a) sexual feelings; (b) sexual behaviors; and (c) romantic feelings. The sequence 120, for example, is someone who scores as somewhat homosexual in sexual feelings, more homosexual in sexual activity, and completely heterosexual in romantic feelings.

What these scores reveal are a variety of bisexualities (Table 4.5). We focus on *groups* of respondents who have similar patterns on the three Kinsey scale scores. For instance, one group of our self-identified bisex-ual men clustered as follows: 1–2, 5–6, 0. This means they scored them-selves as somewhat homosexual in their sexual feelings (1–2), predomi-nantly homosexual in their sexual behaviors (5–6), and completely heterosexual in their romantic feelings (0).

The Pure Type and Mid Type

The "pure" bisexual type would be someone who is a perfect 3 on all three dimensions of the Kinsey scale. Only 2 percent of the self-defined bisexual men fit this pattern. However, 17 percent of the women scored themselves as 3's across the board.

The "mid type" of bisexual refers mainly to those who scored them-selves as a 3 on at least one dimension and anywhere from a 2 to a 4 on the other dimensions. A substantial number of those in our study fell here—about a third of both the men and the women.

The Heterosexual-Leaning Type

Some of our respondents scored themselves as more heterosexual than homosexual on all three dimensions, falling mainly between 0 and 2 on each dimension. More men than women fit this pattern—about 45 per-cent of the men compared with about 20 percent of the women. This cluster tends to represent men who scored themselves as 1–2 in sexual feelings, engaged in less homosexual than heterosexual activity, and had little in the way of homosexual romantic attachments (scored 0–1). On the other hand, none of the women scored a 0 (exclusively heterosexual) in romantic feelings.

The Homosexual-Leaning Type

Another group is more homosexual than heterosexual in sexual feelings, sexual behaviors, and romantic feelings—mainly being between a 4 and a 6 on each dimension. We find only 15 percent of the self-identified bisexual men on the homosexual side of the scales, and about the same percentage of women. This type is therefore almost as common as the heterosexual-leaning type among the women. Men, in contrast, were three times more likely to be heterosexual-leaning types than homosexual-leaning types.

The Varied Type

Some of our self-identified bisexuals were substantially more heterosexual on one dimension while much more homosexual on another one. If they didn't fit into any of the above types, and they were at least *three* Kinsey scale points between any two dimensions, they were classified as the "varied type." About 10 percent of the men and the women fit into this category. The disjunction for the men was a greater degree of homosexual sexual activity than one might predict from their more heterosexual sexual feelings or romantic feelings.

In conclusion, the development of sexual preference shows distinctive patterns. It seems that generally sexual attraction preceded sexual behavior regardless of same- or opposite-sex interest. However, heterosexual development appeared to be completed before homosexual development, suggesting that for many bisexuals, homosexuality is an "add on" to an already-developed heterosexuality.

Our data showed that all bisexuals are not alike, though, even among those who have accepted a bisexual identity. Indeed, we were able to design a typology of "bisexualities" from the possible Kinsey scale combinations. We also find that gender played an important role in the patterns people reported. Both men and women thus may have a bisexual sexual preference, but they are men and women first, subject to the effects of socialization and of the opportunities that befall any man or woman in our society, which shape any type of sexual preference. Thus, compared with women, men seem less likely to experience or explore homosexual love. Although other men may meet their needs for sex, it is women to whom they turn for romance. On the other hand, women seem more likely to experience love for a same-sex partner than do men.

Few bisexuals in our study were consistently at the midpoint on all three dimensions of sexual preference, which suggests that bisexuality represents a preference rather than indifference toward the sex of the sexual partner. If may also reflect a difficulty in achieving and maintaining such a consistency. In addition, the tendency for a greater heterosexual than homosexual weighting is consistent with our interpretation of sexual development: that for many, bisexuality involves an "add on" of homosexuality to an existing heterosexuality.

We tentatively conclude that bisexuals experience a mix of feelings, attractions, and behaviors during their sexual development that they cannot satisfactorily understand by adopting the identity of "heterosexual" or "homosexual." This means that they are open to the effects of further sexual experiences that these exclusive identities would tend to deny. These experiences differ for men and women due to the ways gender is constructed in our society. These experiences can also be tremendously varied for other reasons as well, which suggest various routes to the same destination—the adoption of a "bisexual" identity. In the next chapter we examine more closely what the experience of dual attraction involves.

5

The Nature of
Dual Attraction

Bisexuality consists of a mingling of sexual feelings, behaviors, and romantic inclinations that does not easily gel with society's categories of typical sexuality. But just how different is bisexuality from other forms of sexual attraction? What is sexual attraction, and how do we learn to be attracted to one sex or the other, or both?

Ideas about gender and gender roles are crucial factors in how we think about our sexuality. As we develop sexual identities, our frame of reference generally is not sexuality *per se*. Instead of learning *directly* to eroticize one gender or the other, we learn to act as a woman or to act as a man. A woman learns that to be an adequate woman in our society her sexual feelings and behaviors should be directed toward men, because this is "what women do," while a man is socialized to direct his sexual feelings and behaviors toward women, because this is "what men do." Gender, not sexuality, is the encompassing framework through which we learn and process this information. This framework—this cognitive map—has been referred to as the "gender schema."[1] The gender schema becomes the foundation for the meaning we give to our sexuality.

Yet this learning is by no means consistent or uniform. People clearly do not automatically develop a heterosexual identity, with an exclusive preference for the opposite sex. Obviously those who develop a bisexual sexual preference are not exclusive in their sexuality. Either they have not learned, or have learned and rejected, the traditional gender schema. The latter seems more likely the case, since bisexuality often follows an initial heterosexual development.

If sexual preference is linked to the gender schema, then a dual attraction arises to the extent that the traditional gender schema gets replaced with what we call an "open" one. This involves a disconnection between gender and sexual preference and makes the direction of a person's desires (toward the opposite or same sex) *independent* of whether the person is a man or woman. Put simply, one's *own* gender becomes theoretically irrelevant to partner choice.

We are unable to say exactly *why* some people adopt open gender schemas and others do not. We can, however, get at this issue by examining the nature of dual attraction as it was described by the bisexuals in our study. What is it about each gender that bisexuals like or find pleasurable? Do bisexual men and women view each gender the same way? How do they make choices between genders if it is pleasureful to relate to both? In other words, how exactly does an open gender schema work?

The Sexual Experience

In our interviews we asked bisexuals to describe whether the experience of being sexual was different with a woman compared with a man. Just over four-fifths said there was a difference. We then asked them to explain what it was that was different. Three themes stood out: differences in actual behavior, in physique and body texture, and in the emotional tone of the experience.

Behavioral Differences

The most commonly noted differences about being sexual with men versus women had to do with the kinds of behaviors reported and the ways in which they were experienced. Both women and men, but especially women, emphasized that there was more touching, hugging, and caressing with female partners, along with an altogether gentler, slower, and softer quality to the interaction.

> With a woman the sex is more tactile. There is more touching and kissing. Just that is satisfactory enough for me.   (F)

> I love to kiss women. I love the total body touching and caressing. . . . And with women I really enjoy the cuddling and being close.   (M)

Men were seen as less gentle, more urgent and rougher, and more dominant. Men seemed to get turned on by the sheer physicalness of the experience with other men. Women, however, often resented it.

> With men, there is very little emotion involved. But I find I can go to the limits with the person, physically. Men can endure much heavier

physical activity than women can. Sex with men is harder, stronger, and longer than with women. There is more physical than emotional excitement.   (M)

Men are usually a lot rougher in sex. Most women like a lighter touch. Men kind of tackle you.   (F)

Sex doesn't seem quite so urgent with women. It's easier to be a woman and just do little things like touching fingers.   (F)

These differences, summed up as the "sensual" versus the "sexual" by one respondent, also underlie another commonly perceived difference— that men are more genitally focused, a response given mainly by women about men.

With men it's a hop, skip, and a jump from lips to nipples to pussy. I think that for men sex revolves around their cocks.   (F)

With a woman I feel sex is more of an option. I don't have the burning desire to go to orgasm with a woman. . . . I don't feel the need to have intercourse or oral stimulation to feel complete.   (F)

Many of those we interviewed also felt that a person of one's own sex was more knowledgeable about their sexual responsiveness than a person of the opposite sex. This was mentioned by men and women, but again was most frequently mentioned by women about their sexual experiences with other women.

It is like making love to yourself. All those things you know you want to have done to yourself sexually—you do that for the other woman.   (F)

In terms of physical pleasure men are much more tuned into what turns me on in a way a woman never could. I guess because she does not have the same anatomy.   (M)

Men also emphasized that with women they felt locked into a traditionally dominant role, one that they preferred to do without. With men, they had a sense of greater role freedom. They said that there was less role rigidity in sex with men and that they could be submissive or dominant as they chose.

A few men and women noted that same-sex relationships were more reciprocal and responsive. The same sex was said to make love *with* you, participate *together,* "fuck *back*" more.

Both men and women noted that sex with a man tended to be hotter, more intense, exciting, and passionate. They said that men "turn on" more, have more sexual energy, provide more sexual sparks and more passion.

Bodily Differences

As one would expect, participants in our study said that basic differences in the bodies of men and women had an impact on the meaning of sexual experience. The most frequent stated difference, consistent with the difference for behavior, had to do with touch or the general feel of the body. When referring to female partners, both men and women emphasized the smoothness and softness of the female body and how it was smaller in size. When referring to men, both men and women referred to the muscularity and hardness of the male body, as well as its greater overall size.

> Physically, I enjoy the differences between the way that men and women feel. . . . Women are softer, smaller, curvier. . . . I enjoy men because they are stronger, more muscular, and have a greater presence.   (F)

> Males are harder, bonier, and females are softer, smoother, and I feel their bodies are different from mine.   (M)

> Females are softer, less hairy, and I like the fact they are closer to my size.   (F)

When it came to women, men seemed to especially focus on how the anatomical fit was better or more natural, while women did not mention this about sex with men.

> It's physically more comfortable to have sex with a woman. My feeling is that the bodies fit better together, a better physical fit.   (M)

> With a woman, I subconsciously feel that the act is somehow more "normal." . . . Physically, vaginal sex is more satisfactory than anal sex in that it self-lubricates.   (M)

Different genitalia was mentioned by some respondents. Men and women stated equally that they were turned on by penises, loved penises, etc. While vaginas were also mentioned, there were no remarks of adoration.

Emotional Differences

When describing the emotional character of the sexual act, both genders stressed that men are more impersonal and less sensitive and women are more person-centered and caring during sex.

Women care about satisfying you sexually *and* emotionally.   (F)

For men that is OK. . . . [B]ut women need more of this [caring] than men do. . . . Men would rather fuck.   (F)

Both men and women mentioned that women were more affection-ate, personal, tender, caring, nurturing, comforting, giving, emotionally involved, loving, genuine, and mutual. But men and women thought that their same-sex partners were more open (men to men and women to women) and that it was more difficult to understand what the opposite sex was thinking. Men thought men treated sex less seriously, more casually, more as fun or recreation, as having sex first and friendship later, and as having less at stake than women do. Men also felt more protective of female partners than male partners. Women saw male part-ners as more lustful and hedonistic and viewed sexual attraction as men's main bond with women.

The Nature of Relationships

The emotional differences that accompany the sexual act were described as part of a more general difference in emotional tone projected by men and women. Two-thirds of the bisexuals we interviewed said there was a definite difference in a close relationship with a woman compared with one with a man, especially in three areas: expressiveness, power, and the strength of the interpersonal bond.

Ability to Express Emotions

The most glaring difference about relationships had to do with the ability to express emotions. Both the men and women thought it was more difficult for men to reveal their feelings. Ease in sharing emotions was seen especially by women in their relationships with other women.

Men seem to have a compulsion to invalidate their own feelings, whereas women are more open and honest about their feel-ings.   (M)

Women like to talk about their feelings much more. Men talk about impersonal things—cameras, sports; women talk about their fami-lies and their children.   (F)

With women I feel more able to talk about things that trouble me, to reveal weakness.   (M)

Power Imbalances

The second most mentioned difference in close relationships with men and women points to the inequality between males and females, the

power imbalances, and the sort of role playing that often occurs. Feelings
of social inequality were reported by both men and women, but most
frequently by women about men. Women believed that their relationships with female partners were more egalitarian:

> Women were more cooperative. . . . Men want to control the situa-
> tion, which creates a lot of problems. . . . I need to be in control of
> my own life. . . . [A] lot of men do not let you do that.   (F)

Some men also saw men as playing dominance games:

> With a man, there is always the question of role playing—who dom-
> inates. With a woman, there is less of this, although there is the
> question of achieving total equality of the genders.   (M)

Other men saw it the opposite way:

> Men are more like equals. With women it is more like traditional
> role playing. You can touch and caress men more easily; when you
> do that to women they say, "All you want is sex." They see it as a
> lead to a fuck. Men are more playful and accepting. Women are a
> pain in the ass. They are loaded with psychological game playing. It
> is a power struggle; they want to have control.   (M)

Common Bonds

Many bisexuals emphasized that intimacy was enhanced and was easier
to achieve in a relationship with a member of the same sex. They pointed
out that people of the same sex had more in common. Consequently,
their bond was stronger or more complete.

> It's fuller. There are more levels because of the *sisterhood*. Our condi-
> tioning, our struggles, our comparison of our lives to each other are
> more similar. . . . I do not know as much instinctively about a man
> as I do a woman.   (F)

> I have more in common with men. With a man I have fantasies of
> being buddies, hiking and running together.   (M)

Making Choices

It seems apparent from these responses that the bisexuals we interviewed
experience men and women in much the same way that other people do.
We suggest, then, that the perceived differences in sexual contact and
intimate relationships reflect institutionalized gender scripts in our society rather than something peculiar to bisexuality as a sexual preference.

The differences between bisexuals and others lie more in the fact that they *eroticize qualities of both sexes* to varying degrees.

So why do they choose one sex or the other? To answer this we asked a hypothetical question: "Many people have asked us, when bisexuals go out, for example, to a party, how do they decide whether they want to relate sexually to an equally attractive man or woman?" We found that our bisexual respondents usually did not decide ahead of time whether they wanted a male or female partner. Instead, they tended to remain open to whatever situation happened to arise.

> It is just chance and circumstances. What they are like intellectually, culturally, etc. I do not think there is any logic involved. It's just an impulse or instinct. Rarely do I decide I'm going to pick up somebody. (F)

> I usually do not decide, I just watch things as they develop. (M)

Spontaneity, however, was sometimes mitigated by the sexual preference of other people—whether they were thought to be straight, gay, or bi.

> If I'm with my straight friends I probably won't pick up a man. If I am with my gay friends I probably won't pick up a woman. And my bi friends let me pick up anyone I want to. (M)

Others referred to the "mood" they were in, which also affected the gender they might choose.

> I don't think it has much to do with pitting a good-looking man against a good-looking woman. I think it has more to do with my own feelings of whether I'm attracted to men or women more at a particular point. (M)

A number of people indicated that in certain instances their mood could be quite specific, so much so that it determined who they would approach. Their mood was not necessarily for someone of a particular gender.

> It depends on where I am personally, whether I'm feeling active or passive. I look for active people if I'm feeling passive and vice versa. (M)

Certain "moods," however, did seem to affect the desire for one gender rather than the other.

> It would depend on my feelings about myself at the time. If my need is for emotional involvement, I would want to relate to a woman. If I

had a desire for more immediate sexual excitement, I would want a man. (M)

If I'm feeling good about myself, if I'm feeling attractive, I probably would go toward the woman. (M)

If I wanted intellectual sharing, it would probably be the man. . . . If I wanted a more relaxed interaction, also, it would be easier, less tense with a man. (M)

Note that the last quotations, all from men, reflect gender norms in our society. Since men are expected to initiate heterosexual sex, being in a confident "mood" makes the choice of a woman more likely. Also, interaction is more relaxed with another man. Emotional involvement is seen to be more likely with a woman.

Some respondents did indicate that they made a decision ahead of time to look for either a man or a woman. A common theme was evident in these replies: if they had been relating predominantly to one sex for a period of time, their preference would be more for someone of the other sex.

It depends on what I need to explore more of at a particular time of my life. If I had a long relationship with a woman then I would try exploring a man and vice versa. (F)

I am looking for a primary relationship with a woman. I have had numerous male primary relationships. . . . Since I feel a need for a greater balance in my sexual relationships, I feel "ready" for one with a woman now. (M)

A common set of responses to the question of choice dealt with how attractive or desirable another person made the respondent feel. For many bisexuals, the sex of the partner took a back seat to this factor.

I try to relate with anyone I feel compatible with or who shows an interest in me, regardless of gender. (M)

Finding a woman who is cooperative and attracted to me is difficult. There are not many women around who are willing to relate to me personally. (F)

Finally, many bisexuals referred to the shared values or personal attributes of people who they found appealing. Again, the particular sex of the partner did not seem to be important, making their sexual preference a fluid one.

Compatibility, shared SES, religion, philosophy, educational level, and who finds me attractive as well. (M)

I think it is the person regardless of their sex. It is kind of an energy attraction, a personality attraction. How comfortable I feel with that person.   (F)

It depends on the personality of the people involved. Warmth, responsiveness, a considerate feeling, intelligence. Physical looks are an immediate attraction, after that, those other qualities become much more important.   (M)

As with anyone, mood, situation, and personality factors entered into the choice of partners. But because bisexuals are unconstrained by gender boundaries, they have a larger field of eligible partners from which to choose. With an open gender schema, a need—feeling vulnerable or needing support, for example—can be met by accepting a strong supportive man *or* woman as a partner.

In a group that often sets itself against societal norms, we were surprised to discover that bisexual respondents organized their sexual preferences along the lines of traditional gender stereotypes. As with heterosexuals and homosexuals, gender is the building material from which they put together their sexuality. Unlike these groups, however, the edifice built is not restricted to one gender.

How does this come about? As we mentioned at the beginning of this chapter, the traditional gender schema characterizes differences between the sexes that are mutually exclusive. Yet such differences are considered complementary. Men are expected to be rational and women emotional, but these two traits are assumed to complement each other in a paired relationship. But even while persons learn to be attracted to gender characteristics of the opposite sex, they learn what is attractive about their *own* sex. Not only are these traits complementary, therefore "good"; they also are held up as traits to which they should aspire to in achieving "gender" for themselves.[2]

As traditional gender socialization plays itself out, then, considerable efforts are made to prevent the eroticization of one's own gender. There is the threat of penalties for those who would step out of line. Learning bisexuality can therefore be seen as completing one's heterosexual development but failing to unlearn the desirable aspects of one's own gender, which thus makes the eroticization of the same sex possible. This is what we mean when we suggest that bisexuals' initial sexual socialization makes them more responsive to subsequent experiences: an open gender schema allows them to react sexually to a wider range of stimuli.

Developing an open gender schema seems to involve a "discovery" of attractions to the same sex. Some bisexuals experience this as the discovery of "something that has always been there," now recognized and celebrated. Others see it as a complement to their growing personalities, a new potential they have discovered—for example, in the case of exclusively heterosexual women who adopt feminism and become open to lesbianism. The "add on" nature of bisexuality may be experienced differently by various people. We suggest that for men, more widespread involvement in early same-sex behaviors makes the "add on" experience more likely to be one of continuity. For women, more early emphasis on emotional exploration makes the experiences more likely to be one of personal growth.

However, developing an open gender schema is a necessary, but *not* a sufficient, condition to explain the development of a bisexual identity. We know that most persons engaging in bisexual *behavior* do not take on the identity of "bisexual." Such persons somehow temporarily set the gender schema aside to avoid the implications of their bisexual behavior. Thus we find the "heterosexual" married man who frequents public toilets for homosexual sex, the male hustler who has a girlfriend, and the "heterosexual" married woman who engages in homosexual sex at swing parties.[3] It was coming to refer to themselves as "bisexual" that made our group of bisexually behaving persons unique. Adopting such an identity made sense of their feelings, attractions, and behaviors and helped them form a community with others who identified the same way. In the next chapter we examine just how intricate the relationship between gender, sexual preference, and sexual identity can be.

6

Transsexual Bisexuals

Was I "heterosexual" or "homosexual" or what? I had to find out *who* I was before I could identify *what* I was.

A Transsexual Respondent

In the last chapter, we argued that people become bisexual by developing an "open gender schema" that allows for the eroticization of both sexes. Further, for some persons, this form of sexual attraction can be organized around an identity—"bisexual."

But the *links* between the gender schema, sexual preference, and sexual identity are still not completely clear. In helping us to understand these connections, we were fortunate to have among the people we interviewed a number of transsexual bisexuals, people for whom gender is a major life concern. They had dealt with gender in ways that most people had not. Because of their transsexuality they could more clearly articulate the relationships between gender, sexual preference, and sexual identity, illuminating how the open gender schema operates.

Transsexuals in the Study

Transsexuals claim that their "true sex" is opposite to their genetic sex. Of the eleven transsexuals in the study, eight were male-to-female *post*operative (i.e., had had their male genitalia transformed into "vaginas" through surgery), two were male-to-female *pre*operative (i.e., had been undergoing hormonal [estrogen] treatments, but had not yet had genital surgery), and one was a female-to-male *pre*operative transsexual (i.e., had had a double mastectomy, but no genital surgery or hormonal therapy).[1] They ranged in age from twenty-eight to fifty-four years old (most were in their late thirties and early forties). Their educational background was fairly advanced: seven of the eleven were college graduates and three had postgraduate degrees as well. Incomes ranged from $15,000 per year to over $70,000. Three of the respondents worked in some form of commercial sex. All of the others were in conventional white-collar occupations.[2]

The pronouns we use to refer to the transsexuals, as well as the heterosexual/homosexual direction of their sexual preference, are based on the gender they claim, not their genetic sex.

Transsexuality

Persons who define themselves as "transsexual" experience what is called "gender dysphoria," the feeling that they are trapped in the wrong body, that their physical appearance (masculine or feminine) differs from their psychological feelings of what gender they are (man or woman). Because of this discordance, they may strongly desire hormonal treatments and genital surgery to make their bodies fit their gender identity. From their point of view, transsexuality is a *transitional* status that should disappear once nature's mistake has been rectified through surgery.

Paradoxically, despite their desire for a "sex change" operation, many transsexuals and their doctors support a very conservative notion of gender.[3] This, in effect, says that there are only two genders and each gender is marked by certain stable signs. These signs are not only physical ones like genitalia but also include notions of gender roles—how men and women are expected to appear, feel, and behave. Among these notions are beliefs that impinge upon sexual preference.

Sexual Preference

Sexual preference is clearly related to gender because terms like "heterosexual" and "homosexual" are gender-based categories, referring to a choice a person makes about the sex of his or her sex partners. The traditional gender schema directs the choice exclusively toward the opposite sex. Men are expected to choose women as sex partners and vice versa. Such social expectations affect the transsexual. A man who claims to be a woman can support this claim by expressing an exclusive interest in men as sexual partners. This interest in the "correct sex" is what the medical profession has usually required (among other things) as proof of transsexual status and a prerequisite for genital surgery. The "true" or classic transsexual has been defined as a person who moves to exclusive heterosexuality as a sexual preference.[4] All of the above is based on the central assumption of the traditional gender schema—*that gender identity and sexual preference fit together in some necessary way*. But this assumption is breached by the transsexual bisexual. We can understand the relationship between gender and sexual preference by examining this breach.

Gender, Sexual Preference, and Self-Labeling

Knowing a person's gender does not allow us to say much about that person's sexual preference. Just from knowing that someone claims to be a man, we cannot automatically know whether he is a heterosexual man, a homosexual man, or a bisexual man. Conversely, knowing that a person claims to be heterosexual, homosexual, or bisexual obviously gives no clue as to that person's gender. Gender and sexual preference should be considered *independent,* separate entities that are linked together in ways that must be empirically examined.[5]

For most persons who identify themselves as "heterosexual," "homosexual," or "bisexual," their self-definition is based on a stable sense of gender identity. But for those who define themselves as "transsexual," gender identity is problematic. They must sort out their gender identity before they can apply any label about sexual preference. It was in dealing with these problems that our transsexuals ended up adopting what we call an "open gender schema." That is, they came to understand that gender and sexual preference are *not necessarily* tied to each other. Some did try, at some point, to choose sexual partners only from the "opposite sex," in order to confirm their claimed gender (i.e., as male to female transsexuals they chose genetic males). For reasons we will detail shortly, they found the question, "If I am a woman, what sort of woman am I— one who likes men, or women, or both?"—one that did not admit of an easy answer. They became acutely aware of the open gender schema— the independence of gender and sexual preference. They could, for different reasons, relate sexually to men, women, or both. Even after settling the question of gender and deciding what sexual preference label fit them best, they were often denied social support for these decisions. Not only did society at large refuse to see them as "real" men or women, but other groups did not accept them as "real" heterosexuals or homosexuals because of their violation of gender beliefs. Thus gender *and* sexual preference are often provisional for the transsexual in ways not experienced by other groups in our society.

The Social Basis of Sexual Preference

If transsexuals' life situations produced in them an open gender schema, then it was possible for them to be heterosexual, homosexual, or bisexual in sexual preference. It is at this point that we can clearly see the influence of environmental factors that affected their choices. Some of these are detailed below.

Preference in the Heterosexual Direction

As previously mentioned, the medical definition of transsexualism emphasizes heterosexuality as a crucial signifier of the diagnosis. Male-to-female transsexuals who seek surgery learn that their appropriate sexual partners should be *males*. Even if they are not sexually active, a definite "heterosexual" interest is required as confirmation of their gender identity as a woman. But this can present a number of problems.

First, despite their gender identity as women, most of these genetic males have been socialized toward eroticizing the opposite sex (i.e., females). Indeed, some of our transsexual respondents had been married and had children. Whatever their success as heterosexual men, they certainly had had little direction and experience in playing the role of heterosexual women.

If they are able to eroticize males, the question of whether they are "really homosexual" rather than "really transsexual" inevitably arises, especially in their preoperative state when they have the same genitalia as their partners. Thus, they may lie about their same-sex behaviors or report revulsion toward experimenting with males because of their "homophobia." This has led some sexologists to claim that the desire for sex reassignment may be due to the person's unwillingness to accept that he is "really homosexual."[6] Even though this construction of "transsexualism" is less often accepted than in the past, some of our respondents avoided actual sexual experiences with men during their preoperative stage—yet were still able to claim that their feelings and affections were *heterosexually directed* (i.e., toward males) at that time.

There are also problems in pursuing "heterosexuality" for the male-to-female transsexuals. For *pre*operative transsexuals, sexual interaction with men is difficult. They may feel the need to stop sexual intimacy from escalating, lest their male genitals be discovered, which could lead to a violent response from their partner. Limiting intimacy can mean a constricted time span in their relationships with males, who often leave them for "real" females. The frustration, humiliation, and sometimes fear they experience with men in their preoperative state may also spill over to their postoperative lives in the form of a continuing distrust and fear of men. Even *post*operatively, and despite having functioning "vaginas," some of our transsexuals were not always able to get male partners. Some are still "masculine" in appearance and make unattractive women. Others did not have surgery until late in life and are "older women" in a society that puts a premium on young women. Even those who do get male partners can be disappointed. They usually have a romantic notion of male-female relationships, yet become involved with men who are chauvinistic or abusive or who are only interested in having sex with a "transsexual." Two of our respondents became involved in prostitution to pay for their surgery. Commercial sex is not conducive to the kind of male-female relationships they are seeking. Obtaining money this way was not uncommon. But many transsexuals never obtain enough funds for surgery and thus remain in a preoperative state. (Such persons may be labeled "she-males.")

Finally, with the growing awareness of AIDS, the transsexuals see male sex partners of whatever sexual preference as being a higher risk.

Thus a further factor directing them away from a "heterosexual preference" (having men as partners) and toward becoming "lesbians" (having only women as partners) has appeared in recent years.

Despite these problems, some transsexuals regard heterosexuality as an important source of identity confirmation and pursue men to support their claim that they are *really* women. Others, however, learn from these experiences that their gender identity and their sexual preference are separate issues. Thus, for some transsexuals, their gender identity as a woman does not necessarily mandate a heterosexual preference and the problems this seems to involve.

Preference in the Homosexual Direction

A number of factors push the transsexual in a homosexual direction, i.e., a *male-to-female transsexual* seeking genetic females as partners.

First, during their upbringing they have been socialized as males to be sexually attracted to females. Eroticizing women then is not something that they have to learn *de novo*.

Then, preoperatively, they may have tried relationships with male homosexuals. These appear to be uniformly dissatisfying, as male homosexuals treat them as "men," or specifically men who cross-dress. Transsexuals know that they are not homosexual "drag queens" or heterosexual "transvestites" but *women,* so that relationships with homosexual men are not identity-confirming. Thus, their homosexuality centers around their self-identification as "lesbian" rather than "gay." This is made easier because homophobia is directed less at female than at male homosexuals.

Relationships with heterosexual men are also problematic, in that heterosexual males may have posed a threat to them when they were children, or as adolescents they may have related better to girls than boys. Such fear of men, as we have seen, and a feeling of greater comfort with women, may continue to be reinforced in adulthood. Thus it may be difficult to enter into or maintain a sexual relationship with a male; women, on the other hand, may be easier to get along with and may have those gender attributes that transsexuals admire and with which transsexuals are already comfortable. Some women may therefore relate to male-to-female transsexuals in ways that can confirm the transsexual's identity as a woman.

Finally, it should always be remembered that for all transsexuals, merging into the heterosexual world involves a complete identity transformation and not just a change of genitalia. Birth certificates must be changed, driver's licenses altered, and one's prior history hidden or reworked. "Personal identity" resolution can be completed in ways that "social identity" may never be—so the transsexual always remains "discreditable," a further inducement *not* to adopt a heterosexual role. On the other hand, it is not always easy to be accepted as a lesbian. Male-to-female transsexuals report difficulty with some lesbians who refuse to accept them as women, defining them as really men or, at worst, female impersonators. Links with the lesbian subculture, however, can lead them to consider "lesbianism" as a valid sexual preference.

The Transsexual Bisexual

Since transsexuals by necessity develop an open gender schema in which gender and sexual preference are separated, they may not choose a completely heterosexual or homosexual sexual preference, but rather a bisexual one. This would appear to be unremarkable given that the transsexual has not only experienced *being* both genders, and possibly the sexual socialization associated with each, but also the problems associated with trying to be *exclusively* one or the other. In addition, the transsexual can be exposed to a variety of sexualities from heterosexual marriage to prostitution to a range of different types of partners. Thus bisexuality may appear to them to be a viable option. Given too that attracting a partner, any partner, presents a problem for the transsexual, adopting a bisexual identity widens the range of potential eligibles. Many of the transsexuals we studied reported great loneliness and longing for affection and love. Ideally, male-to-female transsexuals may want a heterosexual male partner who will validate that they are really women, but they may come to meet those needs by accepting a partner from some sexually marginal group, e.g., transvestites or other transsexuals, or bisexual partners of either sex.

Similarities and Differences Between Transsexual and Nontranssexual Bisexuals

There were both general similarities and differences between the transsexual bisexuals and the nontranssexual bisexuals in our study. The most important *similarity* is that both were aware of the independence of gender and sexual preference that marks an open gender schema. This is true for the nontranssexual bisexual who has learned to choose sexual partners regardless of their sex. And it is true for the transsexual bisexuals, who may feel they should be heterosexual, but who have difficulties achieving this and may come to entertain other options. Another similarity is that in adopting an open gender schema, both have sexual preferences that are very responsive to environmental changes such as discovering new sexual opportunities, finding or losing a partner, or entering a stable relationship.

The major *difference* is that the nontranssexual bisexuals do not worry about their gender identity. They are secure in the knowledge that they are either men or women. For them, bisexuality is a matter of their sexual preference, and they are more likely to adopt the term "bisexual" as a self-label that indicates a particular sexual lifestyle. The transsexual bisexual, however, is more likely to view bisexuality in terms linked to gender identity. For some, bisexuality is a transitional stage, and they may utilize sex not so much for pleasure (as do other bisexuals), but as a way to validate their gender. They may be slower to adopt a self-label that indicates a sexual preference ("bisexual") than one involving gender ("man," "woman," or "transsexual"). For those who do refer to themselves as "bisexual," this may come only after they feel secure in their gender identity.

In conclusion, then, transsexual bisexuals highlight some of our conclusions about the nature of dual attraction. First, they demonstrate the importance of the gender schema as the basis for sexuality. Their ideas about gender, moreover, are quite traditional. Second, they have developed an open gender schema as a result of lifelong conflicts with society over their gender identities. Third, this open gender schema allows for a disconnection of gender and sexual preference. Finally, specific environmental factors—sexual opportunities, negative social reactions, support from other sexual nonconformists, etc.—have pushed them in a heterosexual or homosexual direction, making bisexuality the most viable alternative for them.

7

Sexual Activities

We have seen how bisexuals give meaning to their dual attractions. In this chapter, we examine how this relates to their sexual behaviors. We look at a variety of issues. First, what is the mix of opposite and same-sex patterns in bisexuals' sexual lives. This will tell us just how loose the link between sex and gender is for them. Also, by looking at the types of partners they have (from significant to impersonal relationships), we can learn something about the opportunities they have for sex. Moving from types of partners to the incidence and frequency of particular sexual acts can illuminate what the boundaries of eroticism are for them. Just how adventurous is our group? We then look at the sexual problems that can be attendant on a lifestyle that often seems to center on pursuing the purely erotic. To what extent are there problems with sexual dysfunction, communication between partners, sexual satisfaction, and the like. A crucial consideration for our later concerns is bisexuals' experience with sexually transmitted diseases.

Finally, throughout we will look for differences between bisexual men and women. Does becoming bisexual erase the impact of gender differences on sexual patterns, or does being a man or a woman transcend sexual preference? Do fewer women than men seek sex for its own sake, and are women more likely to predicate their sexual behavior on the establishment of a relationship? Both the nonsexist ideology of the Bisexual Center and our own observations suggested that a gender bias like this would be more limited among the people in our study than in the larger population. Many of the women did seem to enjoy sex for its own sake and had been actively involved in swinging, threesomes, and other forms of impersonal sex.

By the same token, much same-sex activity among men may not be as impersonal as is often thought.[1] In San Francisco, locales for impersonal sex such as movie theaters and glory-hole clubs, which were frequented by some of our male subjects, involve a great deal of courtesy, reciprocity, and camaraderie that is far from impersonal. There is anonymity and silence, as in a bathhouse, but these do not rule out other forms of intimacy. Perhaps men and women are intimate in different ways, just as they may be sexual in different ways.

Gender differences become even more fascinating given what we know about the differences between and among heterosexual and homosexual men and women. For example, although gay men have less sex in their primary relationships than heterosexual couples do, they have the highest frequencies of "extramarital" sex.[2] Gay men it seems are able to sustain primary relationships because they do not fuse sex and love. Lesbians, on the other hand, are less likely to disconnect sex and love, which seems to explain why lesbian relationships have the highest rate of breakups. Lesbian courtships are often brief, they "marry," and if "extramarital" sex occurs it is more likely to be an affair than a one-night stand.[3] If male-female differences of this kind appear among homosexuals, do they apply to the dual attraction that typifies bisexuals?

We now turn to the data that describes what the people in our survey were doing sexually in 1983. We describe this with respect to the sex of each person we interviewed, the sex of his or her partner, and the character of their sexual behavior.

Sexual Partners

One of the intriguing questions about bisexuality is: How do bisexuals divide their sexuality between the two sexes; what kind of heterosexual/homosexual mix occurs? Given our previous suggestion that homosexuality is often an "add-on" to heterosexuality, are partners and behaviors primarily heterosexual? Given the vagaries of temperament, attraction, and opportunities, is it likely that there are more same-sex partners or a more equal balance? Finally, since bisexuality is different for men and women, how does gender interact with sexual preference to produce different patterns of sexual partnerships and activities?

Number of Sexual Partners

The bisexual men in our 1983 study reported having had an average (median category) of fifteen to twenty-four female sexual partners over their lifetime, and a median of one to two over the last twelve months (Table 7.1).

The median number of male partners for the men was twenty-five to forty-nine in their lifetime and five to nine in the last twelve months.

This is greater than their number of female partners for both time spans (although these differences do not reach statistical significance).

Compared to the men, women were more likely to have had more opposite-sex and fewer same-sex partners during their lifetime. In 1983, the bisexual women reported a median of twenty-five to forty-nine opposite-sex partners over their lifetime, (and three to four over the last twelve months).

For same-sex partners, the median was in the category five to nine in their lifetime and one to two in the last twelve months.

Casual and Anonymous Partners

About half of the men said that in the last twelve months they engaged in *casual sex* once a month or more (sex with someone the respondent knew but was not significantly involved with). On the other hand, close to a quarter said not at all. Only one-fifth of the men engaging in casual sex said these partners were more likely to be or were always women rather than men. About a quarter of the men reported engaging in *anonymous sex*—sex with strangers—once a month or more in the last twelve months. Some 45 percent said not at all. Only a few of the men who were engaging in anonymous sex said that these partners were more likely to be, or were always, women rather than men.

A third of the women said that in the previous twelve months they engaged in *casual sex* once a month or more; on the other hand, a little less than a quarter said not at all. More than half of the women said that their casual sex partners were usually or always men. Twelve percent of the women said they engaged in *anonymous sex* once a month or more. Nearly 60 percent of these women said that their anonymous partners were more likely to be men or were always men. Thus for both men and women, casual and anonymous partners were more often men than women.

Multiple Partners

Bisexual sexual activity is not necessarily one-on-one, but can involve "threesomes," "swinging," or group sex. The combinations of partners is often quite complex. Roughly half of the men and women who responded to our questionnaire had engaged in "threesomes" in the last twelve months. For these, the median frequency was two to three times a year. The two most common gender combinations were primarily male biased: either three men or two men and a woman. About 40 percent of the respondents had engaged in swinging or group sex in the last twelve months. Of those who did, the median was two to three times a year. Sex with multiple partners was thus a current or recent practice for many of our respondents, but not a frequent one.

The overall heterosexual/homosexual mix shows men reporting more *same-sex* partners and women more *opposite-sex* partners. In both cases, the greater number of partners are men. We argue that these findings reflect the different access to sexual opportunities available to men and women. That is, bisexual men find it easier than bisexual women to "add on" *same-sex* partners because there were more opportunities for pursuing homosexual sex in San Francisco (e.g., bars, baths, cruising areas). Men not only have more resources to establish and sustain institutions (like gay bars), but also enjoy the safety (relative to women) of pursuing sex in public places (like parks). Within such locations, as a result of gender socialization, men are able to obtain and enjoy impersonal sex. Thus men have a higher proportion of casual and anonymous sex partners than do women, and these partners are likely to be other
men.

Bisexual women find it easier to find *opposite-sex* partners. One reason is there are far fewer lesbian bars than male gay bars in San Francisco. On the other hand, the women partake in many of the traditional heterosexual opportunities most large cities provide. In addition, the "sexual underground" in San Francisco provides other heterosexual opportunities— like swing parties and SM clubs (SM is a colloquial term that stands for sadomasochism).

Gender socialization plays a role in the reported patterns. Women are the "gate-keepers" of heterosexual sex, with the power to refuse or accept sexual propositions from men. Their socialization—which conjoins sex and intimacy more than men's socialization does—may make it more difficult for them to accept and act on the opportunities they are offered. Their casual and anonymous opportunities are more likely to involve men not only because of the lack of opportunity structures for such behaviors with women, but also because women are less likely to want such forms of sex.

Whatever the cause—gendered socialization or gendered opportunities —there are differences in male and female bisexuality, and these differences parallel the gender differences found among heterosexuals and homosexuals. The link between sex and gender among bisexuals is not as loose as it may seem.

Sexual Frequencies

Bisexuals are often considered "oversexed" because they have partners of the same and the opposite sex. But a greater number of sexual partners does not necessarily mean a greater frequency and range of sexual activities. By investigating the actual frequency of various sexual behaviors, we may get some idea of just how far our respondents extend the boundaries of their eroticism. Specifically, we can ask whether the boundary for men is different from that for women, further extending our analysis of male and female bisexuality.

We asked the bisexuals in our study a number of questions about their sexual behavior and the frequency with which they engaged in a variety of sexual acts during the preceding twelve months (see Table 7.2). The findings show that, overall, women had a relatively lower frequency of *all* homosexual than heterosexual behaviors when compared to men. This replicates previous studies,[4] as does the finding that men had higher frequencies of oral-genital sex (but not masturbation) in their same-sex relationships than did women.[5] We should note that in 1983, over a quarter of the men were receiving anal intercourse once a month or more. This indicates an acculturation to homosexuality for the bisexual men and it indicates that an awareness of the risk involved in such behavior had not become widespread at that time.

Finally, with regard to the relative distribution of heterosexual and homosexual acts *within* each gender, both men and women showed the greater weighting toward heterosexuality that we have seen in earlier chapters. For example, both men and women were more likely to engage in masturbation and oral genital sex with opposite-sex partners than with same-sex partners.

SM and Other Unconventional Activities

In our fieldwork we observed that some of the bisexuals were involved in other aspects of what we have called the San Francisco sexual underground. One activity that stood out was sadomasochism. Thus people reported going to SM parties or SM clubs. Basically, SM referred to behaviors involving dominance and submission and the appearance of psychological or physical pain.[6] There were also clubs for people interested in urine play, use of enemas, anal fisting, and feces play. We asked an omnibus question as to whether they had engaged in SM or these other unconventional sexual activities in the past twelve months (Tables 7.3 and 7.4). A third of the respondents said they had.

Sadomasochism was the most common of these activities. Just under 30 percent of the men and women had engaged in SM in the last 12 months. While both the men and women were more likely to have engaged in SM with opposite-sex partners, overall, playing dominant or submissive roles was not linked to the sex of the partner. Virtually all of the participants engaged in these behaviors less than once a month. In addition, about three-quarters of the men who had engaged in SM with a woman in the past 12 months had also engaged in it with men. Likewise, over half of the women who had engaged in it had done so with both sexes.

The incidence and frequencies of other unconventional behaviors were very low. None of the respondents had engaged in "feces play" during the twelve-month time period. All but one of the small number who engaged in urine or enema play also engaged in SM. Anal fisting was considered one of the most "advanced sexual techniques" in the homosexual male subculture in 1983. This practice had also found its way into a sector of the lesbian subculture. Through contact with these subcultures as well as contacts between bisexual men and women, it had become part of the repertoire of twelve of the bisexuals. The most common frequency for all of the above had been "less than once a month."

Thus, a considerable number of the interviewees had experienced unconventional sexual patterns, but the overall frequency was relatively low. Despite the existence of the "sexual underground" there was a reluctance to make these practices an integral part of their sexual repertoire.

Bisexual men and women, then, seem to differ in the frequency and composition of their heterosexual and homosexual acts. Although both are more heterosexual than homosexual, the men seem to have expanded their sexual boundaries more to include a greater frequency and range of same-sex behaviors.

Sexual Problems

A number of possible sexual problems would seem to confront the bisexual—for example, the problem of finding the desired mix of male and female sex partners. We asked a number of questions about sexual problems to discover, first, whether bisexuals have different problems in their heterosexual relationships than in their homosexual relationships and second, whether men and women experienced problems in their bisexual relationships that seem to be different from problems found in more conventional relationships (Table 7.5).[7]

Men's Problems

Over half of the men reported a problem in finding a suitable opposite-sex (female) partner and about three-quarters in finding a suitable same-sex (male) partner. About a quarter reported *serious* problems in finding a suitable opposite-sex partner but almost 40 percent in finding a suitable same-sex partner. Thus men found it harder to obtain a suitable same-sex than opposite-sex partner. This may seem odd given the earlier finding that men reported having more homosexual than heterosexual partners. Many of these same-sex relationships, however, were casual or anonymous in nature, whereas many of the men also wanted to find same-sex partners with whom they could have a more substantial relationship.

Over two-thirds of the men wanted greater heterosexual *and* homosexual sex. Thus, although bisexuals have an expanded pool of eligible partners (males *and* females) they still may not obtain the sexual frequency they want.

A number of other problems were reported, but none were seen as serious by a large number of men. Approximately three-fifths to two-thirds of the men reported a problem in discussing their sexual needs with their male and female sex partners, and a similar degree of problems with their partners' disclosure of their sexual needs to them. However, sharing sexual needs was not seen as a serious problem, and it was unrelated to the sex of the partner.

Almost 60 percent of the men reported failing to meet their partners' sexual needs (again regardless of whether the partner was a man or a woman). When it came to the partner failing to meet their needs, though, they said that it was more likely to be their same-sex partners who failed—around 85 percent—as compared with about 60 percent who said this was true of their opposite-sex partners. Very few said this was a serious problem.

Various questions were asked about sexual performance—problems with arousal, lack of orgasm, speed of orgasm—with respect to themselves and their partners. Roughly 30 to 40 percent reported that they themselves had experienced one or another such problem at least "a little." Roughly forty-five percent reported one or another such problem as characteristic of their partners. Again, few saw these as serious problems.

Around half of the men felt sexually inadequate with female partners and two-fifths reported this problem with male partners; very few defined this as a serious problem. In addition, around 40 percent reported a problem in maintaining love for their male and female partners. Again, few reported this as a serious problem, and overall it was unrelated to whether their partner was a man or a woman.

In sum, for the men, two *serious* sexual problems existed: difficulties in finding suitable partners (more so with male partners) and a desire for more frequent sex (with both men and women). Although other problems existed, they were not defined as serious by many men. What is of interest is the overall *lack* of a relationship between these other problems and whether the partner is male or female.

Women's Problems

Just over half of the women reported difficulty in finding a suitable opposite-sex (male) partner, but almost 90 percent had trouble finding a suitable same-sex (female) partner. A quarter reported *serious* problems in finding a suitable opposite-sex partner and 45 percent in finding a suitable same-sex partner.

About two-thirds of the women indicated an inadequate frequency of sex with opposite-sex partners and over 90 percent with same-sex partners. Over twice as many reported this to be a *serious* problem in their *homosexual* sex life compared with their heterosexual sex life. An inadequate frequency of sex with same-sex partners was more of a problem for the women than the men.

About three-quarters of the women reported difficulty in discussing their sexual needs regardless of whether the partner was a man or a woman, but only about 15 percent defined this as a serious problem. Similar results appeared when they reported their partners' disclosure of needs.

About the same percentage of the women reported not meeting their partner's sexual needs and their partner not meeting their sexual needs. Generally, this was not considered to be a serious problem, but overall about 20 percent of the women saw it as such. In general, it appeared to be unrelated to the sex of their partner.

Over 60 percent of the women reported difficulty in becoming or staying aroused with male partners; 45 percent, with female partners. For over 10 percent this was reported to be a serious problem. A similar proportion of the women reported problems with a failure to reach orgasm and this was defined as serious by about a fifth of them. Again, it was not related to whether the partner was a man or a woman. Seventy percent of the women, though, reported reaching orgasm too slowly with men as compared to 40 percent with women. This was defined as a serious problem for 21 percent and 15 percent of them, respectively.

Reaching orgasm too slowly with an opposite-sex partner was also reported more by women than men, but this was not the case with same-sex partners. Overall, about 55 percent said their male and female partners failed to reach orgasm or reached orgasm too slowly, but only a few defined this as much of a problem. On the other hand, 58 percent complained that their male partners reached orgasm too quickly compared to only 3 percent for their female partners. This was defined as a serious problem by 14 percent of the women with respect to their male partners. In this regard, the women defined this as more of a problem in their heterosexual sex life than did the men in their homosexual sex life. In both cases the partner was male.

About two-fifths of the women reported feeling sexually inadequate with both men and women, but very few defined this as much of a problem. This is similar to what we found for the men. Nearly twice as many women, however, said they had greater problems maintaining love for their male versus their female partners, with a fifth saying that this was a serious problem with their male partners, but only 5 percent with their female partners.

In sum, for the women, five *serious* sexual problems were reported, two of which they shared with the men: they had difficulty finding suitable same-sex partners and they wanted more frequent sex. They especially wanted more frequent *homosexual* sex, and this was defined as far more serious by the women than was the lack of heterosexual sex by the men.

A third and fourth problem for the women concerned sexual performance in their *heterosexual* sex: they complained that they reach orgasm too slowly and that their male partners climaxed too quickly. Fifth, women reported a serious problem in maintaining feelings of love for their male partners.

These reports of sexual problems further clarify the relationship among gender, sexual patterns, and the opportunity structure. Both men and women complain of a difficulty in finding suitable same-sex partners. For the men, it appears that the unsuitability lies not in getting a partner, but in getting one with whom some kind of intimate continuing relationship can be established. Not meeting such a person could be due to the bisexual man's lack of integration into that part of the homosexual subculture that is not solely directed toward sexuality (e.g., gay organizations, social clubs, self-help groups). Thus, their social contact with homosexual men may be somewhat limited to sex. As for their social contact with bisexual men there was the Bisexual Center, but only limited numbers of men attended the organization with the purpose of establishing partnerships.

For the women, suitability seems to revolve more around finding a woman with whom to have sex. Whereas both men and women complain of a lack of sexual frequency, women are more likely to desire more *homosexual* sex—and this deficiency is defined as much more serious by the women than lack of heterosexual frequency is for the men. Bisexual women's limited entrée to the lesbian subculture (their rejection by many lesbians on account of their bisexuality) reduces opportunities for meeting same-sex partners. Even when they do meet them, the level of sexual frequency they experience can be unsatisfactory; they may confront the relatively low level of sexuality that often characterizes lesbian relationships. Given the high sexual frequencies in their heterosexual relationships, this could create a sense of relative deprivation.

These findings deepen our understanding of the "add-on" nature of bisexuality for many of our respondents. Both men and women are dissatisfied with exclusive heterosexuality. Men are more often dissatisfied because of their desire for more frequent sex in general; women are more often dissatisfied because of the emotional quality of their heterosexual relationships. Thus, men may increase their sexual frequencies by turning to other men as sex partners, but the kind of sex they get may be impersonal. Women, on the other hand, may turn to other women for the emotional quality of relationships they aren't finding with men, but they may find that these relationships are lacking in the sexual frequencies they have come to expect.

Bisexuality, therefore, for both men and women, does not guarantee a successful sexual and emotional life. It does, however, offer enough of a promise for many persons (and notable successes for some) that it remains a viable alternative to other sexual adaptations.

Sexually Transmitted Diseases

Because of their relationships with both sexes and with many partners, bisexuals are often seen as likely carriers of sexually transmitted diseases (STDs). This belief is especially prevalent among lesbians, who fear the possibility of disease transmission from the male partners of bisexual women. In general, the belief that "bisexuals" are promiscuous and seek group sex and multiple partners sustains the ancillary belief that they are disease carriers.

Approximately 60 percent of both the male and female respondents had contracted a sexually transmitted disease at least once in their lives (Table 7.6). At the time of the 1983 interview, a small proportion were infected with such a disease—the most common being inactive herpes, from which 6 percent of the men and 14 percent of the women suffered. The other cases included gonorrhea, syphilis, active herpes, and several other less serious STDs. The lifetime median frequency of having contracted a sexually transmitted disease was once. Around a quarter of each sex reported four or more STDs in their lifetime. The most common diseases were gonorrhea (experienced by a quarter of both the men and the women), urethral infections (a quarter of the men and 16 percent of the women), and venereal warts (a quarter of the men and 11 percent of the women). The differences among the men and women were not significant. The apparent increase in STDs, though, had affected men's sexual patterns significantly more than women's.

In 1983 no one reported having AIDS. At that time, herpes was the most feared STD, and awareness of the catastrophic significance of AIDS was just beginning to dawn. When we returned to San Francisco in 1988, however, AIDS overshadowed everything else as a sex-related problem and significantly affected the sexual opportunity structure and the bisexual patterns reported in this chapter. We explore the incidence and impact of AIDS in the last portion of this book.

8

Significant Others

The way bisexual desires are realized raises interesting questions, especially because our society places great value on heterosexuality and monogamy, and the homosexual subculture embraces gender exclusiveness. Can bisexuals be truly committed to more than one partner or to partners of different sexes? How does the knowledge of same-sex relationships affect opposite-sex partners and vice versa? How are multiple relationships organized in terms of time, living arrangements, and emotional investments? The widespread belief that bisexuals are untrustworthy, promiscuous, and uncommitted suggests some of the potential problems. We now look at the social arrangements of bisexual relationships in general. Their complexity is illustrated from our field notes taken at a "rap group" on relationships at the Bisexual Center.

> My ideal situation is one with a primary partner with each of us having outside partners. I don't believe that one human being can satisfy my needs. I'm too multifaceted.   (F)

> Bill said he spent his time looking for *both* a man and a woman to start simultaneous relationships with. Finally he gave up looking for this, thinking it was impossible. Then he said the relationships he wanted started happening.

> One woman said that in her sexual network, relationships were constantly changing; new ones were starting up, old friendships were becoming sexual, old sexual partners were becoming friends instead, and so forth.

The Ideal Arrangement

We began our examination of bisexual relationships by asking the bisexuals what their *ideal* sociosexual arrangement would be. Would it be organized as a *ménage à trois,* group sex, communal marriage, or some other unconventional arrangement? Would it be a heterosexual marriage with outside same-sex partners? A core homosexual relationship with outside opposite-sex partners? Some other arrangement?

The answers (Table 8.1) showed that in 1983 the most common ideal arrangement was to have two core relationships—one heterosexual and one homosexual—that could be organized in different ways. Men and women were equally likely to construct their ideal in this way.

An ongoing relationship with both a man and a woman. The three of us would be in a threesome, unmarried, open to outside relationships.   (M)

A deep relationship with a man and a woman. . . . It would not be a live-in situation. I would not want the man and the woman to be sexual with each other. . . . I would want them to be friends, but I am not looking for them to be bisexual.   (F)

Most often, part of this ideal construction was a relationship that would be open or free of boundaries about possible sexual partners.

An open relationship with a man and a woman, all living together and all having sex with each other, with sex outside being possible.   (F)

I would like to live with a male and female who each had their own bedroom in a large house. We would all have sexual and emotional relationships with each other, as well as others outside of the household.   (M)

The second most common ideal arrangement was to have a core heterosexual relationship (which may or may not be a marriage) with additional homosexual relationships on the outside. People often used the term "primary" or "secondary" to distinguish the relationships. This arrangement option was desired more by men than women.

I'd like to have a long-term committed relationship with a woman, and I'd like to have outside relationships with men.   (M)

Ideally, I would like a marriage with some woman. And perhaps, having a man involved in some way with the two of us. But in a secondary way to the marriage.   (F)

A good wife, a family, and a younger man for an athletic companion
—it would include sexuality.   (M)

The opposite arrangement—of one core homosexual relationship
with one or more additional heterosexual relationship(s)—was not a
common ideal, but when it was, it was much more popular among the
women.

Finally, the third most common ideal arrangement—though not
mentioned very often—was some kind of communal group.

A group of two males and two females where all had equal attraction
to each other, who were all bisexual, and who all functioned as a
family.   (M)

A closed group of perhaps eight or ten men and women who would
live together—sharing work, emotions, sexuality, meals. . . . There
may be primary relationships, but everyone could relate sexually and
emotionally in whatever groupings they choose.   (M)

Regardless of the structural form of the relationship, the *sine qua non*
for the ideal situation was for the person to be nonmonogamous. The
major reason for this was that respondents preferred sexual variety, including the freedom to be with both sexes. A frequent remark was that it
was unrealistic to expect one partner or one sex to meet all of a person's
sexual needs.

I get love with one person. There are things I still want to learn about
sex and relating to people and I can't do that if I don't have the
opportunity to relate to both males and females.   (F)

Other responses did not specifically mention sexual freedom, but the
personal independence gained by not being confined to an exclusive
relationship.

I like a lot of space. When I leave my home and return, I like to know
that it is just the way I left it. And, if I choose not to go home, I do
not have to.   (M)

At the same time, however, and almost as important as freedom, the
respondents mentioned the emotional intimacy and security that a relationship can bring. They believed that significant relationships need not
be based on monogamy and that additional partners could be incorporated into their lifestyle.

I'd have the best of all worlds. I'd have the stability and deep emotional satisfaction of a primary relationship and the spontaneity and
fun of brief affairs.   (F)

The fact that my wife and I would choose to be together would make our marriage stronger—the sense of being tested with a third party and not wanting to separate makes our tie stronger. . . . There would be someone else in our lives who could be there.   (M)

Many of the respondents never achieved their ideal arrangement. Only about one-third felt they actually had, and half of these respondents said it lasted for just six months or less. Since such ideals are not easy to establish or maintain, bisexuals, like anyone else, often settle for something less.

Current Relationships

Two-thirds of our respondents reported currently being in what they defined as a "significant relationship" (Table 8.2). Of this group, about a half had one significant relationship while a third had two of them. The remainder had three or more. (Nineteen percent of all the respondents were in heterosexual marriages—the focus of the next chapter—while 61 percent of those who were unmarried had a significant relationship.) In this chapter, married and unmarried people are combined.

The Most Significant Relationship

In exploring the dynamics of involved relationships, we asked our bisexual respondents who reported only one involvement to tell us about it. We asked those who had more than one involved partner to first tell us about the partner they considered most significant.

Among those who described their only or most significant relationship, more than three-quarters said their partner was of the opposite sex (Table 8.3). With regard to sexual preference, over 40 percent of the most significant partners were said to be "bisexual;" a similar percentage, "straight" (heterosexual). The remainder were described as "lesbian" or "gay" (homosexual).

We were especially interested in how bisexuals organized and experienced their relationships: in terms of their importance, whether they lived together or apart from their partners, whether they had an open or closed sexual arrangement, their emotional feelings for their partner, how sexually satisfied they were, and how long they thought the relationship would last. We begin by describing what respondents referred to as their most (or only) significant relationship. For some, their relationships were new and uncertain:

He is the newest one. He is thirty, attractive, intelligent, and bisexual. We have been together maybe six times. I am a little nervous about the age difference, and I do not know how profound the relationship will become yet. It is too new. I would like it to go on forever, but that is probably not going to happen. I do not know whether I know him well enough to really be sure. I do not know his feelings well enough to know what he wants over the long haul.   (F)

Others noted their relationships were more well-established:

The guy I've been living with for twenty-five years. It's a very committed relationship with a lot of freedom on both sides. We're there for each other emotionally and otherwise. He's my best friend. . . . Whatever comes up, no one's going to run out, that kind of thing. Some kind of problem or trauma, accident, whatever. We try not to put restrictions on each other's behavior both sexually and otherwise. We don't demand monogamy of each other. We are supportive of each other's outside relationships.   (F)

More respondents indicated the relationship had been of limited duration:

We have been hooked up—together—for two and one half years. It is the most important relationship in my life currently. We have been through a lot of struggles, and I feel good about the understandings we have come to. I get a lot of support from her, and I have learned a lot being with her. I can be as open and honest with her as with any woman I have ever been with. Emotional support about personal growth, about dealing with feelings, about communication. Also, she is very hot sexually. She likes the sexual activities that I like to indulge in, and likes to do them as frequently as I do. I also feel equal with her intellectually.   (M)

They usually were also said to involve serious commitment:

We are very committed to each other's well-being and happiness. That means that we would not just stop seeing each other. We would do our best to work out a compromise given any conflict. We would not just walk out on each other, and both of us would be very upset if the relationship were to end.   (M)

Still, serious commitment was not always lasting:

We have been in a primary relationship for two years and involved sexually with each for two and one half years. A very important component in our relationship is the sexual aspect of it. It is just a real hot connection. Our sexual energies and interests are just very compatible. We are also very socially and intellectually compatible. We like the same kind of thing, the same kind of people. The nature of the relationship now is undergoing change because he recently moved to Monterey. Now we live three hours away from each other instead of just ten minutes. . . . We do not have any long-term commitment, goals, or plans for the relationship. It just sort of goes from week to week, or month to month.   (F)

Sometimes people mentioned that they and their partners lived very separate lives:

We live separately. We are both bi and interested in having other lovers. Occasionally we experience other people. We care about each other, but don't own each other. We live very separate lives. We have many different interests and social groups that we don't share.   (F)

The most significant relationship was also sometimes homosexual:

I have never loved a man before. And I never thought I could love a man as I have women. My feelings are unsure. I have love feelings for him, but I have not allowed myself to say those three little words, "I love you." I am not as attracted to him nor in love with him as much as he is to me.   (M)

Other Aspects of Their Most Significant Relationship

About three-quarters of those who said that they were in a significant relationship reported making a distinction between "primary" and "secondary" relationships. Nearly all who did so labeled their most significant relationship as "primary."

Where did they meet their most significant partner? Half met them at school, work, or in a recreation group, close to a third in bi/gay/lesbian support groups, just over a tenth through friends or at conventional parties, dinners, or get-togethers, and the same proportion at bars, sex-related clubs (e.g., an SM club), swinging or other sex parties or get-togethers. In sum, about three-quarters of the respondents had met their most significant partner at school, work, or in recreational or support-type groups.

We asked them to describe the emotions they felt toward their most significant partner. Many replied in terms of romantic love—to paraphrase their words, passion, a holding hands kind of love, being very much in love, feeling in love every time they set eyes on each other. There were feelings of a unique love: one that feels especially deep, the partner making life worth living, having never loved anyone as much. Some spoke of tenderness, warmth, affection, delight, playfulness, cuddliness, lightheartedness, and laughing and crying together. And, of course, there were feelings of lust and intense sexual feelings and sexual excitement. Many described friendship, companionship, connectedness, enjoying doing things together, a soulmate, sharing a lot of their lives together. Empathy was also mentioned: caring, being concerned about the other's well-being, supportiveness, and sharing their lives. There were feelings of compatibility, an intuitive understanding of each other, liking each other, feelings of permanence and commitment, and many more positive emotions.

There were less positive emotions as well: ambivalence, not sure what they felt emotionally, having trouble identifying the feelings that they had for their partner, a "too soon to tell" attitude, not being sure they wanted to be with a partner any more, or that it could "go either way." And for some: the romance is dead, it's a worn-out love, the feelings of love are fading. For others there was anger, frustration, exasperation, enough is enough, feeling too crowded, feeling trapped, feeling restricted, annoyed, feeling the relationship was too hard emotionally, and feeling bored. In short, these primary relationships sounded much like many others in our society.

The people we interviewed spent a substantial amount of time with their most significant partner. About half lived with their partner and those who did almost always shared the same bed or bedroom. Among the group that did not live together, over three-quarters saw their partner once a week or more. Usually this involved overnight visits, weekends, or a string of days and nights together, as well as evenings out, etc.

Most people said that sex was an important aspect of their relationship. About three-quarters reported engaging in sex with their partner once a week or more. Two-thirds rated the sexual relationship as positive. For example, various respondents described their sex as intense and mutually satisfying, spoke about sexual compatibility, indicated their ability to talk together openly and frankly about sex, mentioned the willingness of their partner to engage in diverse or unconventional activities, or described it as simply exciting and fun.

Among those who described their sexual relationship in less glowing terms, most stated that this was due to a lack of frequency. Some mentioned situational factors (geographic distance, conflicting work schedules, a lack of time and opportunity, living arrangements not permitting it, and the need to spend more time together). Others cited personal factors, one or both of them no longer sexually attracted, other sexual incompatibilities, or the partner being uninterested in sex generally.

For most of our respondents, the relationship appeared to be quite stable. The average length of time in the relationship was just under four years. About a fifth of the women and a third of the men, however, had experienced a temporary breakup during this period of time. Some of these temporary breakups were linked to the open nonmonogamous lifestyle the respondents wanted:

A breakdown in communication. The argument was over the amount of time we would spend together and how other relationships got in the way of us. I could not manage all the relationships I had. I felt overloaded, so I started to pull back. And, I decided I just wanted to spend some time alone.   (M)

Other breakups were akin to those in more conventional lifestyles:

> I get tired of her bullshit, or she gets tired of mine, and we part company for a short period of time. Phone calls call a truce. "I will be nice if you will be nice."   (M)

> Opposites sometimes attract and sometimes conflict. Jane is a neat and orderly person. I am very messy and disorganized. Jane is very thrifty; I spend money like it is going out of style. Jane is very romantic and likes to be very romantic with the mood, lights, and everything perfect. To her it is always making love, never having sex. I do not believe I make love; I just have sex.   (M)

Most respondents, though, saw the relationship as more or less permanent. Two-thirds expected the relationship to continue for more than ten years, even given the complications of being bisexual.

The most frequently cited potential cause of their relationship ending was relationship stagnation—to paraphrase their words, not growing together any more, growing apart in general, not meeting each other's emotional or sexual needs any more, emotional, intellectual, spiritual lethargy, just not making each other happy, interests waning and changing to other things, no excitement in the relationship, not being fun any more, being tired of each other.

The second most common potential cause of breakups was said to be finding someone else more satisfying, someone who had more to offer, someone more ideal, someone who met more of their needs, or finding a more permanent replacement.

A third potential reason was geographic: one partner moving out of the area for schooling, a career move, or for a financial reason.

The fourth was another relationship getting in the way, a competing primary partner, another relationship becoming too involved, another partner becoming more primary, falling seriously in love with someone else, becoming involved with someone the primary partner doesn't approve of, putting more time into another relationship, the primary partner not liking the idea that someone else is becoming equally primary.

Finally, another potential reason was conflict over the openness of the relationship, the partner wanting to be monogamous, the partner wanting to be nonmonogamous, jealousy over outside sex, wanting fewer sexual restrictions than the partner does, or the partner getting too possessive and wanting them to cut back on outside sex.

The temporary breakups that had already occurred were also said to be a result of the above. When asked hypothetically how they would feel if the relationship were to break up permanently, two-thirds said they would be very upset, and only five percent not at all upset.

Three-quarters, however, said that they did *not* feel that being bisexual affected their ability to sustain long-term relationships. They gave two reasons for this. First, they said that things other than sexual preference were more important for keeping a relationship going. For instance, in the words of one woman: "There's more to a relationship than sex—things like maturity, compatibility, etc." Second, to avoid rejection, they selected persons they considered sexually liberal:

> I don't relate to people who are uncomfortable with my bisexuality. Before getting involved I tell the person I'm bisexual which seems to screen out those bothered by my bisexuality.   (F)

Of those who said bisexuality did affect their ability to sustain long-term relationships, the most frequently mentioned reason was the strain of nonmonogamy. This was most often mentioned by the bisexual women.

> The monogamy issue in my present relationship may end it prematurely, because it's a lot to deal with.   (F)

> Open relationships strain the relationship. I wouldn't want an open relationship if I were not bisexual.   (F)

But some said that problems with monogamy existed for heterosexuals and homosexuals too—that a desire for sexual variety was independent of sexual preference.

> It is the desire for a variety of sexual experience. I might have that same desire if I were not bisexual. I can't really pin anything on bisexuality.   (M)

We now turn to those who also had a second significant relationship including a discussion of what were said to be its effects on the first one.

Second Significant Relationship

About half of those who reported having a "significant relationship" also reported a second one (Table 8.4). This relationship tended to be quite different in character from the most important one. Of the women who had a second significant relationship, about 20 percent said their partner defined herself (or himself) as "straight," over 70 percent as "bisexual," and 6 percent as "lesbian." For the men who had a second significant relationship, 14 percent said their partner defined himself (or herself) as "straight," about 60 percent as "bisexual," and about 30 percent as "gay." Compared with the most important relationship there was a smaller percentage of *heterosexual* partners in the secondary relationship (42 percent versus 19 percent). Also there was a larger percentage of *bisexual* partners (43 percent versus 66 percent). Some of these arrangements were described as follows:

> Karen and I were close friends for three or four years before we became lovers—she was living in our communal household. We are very much in love with each other and have a very high degree of commitment and intend to stay together for a long time. It is also a stable relationship, although not as stable as with Eric—because she's gone through a lot of emotional trauma and she's not as stable.   (F)

> We are best friends. We are like sisters emotionally. We see each other just about every day. We both expressed we love each other as friends. We are involved sexually. I initiate things. She enjoys sexual experiences. She likes being sexual with me. She said she was in love with me, but not "in love with me." The love is not the same as with her husband. . . . I enjoy spending time with her. I enjoy seeing her every day. We have been working at the same job. I enjoy her intellectually and how she views the world, her sense of humor. I enjoy her sexually. She is one of the best human beings I've ever met.   (F)

> It is casual. He is very easy to be with. There is not a lot of doing other things together, social events, etc. Usually what we do is get together and talk for hours into the night and then fuck. He preaches being mellow, teaches yoga, that kind of thing. In three years, we have probably gone out to do something only two or three times.   (M)

Other Aspects of Their Second Most Significant Relationship

Half of the respondents thought of their second relationship in terms of the categories "primary" and "secondary." Of those who thought of it in these terms, three-quarters labeled it as "secondary." Some of the reasons given for this classification were, to paraphrase: they already have a primary relationship, a person can only have one primary relationship, they love someone else more, the person isn't a regular but rather an occasional sexual partner; they don't live with this person, their priorities lie with somebody else, the person comes behind other concerns, they don't share sex together, a person can't have a nonsexual primary relationship, the partner is not very close geographically, they don't share economic responsibilities with this person, they don't expect the relationship to last very long, not as long as the other relationship.

When respondents were asked to describe the second relationship, they often mentioned things that were similar to their most significant relationship. One major difference, though, was that the emotional attachment was less intense. For example, fewer described the relationships in terms of strong feelings of love.

> Warmth, tenderness, desire, respect. I do not drool at the thought of her, but we do have good sex when we are together. We do not have the depth of emotional involvement, though, that Jay [her primary partner] and I have. It is partial intimacy, I guess. I do not share everything with her like I do with Jay. We have serious discussions at times, but we do not share economics or living space, those kinds of things. She is more secondary. We enjoy walks together and being together. All of the emotions that I feel for Jay I feel with this woman, but they are more faded, less intense.   (F)

> I do not feel romance but do feel companionship and empathy. I see it more as fondness and having fun.   (M)

While work, school, or a recreational group were the most common places to meet their most significant partner, more met the second partner through friends or at a conventional party, dinner, or get-together (about a third). Another third met their second significant partner in bisexual, gay, or lesbian rap-type groups, a quarter at school, work, or in a recreational group, and a tenth at a swing party or gay bath.

The second relationship was described as less involved and more casual than the most important one. It was less likely to involve living together. Partners got together less often and spent less time together. In the second relationship, the partners were more likely to get together for a specific rather than a more global reason—e.g., for dinner, a movie, to have sex. Sexual activity was also less frequent: e.g., about a third had sex with this partner one to three times a week or more compared with just over three-quarters who had the same frequency with their most significant partner. They did not, however, evaluate sex any less positively with this partner. But while there were some complaints about the frequency of sex with regard to the first relationship, this was voiced more often with regard to the second relationship.

> It [sex] isn't enough, but it is . . . getting better and I'm inclined to be patient at this point. When it does happen, our emotions are serene, harmonious . . . joyful, hot.   (M)

And problems from the first relationship sometimes fed into the second one:

> I feel very emotionally satisfied and sometimes very sexually excited and satisfied. Because of problems she is having, it is not as exciting sexually as it has been in the past. She has had lots of problems with the guy she is living with, with her job, and with finances.   (F)

The second relationship also seemed less stable and enduring. Compared to the first relationship, it was more likely to be new and seldom extended beyond 10 years. One third said it was less than 6 months old. In addition, about twice as many of those we interviewed reported temporary breakups with this partner compared with their most significant relationship.

Reasons for this separation were different from those given importance for the breakup of the first significant relationship. Stagnation of the relationship was half as likely to be mentioned, and conflict over openness was not mentioned at all. On the other hand, rivalry with the other partner was more likely to be reported as the cause of the breakup. To paraphrase our respondents' reasons: the primary relationship getting in the way of their relationship, feeling in competition with the primary partner, its becoming problematic for the partner's other lover, a fear that the primary partner would be hurt by it all, the relationship becoming too difficult to manage, their or their partner's other partner requesting a stop to the relationship, their failure to support the relationship, one of the primary partners not letting them see each other, their being too jealous.

Even given nearly double the breakups, about two-thirds of those we interviewed who had a second significant relationship thought that the relationship would last more than ten years. Overall, this is approximately the same proportion that made such a prediction for the most significant relationship. Reasons projected for a permanent breakup were similar to those given for their actual short-term breakups, which we have discussed above. Half said they would be very upset if there was a permanent breakup (two-thirds for the first relationship), a third moderately upset (a fifth for the first), about a tenth a little upset (similar to the first relationship), and a few not at all upset. Thus, the respondents saw the relationship as very meaningful, if not quite as meaningful as the most significant relationship.

We also asked whether this second significant relationship affected their most significant relationship. Over half said "yes," but a majority reported a positive effect. The most common such effect was that having an additional relationship put them in a better frame of mind—made them less angry, resentful, frustrated, etc., with regard to their bisexuality. In other cases, a second relationship eased the expectations those involved had about their first relationship:

It takes pressure off of our [the primary] relationship, off of Susan— needing to be everything, of her needing to meet all of my sexual needs. (M)

It was said to possibly decrease the first partner's fear of the respondent becoming involved with someone else:

> I think the effect is that my married partner can be secure in the durability of the relationship even when I am emotionally and socially and sexually involved with others. I think it gives my marriage relationship a lot of strength by removing the fear of the relationship breaking up if I should become attracted to someone else. We go beyond the concept that you can love only one person.   (F)

Sometimes it led the primary partner to be more circumspect and appreciative:

> My having another relationship keeps him on his toes. He is concerned that I will leave, but I choose to be with him. He is learning how to be more generous in order to keep me around for a long time.   (F)

Negative effects were, for some, that the second relationship created problems of time allocation and feelings of insecurity for the first partner:

> Time problems are built into a bisexual lifestyle!   (M)

> It led to jealousy. She very clearly wants to deal with this on her own. It is something she has to deal with, and we both know that.   (M)

Finally, there was the possible effect of sexual satiation:

> I am probably having less orgasms with her and him versus if I was with only one of them. I only have so much orgasm energy per week.   (M)

Despite the fact that the *most* significant relationship for the bisexual was more likely to be a heterosexual one, and the *second* significant relationship was more likely to be a homosexual one, both relationships were "significant." Even though they differed in their sexual, social, and psychological character, such arrangements allowed them to express their bisexuality in the context of meaningful relationships. At the same time, however, regardless of the sexual preference of the partners, such arrangements could create problems with the most significant partner.

Other Significant Relationships

A tenth of the respondents reported having more than two significant partners (Table 8.5). They get together with these other partners and had sex with them an average of one to three times a month. At the same time, nearly three-quarters said that sex was not the main reason for getting together, but rather a strong sense of friendship, affection, or mutual caring. These relationships were about as likely to be with a man as with a woman.

Over half said that these additional relationships affected their most significant relationship, and about half as many said that these relationships affected their second significant relationship. All but one person said this effect was a positive one. Sometimes a network of effects among the different relationships was described:

> Everybody knows everybody. When I have a bad time with Doug [her second significant partner], I come home to Gary [her most significant partner], angry and depressed and frustrated and horny. He likes the latter and none of the former. My relationship with Doug has been an exploration ground for SM. I've taken what I learned and brought it to my relationship with Gary. Gary has learned all the best of SM and never had a bad SM experience. My relationship with Doug has been one that Lynn [her third significant partner] feels most competitive with these days. The difficult times I have with Doug become grist for discussion with Lynn. We talk about what works and what doesn't work. My attention to Doug is also something that Lynn is jealous of; in other words she is jealous of the fact that I am fascinated with Doug. She used to be jealous also of Gary, but it is so apparent that Gary is at the center of my life that she gave up jealousy with a lot of heartache. She wanted hers and my relationship to be the most important in my life.   (F)

And again, there was the point that the significant partners get accustomed to partner sharing:

> The effect has been that I started out my relationship by informing him [her most significant partner] that I was in love with this other male partner. Therefore he has always had to share me with other people. Marion, the second person I talked about [her second significant partner] has had to learn to share me with other people, too.   (F)

Other Sexual Partners

Many of the bisexuals who were in a significant relationship were also currently engaged in casual sex (sex with partners they knew but were not significantly involved with) (Table 8.6). The median frequency was less than once a month for both the men and the women. The casual partners were most often same-sex for the men and opposite-sex for the women—i.e., in both cases the casual sexual partners were usually men.

Where did they most often meet their casual sex partners? Approximately half mentioned conventional settings (e.g., conventional get-togethers, school, work) and half less conventional settings (bars, sex-related clubs, the Bisexual Center).

Over 40 percent said that their casual sex affected their most significant relationship, but only one said that it affected their second significant relationship. About two-thirds reported this effect to be a positive one and one-third a negative one—along the lines described in the above sections. An additional effect, one that was *not* reported for the effects of an additional "significant relationship," was the "fantasy material" casual sex provided:

> My partner is for the most part pleased that I am enjoying myself. I think since we generally talk about it, usually not in great detail, it gives him some interesting fantasy material. If I've had an interesting sexual encounter that has made me feel good, then I feel better about him.   (F)

> One effect has been that it has been a turn-on for her sometimes, hearing about it or imagining. On a couple of occasions when I've been away from home, I think I've returned to her happier and more relaxed because I had sex with someone I was visiting.   (M)

As we will discuss in our jealousy chapter, the positive or negative nature of the effects may be related to the sex of the casual partner:

> Engaging with women has stimulated my primary relationship [which is with a male], and being involved with men has depleted it emotionally. We enjoy the same fantasies, and talking about the sexual experiences I have with women. We do this thing where we put ourselves hypothetically in a situation with other women we have related to sexually in the past. And then we will say the things we would say when relating sexually to them, we say this to each other, and couple it with mutual masturbation, and the use of pornography too.   (F)

Or the effects of casual sex may be negative overall in terms of its impact:

> Jealousy. He was upset that I was having any kind of sexual relationship with anyone else. The last guy I related to sexually he does not know about. I never told him. When I met him, I was seeing two other people in the sexual vein. He was not pleased, but I didn't care because at the time, I was afraid I could not handle being emotionally and sexually tied to one person again. He would pout. He would mope around the room. He would comment, "Who are you going to fuck next?"   (F)

About half of the respondents who were in significant relationships also engaged in anonymous sex (sex with partners they did not know). For nearly all the men, these anonymous relationships were homosexual ones. Among the women, they were equally with men or women or more with men. Most often these partners were found in bars, sex-related clubs (e.g., SM), swing and group-sex parties, parks, an X-rated movie house, on the street, or through an ad in an underground newspaper or sex magazine. About a third felt that their anonymous sex affected their most significant relationship and a fifth their second significant relationship. The effect discussed most was their significant partner's fear about contracting sexually transmitted diseases.

> It brings out the fear that I will bring home some disease and I have brought disease home in the past, and have given them to him. But why should he complain? He is not giving me sexually what I want at home; so I have to go out and meet other sexual partners.   (F)

> It was anonymous sex that gave me the VD and that is why we put the rules down.   (F)

Another effect was that a sense of sordidness in anonymous sexual encounters led the respondents to appreciate their significant relationships more:

> It affects me . . . in terms of how I feel about myself . . . my sexuality . . . [and my] relationship. When I have anonymous sex, I often feel very compulsive and at times it feels dirty. This can make me appreciate the intimate and rich feelings I have with my partner and thus bring us closer together. It sometimes makes me aware of how needy I am. And my neediness is better fulfilled within a primary relationship. But I also sometimes feel that anonymous sex can be very liberating when I'm feeling stifled by my partner.   (M)

In conclusion, the bisexuals we interviewed were more heterosexual than homosexual in the direction of their sexual preference. This bias is actualized in the way they constructed their most significant relationships. The overwhelming majority of significant partners were of the opposite sex, with about a fifth of the respondents being married.

An advantage bisexuals have in San Francisco is the existence of the sexual underground: for example, the availability of support groups, rap groups, and workshops centering around unconventional sexual patterns. These create opportunities for meeting others with similar interests. It was the most common place for our respondents to meet their second and third significant partners and casual-sex partners. Anonymous sex partners were most likely to be met in other parts of the sexual underground—at bars and cruising areas such as parks, gay baths, sex-related clubs and parties. At the same time, the most significant partner (usually of the opposite sex) was most likely to be met in the conventional settings of school, work, and recreational groups. This is not surprising, as the most significant partner was likely to be heterosexual, and often a spouse. It was this partner who provided the link with the wider world of conventional sexuality.

Once one has a primary partner, a second partner usually takes a secondary position. Problems in relationships came either from a primary partner who saw the secondary partner taking up too much of the respondent's time and consequently felt insecure, or from a secondary partner who felt short-changed or in competition with the primary partner. It was not so much bisexuality per se, then, that created problems, but rather the nature of a common bisexual social arrangement—that of sustaining a nonmonogamous relationship.

9

Marriage

Under what circumstances can marriage and bisexuality coexist? How can the bisexual partner sustain a marriage without compromising on his/her sexual preference? Can disclosing bisexuality have positive consequences for a marriage or only negative ones? These are important questions, as many of the bisexuals we interviewed were married. Even among those who were divorced, a substantial number expressed a desire to be married again if they could find a suitable partner.

In 1983, around half of the bisexuals we interviewed were, or had been, married. Of these, 14 percent had been married more than once. Approximately a fifth were married at the time they were interviewed (Table 9.1). The lengths of the current marriages were about evenly distributed between less than four years, five to nine years, and ten or more years.

The marriages described to us varied in their basic structure. Some 60 percent of those we interviewed lived with their spouses. Most of the men indicated their wife was heterosexual; half the women indicated their husband was bisexual and about a third that he was heterosexual. Around three-fourths of the respondents said their marriage was non-monogamous and sexually open, and nearly all considered their spouses to be their primary partners. Just over three-quarters rated their marriage as "moderately" to "very" happy.

Disclosure to Spouse

A third of the married bisexuals we interviewed said that they had told their spouses about their bisexuality before getting married (Table 9.2).

The remaining two-thirds did not disclose their bisexuality until after they were married, although three-quarters of this group said that before their marriage they had not yet considered themselves to be bisexual. Usually disclosure was voluntary, but one respondent told his wife he was bisexual after being arrested for homosexual behavior in an adult bookstore. Another felt forced to disclose his bisexuality to his wife because he had contracted a sexually transmitted disease.

Most of the time the method of disclosure was said to be very matter-of-fact. Bisexual spouses usually presented their marital partners with a *fait accompli,* saying that their sexual preference was so important that they could not, or would not, change.

> I told her that I had a gay component that I could not change and that in order to stay married I would have to have the freedom to grow.   (M)

> I told him I was going out to make friends with other women and be sexual. I wanted to explore my bisexual side. I wanted more freedom to choose and have it supported by him.   (F)

In a sense, the partner was given no room to negotiate the question of his or her spouse's sexuality. The partner either had to accept it or end the relationship. A presumption of "acceptance" was supported by the ideology of the San Francisco sexual underground: that all sexual preferences were equally valid and that every individual had the right to sexual freedom. In one case, the disclosure sounded like a condition for doing "business," and an element of power and control was evident.

> I told her I just intended continuing what I am and would not try to cause any embarrassing situations. . . .   (M)

Some said that the reaction they received from their spouse was supportive. The partner seemed to accept the fact of their bisexuality and—*from the respondent's point of view*—their relationship got better.

> He said he wants me to be happy. If this means my having a relationship with a woman, then it is okay. He is very encouraging about me exploring my sexuality.   (F)

> We have a better marriage than we ever had. The one major factor is that I no longer have this big dark secret to hide.   (M)

In these cases, it was the greater openness that appeared to have helped the relationship.

It is unclear whether early disclosure helps or hurts a marriage. A 1983 study found no great differences in the effect of disclosures before or after marriage,[1] but other research shows that bisexual men and women who evaluate their marriages positively are more likely to have disclosed their bisexuality before or soon after marriage rather than later.[2]

Among the bisexuals in our study, the impact of disclosure was mixed. Spouses, we were told, worried about a number of things: whether nonmonogamy would work, the possibility of losing their spouse to another person, what their relationship would be like in the future, and whether they might contract a sexually transmitted disease. A few experienced feelings such as jealousy, betrayal (that they were not told before), and uneasiness about homosexuality.

> Her reaction was mixed—she feels we can develop an open relationship although she is jealous of my outside relationships and does not feel comfortable about me having sex with men.   (M)

No one, however, reported a radical change in the marriage as a result of the disclosure. The most common response of the spouse was either to ignore the issue altogether or to seek out a support group.

> She says to me, "I don't want to know when you go out." She's not one hundred percent comfortable with it.   (M)

Outside support has been shown to be important for bisexual men and their wives, even though the wives may be more reluctant than their husbands to seek out support groups.[3]

Marital Dissolution

Among those who had been married in the past but were *not* married at the time of the interview, about a third said that their bisexual identity and lifestyle contributed directly to the breakup of the marriage (Table 9.3).[4] One of the main reasons was the spouse's nonacceptance of homosexual behavior. The radical sexual ideology of the San Francisco underground apparently had not been accepted by all of the respondents' spouses.

> She knew the man I was having a relationship with and she did not like it. . . . She threatened not to let me see the kids any more because I was too much of a "pervert."   (M)

In a considerable number of cases, respondents said that their spouses struggled more with the issue of monogamy than the issue of homosexuality.

> I was not able to agree to his terms of a traditional marriage because of my interest in other men and women. He wanted me to be a wife

and mother and see my work and other relationships as less important than my relationship with him.   (F)

Some talked about a breakdown in communication:

> Whether I expressed it or not, she was aware that I had leanings and
> attractions, sexually, in both directions, and she would not allow
> discussion of these issues. There was a good part of me going unex-
> pressed, and this led me to build up hostility towards her. It was the
> fact that she would not talk about sex with me. When she did not
> want to discuss something, it was a closed book.   (M)

Other marriages dissolved because the bisexual partner fell in love with
someone of the same sex:

> I was searching. I was unhappy sexually, not knowing exactly what I
> wanted. I was in therapy and met a gay guy that I liked . . . one
> thing led to another.   (M)

As noted, a third of the married bisexuals told their spouses about
their bisexuality before their marriage. Most of those who first thought
of themselves as bisexual after they married told their spouse soon after
that. Overall, this seemed to keep the marriage intact. It also helped if the
disclosure was given voluntarily; if it was not seen to decrease the sense
of commitment to the partner; if the way in which it was disclosed was
seen as reflecting honesty, love, and concern for the partner; if it occurred
within the context of a good relationship; and if a support system was
available. Yet a number of the people interviewed blamed bisexuality as
the cause of their failed marriage.

Also, we should be cautious in coming to overall conclusions about
the lack of negative effects on many of the marriages. Our respondents
were self-identified bisexuals, committed to this identity enough to be
members of a bisexual organization, and living in San Francisco. Whereas
this group may have been able to reconcile bisexuality and marriage,
things may be different for other types of bisexuals and those who live in
environments that are not as liberal as San Francisco. A study of wives of
bisexual men from Honolulu, Portland, and Rochester, N.Y., as well as
San Francisco, found more negative results.[5] But several other researchers found that successful marriages were possible, and that some
couples had found effective coping strategies to preserve their marriages.[6]

Desire for Marriage

The unmarried bisexuals in our sample were fairly evenly divided when
asked whether they would like to marry in the future (Table 9.4). Just

less than one-third answered yes, slightly over one-third said no, and the remaining third were undecided. Excluding those who rejected the idea outright, nearly all agreed that their bisexuality would affect the character of any marriage. Most said the marriage would have to be with another bisexual, or at least with someone understanding enough to accept bisexuality and nonmonogamy as part of the relationship.

> The wife would definitely have to be bisexual, nonmonogamous, and versatile.   (M)

> It is highly doubtful that I would ever marry a man who is not bisexual himself since I find most straight men either to be intolerably homophobic or suspicious of lesbian activity, and I would want an open relationship.   (F)

Among those who were either currently married, who were open to the idea, or who were undecided about it, the main factors affecting their desire for marriage was a desire for love, heterosexual sex, and children. Other reasons for marriage that we asked about—avoiding a homosexual or bisexual label, loneliness, family pressure, or social expectations— were not particularly salient to those we interviewed. Thus, in contrast to the "negative reasons" that forced bisexuals to marry in the past, bisexuals in 1983 were much more likely to marry for "positive reasons" rather than to escape stigma.[7]

Children

While the wish for children may affect the desire to marry, bisexuals must deal with the practical issue of how children and a parent's bisexuality mix. A third of our respondents said they were parents (Table 9.5). Of these, just over 60 percent had more than one child. Three-quarters of all the children were thirteen or older, and two-thirds of them were said to know about or suspect their parents' bisexuality. Parents' disclosure of their bisexuality reportedly had a variety of positive and/or negative effects on the relationships with their children; some mentioned no effects at all.

Some respondents said their children responded to the fact by blocking it out.

> We never talk about it. They know I come to the Bi Center. I would like them to meet more of my friends there, but I think they feel uncomfortable with my being bisexual. What it really boils down to is that we just do not talk about it.   (F)

Others told us that their children found it difficult to explain to their friends.[8]

It affected them because of their own peer pressure. They found it difficult because their friends were not understanding of my sexuality.   (F)

In the extreme, for some the bisexual issue meant a divorce and subsequent loss of the custody of their children.

My bisexuality has seriously curtailed our relationship. It has changed from living with them and being a parent to occasionally writing letters.   (M)

There were also positive effects, however. Some parents said that having to confront their own sexuality improved communication with their children or led to a closer relationship.

I think it has helped to make our communication better. I have always tried to make it clear with them about who I am in the world. And I think my honesty with them has allowed them to be more honest with me about who they are.   (F)

They tell their friends they are glad I am not a regular mom. They are mildly amused by the variety of people who parade in and out of my bedroom. I feel they are more tolerant and understanding than some of their peers. We love each other very much.   (F)

A Rap Session

As mentioned earlier, we attended a rap session in 1983 for married bisexuals at the Bisexual Center. The discussion was unstructured. We listened to people tell their stories and asked a variety of questions of our own. What follows are portions of the taped transcript from the group discussion. It illustrates the various problems and adaptations that married bisexuals and their spouses faced. (The names are pseudonyms, and the order of speech has been altered somewhat to preserve continuity.)

> *Vicki:* John and I have been married about six years. It's my second marriage, his first. I'm not aware of any same-sex feelings within myself and I knew that John had some right from the beginning of our relationship, but it looked to me like he was going to be satisfied with just fantasizing about men. Now I see that that is not so. And so we're dealing with that.
>
> *Researcher:* You knew about John's feelings when you got married. At what point in time was the behavior introduced as an issue?

Vicki: Must be months. When he started talking with certainty about bi feelings. I really guess I'm lucky to have him communicate with me about it at this stage of the game. But I'm not feeling good about it.

Bob: A lot of times what happens is it's not clear whether it's the fact that John wants to be with men or whether John wants to relate sexually with anybody else.

Vicki: Yeah, it's to anybody else!

Researcher: You don't think you'd feel differently whether it's a male or a female involved in the outside relationship?

Vicki: It seems that if it were a woman that it might be possible she'd more likely be someone he'd want to go live with. But I don't even know that that's true.

Researcher: What about with a man?

Vicki: Well, that seems possible also. It is more likely that John might not have enough time or energy left over for me.

Rob: . . . Jill and I have been living together for thirteen years and we've been married over twelve years. We've been in an open relationship for eleven and a half years. That encompasses a whole lot. I was talking to someone else and they said, "You know, you make it sound like it's easy!" One of the things Jill and I talked about was at this point in time it seems easy. But that's after thirteen years of fighting and screaming and shouting and having all sorts of disparities.

Jill: I don't remember us ever being monogamous. I don't want to be monogamous. This is my second marriage and I have two children. . . . Rob and I have raised them, but the two children are from my first marriage. I didn't want to be monogamous when I met Rob. I had a very limited idea of my own sexuality and the possibilities. Monogamy feels like a trap to me. . . . There have been periods during our marriage, however, that I have been monogamous. Periods up to a couple years. I wasn't happy during those periods. Rob was not monogamous, intensively not monogamous, and I found it very difficult. [And] he couldn't handle it [the thought of her being nonmonogamous]. Well, I couldn't handle it [his nonmonogamy] either it turned out. We each spent about five years being crazy jealous. So that means the last two years, I'm proud that we've gotten things kind of together. I have several relationships right now. Actually one, my relationship with Rob, and there's one other man I've been seeing since last June on a regular basis that feels good. There's a woman that's been a friend for a long time that I've been sexual with that I think I'll continue to be sexual with. I find that my interest in women is increasing.

Where I might have been a 1 [on the Kinsey Scale] a year ago, I'm a 2 [on the Kinsey Scale] now, and I seem to be moving. . . . It's taken a long time for me to progress along the [Kinsey] scale. What I had to do first was look at my sexuality and feel free enough and accepting enough of my own sexuality and then find that in addition to men, I was also interested in women.

Researcher: Rob, how do you feel about Jill's other relationships?

Rob: Oh, I feel wonderful. I'm real glad Jill is here. I'm one of the people who tend to use the "bi-monogamous" label. What tends to work for me is to have two relationships. To have my relationship going with Jill and have a relationship going with a man. I only have the intensity to be emotionally and sexually involved with one other person other than Jill. But there are other people I "play with" on a very casual basis that may involve sexuality. I feel fine at this point. It took a long time. When Jill and I first met, I wanted to be monogamous. It was the only role model I had for a relationship. I was not comfortable with my sexuality. I was terrified of it. I still get jealous from time to time. The only difference is I don't feel overwhelmed, whereas at first jealousy kind of like negated all the other feelings I had. At this point in time—okay, that's it, that's interesting. The only thing that feels a little uncomfortable at this point in time, I'm just ending a relationship with a man that I've been seeing for the past three and a half years. So there's—I'm suddenly in this situation. I don't want to start another relationship right now. So I'm kind of in limbo and Jill is involved in two other relationships. I'm aware of the difference in that. But I don't feel jealous.

Researcher: Jill, you have what, three primary relationships?

Jill: No, I do not. I have one primary relationship. My primary relationship is with Rob.

Researcher: Okay, so what are these others?

Jill: The man that I see, I see—oh, every couple weeks. As the sexual intensity has diminished the friendship and the caring has increased. We have more contact over the phone and Valentine cards and that type of thing. He's also living with a woman who's in another relationship with a man, which was a primary requisite for my willingness to be involved with him. I am unwilling at this point to be involved with anybody who is not already either partnered or clearly the kind of person who likes three or four multiple partners at a time and has no interest in getting involved in a primary relationship. I don't have that to offer. I also have eliminated through trial and error people who are deceiving their partners. I got really burned on that one, one time. And I'm not willing to be in that situation again. I'm real clear about those who are possible partners for me. That doesn't mean I wouldn't go, and if I met someone that turned on to me, have casual sex with them. It means that I wouldn't probably continue or entertain it as a possibility for any kind of ongoing relationship. Another issue—and I don't know if it relates to the bisexuality or not. It definitely relates to my sexuality. Motherhood. I'm past that stage in my life. My children are in college. They're grown and raised. I have freedom for the first time since I was in college. I think that is a positive factor, I guess, in expansion of my sexuality and of my bisexuality. . . . I spent the whole last year dealing with my father's illness and death. I spent the year before that dealing with Rob's mother's illness and death and Rob's response to that. The year before that dealing with my son's emerging adolescence. Nightmarish. I'm finally getting the time now to deal with me. Part of what is happening is I am expanding on my social contacts and my people. I've got more room in my life now to be closer to people.

Bob: I've been a member of the center two and a half years. I'm married to a heterosexual, single, monogamous-relating woman—by her choice. I'd have to define myself as bisexual and polygamous.

Researcher: Does your wife know about your bisexuality?

Bob: Yes.

Rob: Is she happy about that?

Bob: Day to day.

Researcher: On the off-days, is there anything special that is involved that make for bad days?

Bob: No, I guess it's just that I think she would prefer if I just related to her.

Researcher: Are there any special strains that occur from time to time? You say day to day, is there something that causes that off-day?

Bob: Yeah, I guess so. Sometimes I haven't made it home right away and that can be upsetting to her.

Researcher: At what point did your wife learn that you were bi?

Bob: Three weeks before we were getting married.

Researcher: Did you tell her?

Bob: Uh-hum.

Rob: At the times that it does not work for your wife, is that because of some behavior that you're engaging in or is it her own insecurities?

     Bob: From her own insecurities.

     Rob: So that you can engage in the same kind of behavior, come home late one day and she'd be very upset, and come home late another day and she wouldn't be upset?

     Bob: True.

    Vicki: Is there anything you do, Bob, to try and help your wife accept this situation?

     Bob: Yeah, as time progresses, she's accepted it a lot better. When I first came to the center two and a half years ago, I used to do once a week rap groups. I did them consecutively for eighteen months. Wednesday night [the Bi Center] was part of my life before I met her and it continued to be part of my life after I met her. She attended a couple of rap groups just to see what it was about. She's attended quite a few social functions here.

    Vicki: You didn't answer my question.

     Bob: I'm sorry.

    Vicki: I asked you do you do anything to help your wife feel better about the situation, about your same-sex interests.

     Bob: No. No, I don't think I do probably.

Researcher: What is the nature of your relationships now? Very casual relationships or relationships over long periods of time?

     Bob: It's been changing. Casual, casual. Yeah, I'd never want to play pots and pans. I'd never want to be married to a man. I'm real clear on a lot of things—what I don't want and what I do want. Oh, the nature of my relationships. I enjoy sex I think in all forms. I think sex is a very, very beautiful thing or can be a beautiful thing. I'm thirty-nine years old and all my sexual experiences have been very pleasurable. I've never had a negative one that I can think of so I've been very, very lucky that way. And to me, I've been involved in group sex, a lot of different styles of sex.

Researcher: These men, were they usually one-night stands?

     Bob: Some one-night stands and some have continued into relationships. Everybody always knows where I'm coming from, that I have a wife and two cats and we don't need any children and I'm not looking to leave my wife.

This rap session reveals some of the dynamics and difficulties of open marriages and the management of jealousy. It also illustrates one person's attempt to deliberately change in terms of the Kinsey scale and how the social context, and perhaps adult socialization, can play a role. Finally, it shows that sexual expression for those who are bisexual, like anyone else, vacillates depending on life circumstances. For example, Rob and Jill had children from her first marriage, and each had to deal with the death of a parent. The net effect was a reduction in their extramarital sex.

Another part of the discussion dealt with the social consequences a couple can face when they try to set up, and engage in, threesomes or swinging:

Frances: Okay, well, I'm Frances. I'm married to a man who is pure-ly heterosexual and we've been together for six and a half years. We have some specific agreements around our sexu-ality. Number one is that I will not go to bed with strangers, and that the purpose of our relationship is to, if possible, in terms of sexual liaisons, bring in a person who we can share if that can work out. We've had several kinds of sexual shar-ing with couples and with women and sometimes with a man who is a friend who is not bisexual. And that has worked out with varying degrees of success and failure. We blew away a lot of friends in the process is basically what happened most of the time. You know it was funny how that worked out. In the beginning of our relationship it was like a lot of people just sort of walked into our lives and after socializing for a little while, somehow we made it known what our ideas were on sexuality. Seems like a whole lot of them said they wanted to try it so we'd give it a try and then there'd always be somebody it wasn't working for in almost every case. But in our own relationship, it seemed like no matter what happened it strengthened our own relationship together. We found that we were able to handle it without being jealous, sharing with other people. It felt real good.

From our viewpoint, it just gets frustrating after a while when other people don't seem to be feeling the same way. Then for the last few years our relationship has gotten more traditional. We've both started working a lot harder. I got a full-time job and his job turned into more than full-time, more like time and a half. So we don't have very much social time available, much less. So it seems that friendships have dropped away and we haven't had time to cultivate new friendships, and that's partly what got me coming to the Bi Center last spring. Because I was really tired of how tradi-tional our lives were getting. It was getting much too boring for me. It was also about, well, it was really about two years ago when one woman that we had had some sharing with who sort of dropped out because it wasn't working for her, then we ran into her again at a party. I talked it over with him [my husband] and found out without ever having thought of it before that it was okay with him if I went out and had a relationship with another woman. I hadn't consid-ered that I could go out and have a relationship with a wom-an before so that opened up a whole new area.

Researcher: How did it get to that point?

Frances: Where the issue came up? Because I ran into this woman at a party that we had had a relationship with and she was interested, we had all had a sharing, in other words, she and I had had sex together. I was interested in getting together with her and so we talked it over and I found out that that would be okay. It didn't have to be with all three of us in order for me to have a relationship with her. So I started coming to the Bi Center to try to meet more people and hopefully women that I could have relationships with.

Reflections on a Marriage

We talked with Rob (whom we introduced earlier) two years later in 1985 during a brief visit to San Francisco. Part of our interest in meeting with him was that we had heard that he and Jill had divorced. Both had been visible and charismatic leaders in the bisexual community. Their relationship was regarded by the many bisexuals they knew as a model to emulate. Rob conveyed that this was a major factor leading to the breakup of the marriage. He said that they were trapped in the situation of having to set an example of an open bisexual partnership for everyone else. What they really needed, according to him, was to be monogamous for awhile. They had been married for fourteen years. Even though he speaks for Jill, as we were not able to reinterview her, this can only be seen as his account of the marriage and its breakup.

Rob: Jill and I differed in our conception of bisexual relationships. I didn't like the distinction between primary and secondary relationships. I think it's dehumanizing to see relationships in this way. Each relationship is unique and special and not "secondary" to another one. She wanted to be considered my primary partner and didn't like what I was saying publicly [in this regard] in talks I gave—that this was degrading her. But she really was my primary partner—so primary that it was insulting to compare her to a relationship with someone else. Even comparing them [Jill and his male lover] was inappropriate. Stresses in our lifestyle also contributed to the breakup and a weakness in bonding. She needed to feel special. I wasn't pointing out how special she was to me. I took the relationship for granted. To be politically correct, I was conscious about how I should present my [sociosexual] relationship to men [when talking] in public. That was stupid. It was a big mistake. By being a role model for bisexuals, my open bi relationship became a prison in some ways. Spontaneity stopped. We had to continue this model open relationship with no partner seen as more primary than the other, and this was at a time when we needed to pull back and be monogamous. We needed to pull back and be monogamous when our son moved in and we had to deal with his drug problems, and when Jill's father got cancer and moved in. A year before my mother moved in for a while with kidney failure. She subsequently died. Three months later Jill's father moved in. He was expected to die in six weeks but lived for about a year. We were *the* bi couple! This meant we couldn't be monogamous when we needed to be monogamous. This created additional stress. One advantage of nonmonogamy is that other people can be there for you—but at times you can use this to avoid marital issues. You can turn to other people. We stopped confronting conflicts because of this. We became different people and grew in different directions. She needed a more traditional relationship. More privacy. It was very stressful to always be under public scrutiny. I was off in one direction and she was off in another direction. She tried harder and harder to have a relationship with a woman, but it wasn't happening. She knows how to initiate and relate [romantically and sexually] to men more than to women. Also she was a 1 or 2 on the Kinsey scale and therefore she didn't need a relationship with a woman as much. Where was her bisexuality leading her? She was getting tired of gay and bi issues. When this was waning for her, she wanted a more exclusive and less public lifestyle. I was relating more and more to the bi and gay community and less and less to her. She wanted a more traditional lifestyle and I was going in the other direction. It was like I was turning my back on her. She got tired of pleading. We didn't know if we could deal with our marital issues any more. The conflict regarding what I said about primary relationships never was resolved. I don't want any longer to recreate the bi fantasy of one male lover and one female lover. Now it is possible for me to have a completely monogamous relationship with a male or female and feel comfortable not using my bi option. I have satisfied my curiosity. I have had intense relationships with men and women. I had to prove it to myself. I've experienced it. I can now limit my options because of the benefits of having an exclusive relationship. A lot of energy goes into making an open relationship work. I don't want to put in the time and energy. I did for fourteen years.

Researcher: Do you think you will ever have an open relationship again?
Rob: No, I won't!

We noted previously that in some cases an open marriage can continue to work. The experience of Rob, who was a leader and proselytizer for such marriages, gives us an example of one that did last for a long time (fourteen years). But even here the social circumstances they faced and the resulting stresses dramatically affected the outcome of the marriage.

10

Jealousy

Although most of the bisexuals we interviewed were in a significant relationship or looking for one, nonmonogamy was a common aspect of these relationships. It took various forms: swinging, sexual triads, group sex parties, multiple involved partners, casual sex with friends, and anonymous sex at such places as gay bath houses or through pick-ups at gay or lesbian bars. These multiple relations were not just for sexual gratification but were often critical for sustaining a sense of one's self as bisexual. Indeed, bisexuals often actively sought partners of both sexes for this reason.

The interest in open relationships was very much a part of the time and place of our research: pre-AIDS 1983 San Francisco. In part, it represented a carryover from the 1960s and the 1970s, an era when many writers began to seriously discuss nonmonogamous relationships. These ideas were an important part of life at the Bisexual Center, where we found the "sexual revolution" very much alive. Indeed many bisexuals expressed pride in their ability to be happily nonmonogamous, proclaiming this to be one of their most enviable characteristics. In this chapter we explore how successful bisexuals were at nonmonogamous relationships. First we look at the incidence of nonmonogamy among bisexuals who said they were in established relationships and at what an "open" relationship meant to them. Then we look at jealousy in such relationships and the ways bisexuals tried to manage such feelings.

Nonmonogamy in the Primary Relationship

Among those who had a primary relationship, just over 80 percent said they were not monogamous, and among these respondents, approximately 90 percent said they had an "open relationship" (Table 10.1). Open relationships meant different things to different people. One type of relationship was very open, permitting emotional as well as sexual involvement. The person was free to fall in love with others and be open to the affectional feelings of others. A second type of relationship was more narrow, permitting only *sexual* relationships with others. A third type was similar to the second in that sex with others was allowed, but there were specific ground rules that defined who were acceptable partners, how much time could be spent with them, etc. Respondents spoke of these relationships respectively as follows:

> We have the freedom to be involved sexually, sensually, and emotionally, and spiritually with other people. (M)

> Open to having other people in our lives sexually and being supportive of that. (F)

> We generally agree that each of us has the freedom to have other sexual relationships with the understanding that it won't happen often. (M)

Jealousy in the Primary Relationship

Even though they characterized their relationships as open, only about a quarter believed that their partners were free of jealousy (Table 10.2). Ten percent said that there was substantial jealousy, about 20 percent that it was "moderate," and just under a half that it existed "only a little." Primary partners were reportedly more jealous of an "outside" partner of their own sex—for example, a man whose primary partner was a woman would say she was more jealous of his relationships with other women. The logic that underlies this was that a person of the same sex as themselves could meet similar needs and thus replace them. A person of the opposite sex would not compete in this way, satisfying a different set of needs for their partner. Thus, a man about his female partner:

> Well, she feels she could be replaced by another female, whereas with men she knows I'm interested just in sex and not any kind of emotional thing. (M)

And a man whose primary relationship was with a male:

> It's okay with him for me to see women. However, with a man he would have trouble understanding that. He would find that more threatening. (M)

> This is not to say that an outside partner of the opposite sex could not be threatening to the primary partner. A few people said that their part-

ners saw a problem in knowing how to compete against such an outside partner.

> I don't know if the term "jealous" is correct here. She just felt at a complete loss having to compete with a *male!* (M)

The other side of the jealousy issue was how the bisexuals we interviewed dealt with their primary partners' other relationships. Sixty percent of the bisexuals believed that their partners were nonmonogamous. At the same time, 14 percent were unsure how extensive their partner's outside sexual activities were. Around three-quarters believed them to be "not very extensive" and 14 percent "extensive."

> I don't know all the details, but he's very attractive, and I'm sure he's getting laid fairly often. (M)

> He sees Tom. He goes to the baths occasionally. He says he's having sex with women too. I just do not know more than that. He does not tell me. (F)

> She varies a lot in her sexual interest in other people. It is really important that she always have the option to relate to other people. There have been times when she has related to a large number of people; there have been times when she was relatively monogamous and I was not. Presently, she is involved with another man and another woman. (M)

Most believed that their partners' outside sexual relationships were more casual and anonymous than "involved." This belief helped to prevent jealousy from arising. Thus, close to half said that they were not jealous at all about their partner's outside sexual activities. (Of those who said they were jealous, a third said they were "only a little jealous," 14 percent "somewhat," and only a few were "quite a bit" or "extremely" jealous.)

What led to jealousy when it occurred? As with anyone, most bisexuals remarked that jealousy was the result of insecurity and the threat of loss. Jealousy was more likely to arise if the secondary partner was perceived to be younger or more attractive, or if their primary partner seemed to give them less attention than the primary partner gave the outside partner. However, jealousy was more easily managed if they were both nonmonogamous, if both had success outside the relationship (or "kept busy in other ways"), and if the outside relationship was not flaunted. Also, it was important that the commitment to the primary partner be constantly assured.

Given the legacy of the "sexual revolution," whose locus was in California in the late 1960s, other remarks had to do with the belief that, in fact, one *did not have the right* to be jealous and thereby intrude on their partner's personal space. Ideally, one should feel secure enough in the relationship so this does not happen. As a result, a majority reported working to control jealousy: wanting to deal with jealousy in a reasonable manner and trying to work through these feelings with their partners, e.g., going to workshops that deal with jealousy.

Ground Rules

To help deal with the potentially divisive issue of jealousy, bisexuals in open relationships established ground rules. About three-quarters of the bisexuals said that they had one or more ground rules with regard to their outside sexual relationship (Table 10.3).

Acceptability of Partners

Just over 40 percent had rules to distinguish acceptable from unacceptable partners. One of the most common rules was that the outside partner had to be in a significant relationship with someone else, therefore making them less of a threat. Others made a rule that the outside relationship be only with a same-sex partner. There were other rules—especially one of a general veto power. One woman commented about her core heterosexual partner:

> Theoretically, he has veto power; but he's never exercised it. He would only veto a person if he felt that it was a bad person, or one who would be hostile to him and try to mess up my head by being negative about him.   (F)

Another woman spoke about her lesbian primary relationship:

> She does not want anyone in the house she does not approve of. She does not want me getting involved with another woman.   (F)

Number of Outside Partners

While there were few ground rules specifically governing the number of sexual partners permitted, those who had such a ground rule mentioned a concern about sexually transmitted diseases:

> We're in the process of talking about this. I don't want him to have a lot of male lovers, nor do I want his current male lover to have a lot of male sexual partners. The reason for this is the disease issue. Disease is more likely as the number of partners he has increases. And I don't want any diseases. So if he continues to want to be sexual with me, I don't want him or his other male sexual partners to be having sex with a lot of other people as well.   (F)

Also mentioned was that the number of partners should not take all the person's energy.

Affectional Involvement

Twenty percent of those we interviewed had ground rules limiting how emotionally involved each could become with outside partners. A number of such rules were mentioned, most of which dealt with the fear of replacement:

> Well Christiana has usually maintained that she wants ours to be the *primary relationship* and she wouldn't accept me having another significant relationship. (M)

> My partner and I have decided that we choose to have the kind of relationship we have *only with each other*. This means that if we become involved enough with another person, so that there is danger of replacing our relationship, then we must choose to limit or eliminate the outside relationship. In this way, we have full security, and trust, that our relationship will come first. (F)

Time

Half of those we interviewed said that they had ground rules for the allocation of time between their significant partners and their outside partners.

> We have ground rules for how much time we spend together. We spend one night a week together, Wednesday, and we spend Saturday night and all day and night Sunday together. We have three nights officially where *we* are together, but we may be with other people socially. Usually, though, we spend six nights a week together. I find it a lot better to know that Steve and I have a particular night together, because I don't worry then about the nights he is with others. (F)

Another formulation was that the time schedule with the primary partner was negotiated on an ongoing basis:

> We need to check in. We consult each other's calendars. We let each other know what nights we will be home, what are our plans for the week, who's coming over for dinner, etc. Just kind of let each other know what is happening. (M)

> We see each other every day right now. If one of us feels the need to have a day off, we do it. Tomorrow, I'm seeing a partner I haven't

seen for six weeks. I told Bruce a week ahead of time I was going out. (F)

For some, not only was it important to have a schedule, but also to be decisive about it and stick to it. It was also important that the time spent together be "quality" time. That is, when they're together they must *really* be together; they must be sober, not watch TV, etc. Most primary partners also insisted that our respondents spend most of their available time with them and that they had their first claim on discretionary time. An open relationship clearly required partners to cope with issues of time and energy management.

Meeting Outside Partners

Two-thirds of our interviewees said that their partners liked to know firsthand the outside person(s) with whom they were relating sexually. The logic behind this rule is stated in the following:

Because an unknown person is very threatening and mysterious. The fantasy and paranoia surrounding an unknown party can cause real problems if the partner is insecure. It's just considerate to put their mind at rest by letting them meet the outside person themselves. (F)

Knowing someone minimizes jealousy. . . . If unknown, the mind can run wild. I tend to make goddesses or gods out of these people and meeting them reduces them to human beings. (F)

By contrast, some of the primary partners did not want to meet or even hear about the outside partner(s), and this became the ground rule. This could cause conflict because it breached the widely held belief at the center that primary partners should get to know the outside partners of their mates. As one woman in a primary homosexual relationship told us:

She really should meet him (the secondary partner). She doesn't want to, although I know she'd enjoy him. (F)

Other Ground Rules

Additional ground rules mentioned by our respondents had to do with honesty: for example, there was to be no lying about outside sexual activities and both partners were to be generally open with each other and not conceal things. Still other ground rules were not to cruise or pick someone up when partners went out together and not to engage in sex with others in the home they shared. Some also wanted to restrict their nonmonogamy to expressing it together—for example, in a *ménage à trois*. These rules didn't always work out, though, for our respondents or their partners.

Honoring the Ground Rules

Most said that the ground rules were the same for themselves and their primary partner. If differences were said to exist it was mainly that they did not apply to the primary partner because he or she was monogamous. Other exceptions appeared where the respondent said the rules were less binding on the primary partner because the respondent was less "uptight." On the other hand, some noted the application of a double standard—they didn't want their primary partner to have the same freedom that they had.

Despite the time and energy given to negotiating rules for open relationships, we found that even when they existed and were agreed on, they were not always followed. Rules concerning time allocation were broken most often. Those with an established time schedule cited certain problems: e.g., when something spontaneous happened or unforeseen circumstances arose or if they forgot or lost track of time. Others simply did not take the rules very seriously. In the words of two interviewees: "If I'm turned on to somebody I go ahead regardless of the rules;" "I break them whenever I choose."

Close to half of those who reported having ground rules said that the rules had changed over time. The major reason cited was that the rules had to be adapted to changing circumstances. Some reported that rules developed in the first place because their primary partner found out about their outside relationships, making negotiations necessary. Others reported a jealousy problem even when ground rules were followed—and additional rules were made to deal with this. Rules changed too when the primary partner became settled in a particular outside relationship. It became commonplace that ground rules changed as the respondent's relationship with the primary partner became closer and deeper. But when there was an increasing desire for nonmonogamy, new rules had to be worked out.

Another pattern was that relationships began with numerous ground rules, but these decreased over time as they appeared to fail or if they were too rigid or too oppressive. Ground rules also diminished when the jealousy they were made to address disappeared.

Finally, some reported that rules were never hard and fast and were constantly under negotiation. For example: "we discuss rules hypothetically before things happen and usually have to make adjustments."

Jealousy and Ground Rules in the Secondary Relationship

Since most people we interviewed implicitly ranked their relationships as primary and secondary, we asked whether there was jealousy in the respondent's relationship with a second significant partner (Table 10.4). Three-quarters of those with a second significant relationship engaged in sex outside of both relationships. All said that the second partner knew about their outside sex and over half that they believed that their secondary partner was free from jealousy. Unlike the primary partner, the sex of the outside partner was not thought to be as relevant.

Almost all the respondents thought that their secondary partner was nonmonogamous. A small proportion of them were unsure of how extensive their second partner's outside sexual activities were: two-thirds believed them to be "not very extensive" and just over a quarter "extensive." Respondents were evenly divided as to whether they believed these relationships to be "significant" or "casual" involvements.

Over half of the respondents with a secondary partner said that they were not jealous of their second partner's outside sexual activities. (More than a quarter said they were "a little jealous," less than a tenth "somewhat," and only about a tenth "quite a bit.") The ideology of the sexual underground and its supporters certainly played a key role in this regard.

Two-thirds of those we interviewed said that there were no ground rules for these second relationships (compared with a fifth for primary relationships). When they existed they referred especially to avoiding sexually transmitted diseases—telling a partner immediately if they might have been exposed to any such disease.

In conclusion, bisexuals were like many other persons in 1983—interested in exploring nonmonogamous relationships. But they had adopted some features that were unique to managing open relationships. First, formally defining partners as "primary" or "secondary" helped reduce jealousy. Primary partners had greater claim over each other, and these relationships were organized by a variety of rules that ostensibly reduced external threats. Secondary relationships were ranked as less important and were narrower in their claims, often being restricted to sexual gratification. Consequently, they were bound by fewer rules. Thus a primary partner need not feel jealous of a secondary partner and vice versa. Although this did not always work in practice, we have seen a vivid example of jealousy occurring in the *absence* of such a ranking. In the last chapter, Rob refused to define his relationship as primary or secondary which made Jill, his wife, very upset. This is not to say that bisexuals do not have more than one primary partner, but that when there is a lack of clarity about where a partner stands, jealousy is more likely.

Second, bisexuals' open relationships are unique because a combination of opposite-sex and same-sex partners can create a different situation vis à vis jealousy. Unlike the jealousy experienced in a heterosexual relationship where the additional partner is the same sex as the established partner so that a fear of being replaced occurs, bisexuals' typical outside relations were different. Most involved partners were of the opposite sex to the established partner. This usually evoked less jealousy, although it was not entirely absent.

Bisexual relationships often share with other types of nonexclusive relationships a set of "ground rules" designed to reduce jealousy. Jealousy is apparently easier to manage when partners can draw boundaries around what the other person can do. But among the bisexuals we studied, the ground rules were often formed as situations warranted, they were sometimes breached depending on the circumstances of the moment, and double standards sometimes operated. Jill (from the last chapter) broke ground rules: She did not check with Rob as to the acceptability to him of a new male partner she had found. She also breached the rule of not having sex in the house that she shared with Rob. This contributed to the breakup of their marriage.

11

Being "Out"

Up to this point we have presented a portrait of how bisexuals cope with the problem of personal identity and how they struggle with intimate relationships. But bisexual people, like others, live and work in a wider community, one that is not particularly accepting of the sexually unconventional. Since what they often consider a private matter can become a public one, it is important to look at the social consequences of living with a bisexual social identity.

We examine first the problems that the people in our study experienced because of being bisexual, specifically the conflicts they had with heterosexual and homosexual men and women. Next we look at the extent to which they disclosed their bisexuality or "came out." We also asked how it felt to have to hide their sexual preference and about the pro's and con's of disclosure versus concealment. And finally, we look at the wider community of other bisexuals and how the Bisexual Center helped its members manage social disfavor.

Social Conflicts

As a generation of homosexual political action has shown, declaring an unconventional sexual preference brings social disapproval and converts a private matter into a public identity. So it is no surprise that our bisexuals received disproval for the same-sex aspect of their sexual preference. What is interesting is that the intensity of the reactions seemed to be mitigated little by their claim to be attracted to the opposite sex as well.

Particularly upsetting to them was the animosity that many felt from homosexuals.

Both gays and lesbians claimed that those who adopted the label "bisexual" did so because they feared the stigma attached to defining themselves as "gay" or "lesbian." Additionally, gays and lesbians saw bisexuality as a transition to becoming homosexual. In other words, they often rejected the bisexual identity in and of itself. Such attacks were said to come especially from politically active homosexuals who deplored the political fragmentation they saw caused by bisexuals who refused to fight the common enemy of "heterosexism." Bisexuals could exercise "heterosexual privilege"—i.e., they could always revert to a comfortable identity rather than suffer the consequences of standing up for their gay rights.

These beliefs affected personal interactions between bisexuals and homosexuals. Bisexuals were accused of being unable to sustain long-term relationships because of their continued desire for and contact with the opposite sex. This criticism was particularly voiced by homosexual women, who complained that they had to compete with men over their female lovers. It was especially anathema to lesbian feminists, who saw any female heterosexuality as "sleeping with the enemy." Generally speaking, bisexuality was equated with "promiscuity." As one homosexual man said to us, "Bisexuals are erotic gluttons." This supposition was shared by many gays and lesbians.

Accompanying this belief was the corollary, that bisexual women spread sexually transmitted diseases to the lesbian community because they had sex with men—especially bisexual men (which is not necessarily the case)—who were having sex with homosexual men (the highest-risk group for sexually transmitted diseases even in 1983).

Bisexuals confronted these accusations with dismay. What they had initially perceived as a potential source of support—the homosexual community—turned out to be another avenue of rejection. Despite seeing themselves as victims of the same type of prejudice as homosexuals, they found themselves victims of further discrimination. A sense of isolation and anger grew toward homosexuals in general.

We asked our respondents to answer the following question: "Because you are bisexual, what problems have you personally faced from homosexual men, homosexual women, heterosexual men, and heterosexual women?" The answers reveal the depth and complexity of our respondents' feelings and suggest the difficult social bind in which they found themselves (Table 11.1).

Relations with Homosexual Men

About half the bisexuals reported hostility from homosexual men over their bisexuality. For many women, these problems arose because of what they saw as the dislike many homosexual men harbored for women in general.

> Some gay males' sexist attitudes are appalling. They describe females using slang, a put down. Also there is a covert sexism that male gays are more important politically than lesbians.   (F)

Some bisexual men resented this attitude toward women too. For example:

> I told [one gay male] I had a wife and asked him if it made a difference. He stated he didn't want to relate to a man who related with fish, which is a put-down of women and I became quite angry.   (M)

Major complaints by our male respondents about homosexual men centered on the way they denied the reality of the bisexual identity.

> They question the validity of my bisexuality and tend to think of me as gay. They feel I can't cope with being gay so I call myself bisexual.   (M)

> They say you are gay—there is no such thing as bisexuality, there is a weakness in that a person can't make up their mind.   (M)

Other common beliefs about bisexuality attributed to homosexual men were not surprising—e.g., the fear of instability or problems with heterosexual privilege.

> Problems are mostly about their feelings of my commitment, whether I'll leave them for a woman.   (M)

> They see me as a traitor, still identified as [in some way] heterosexual for the benefits given to heterosexuals—acceptance, social esteem.   (M)

Relations with Homosexual Women

Twice as many bisexual women as men (about 80 percent compared with 40 percent) said they had experienced problems with homosexual women. They reported that many lesbians complained that bisexual women related more emotionally and sexually to men and therefore could not be trusted. During a rap session one bisexual woman complained of the shame she felt admitting to homosexual women that she was "bi" and that in San Francisco it was harder to "come out" to homosexual women than to heterosexual women or even to claim to be a bisexual feminist. Because the organizations for women in the Bay Area were dominated by lesbians, and since these provided a common place to meet same-sex partners, this was a major problem for women who openly identified themselves as bisexual.

> They see a woman who is still having sex with a man as a traitor. (F)

> They don't think it's politically correct to be bisexual; they don't trust women who are involved with men. (F)

Other complaints that were mentioned also included familiar lists of troubles about bisexuality itself:

> Accusations of sitting on the fence, of being traitorous, of being just in a stage of lesbianism, "You sound just like me when I was twenty-two." Sometimes I don't even dress politically correct. (F)

The men in our sample also described problems with lesbians arising over their gender as well as their sexual preference. These came down to a hatred of men, fear of sexual advances, and the like.

> The relationship is really limited with my lesbian friends—it would be more open if I were gay. But the "bi" is threatening. They are much more guarded and everything has to be clear with no ambiguity. Otherwise sexual overtones are always suspected in expressing affection. (M)

> Mainly I experience a sense of hatred for men. They are not willing to give a man a chance to do anything. I don't like the attitude that some have that they would like the whole opposite sex eliminated from the world. (M)

Thus, cross-gender hostility also affected relationships between bisexuals and homosexuals. This was related to the desire for gender separation among some homosexuals as well as the insecurity the opposite sex posed when one had a bisexual partner. As long as a person was labeled bisexual, he or she faced problems in obtaining consistent support from homosexuals.

Relations with Heterosexual Men

Three-fifths of the bisexuals had personally experienced problems with "straight" men, though these reactions were different from those of homosexuals. Women, for example, mentioned name calling:

> When I have had guys see me with a woman, they would look at me and make snide comments, calling us "dykes." (F)

Or there were comments about the role that women's homosexuality sometimes plays in heterosexual men's fantasies:

> They always want to now what females do sexually. They use my
> bisexuality as fantasy material and don't take my affection with
> women seriously.   (F)

> The last experience I had with a female at a swingers [sex] party, we
> had to tell some straight guys to leave us alone. . . . The females can
> be sexual in the setting and it's a turn-on for the men, but the men
> aren't sexual with each other. Male homosexuality isn't socially ac-
> ceptable.   (F)

Bisexual men appeared to have received much more hostile and serious reactions. In addition to name calling, there were sometimes threats
of violence.

> They see bi's as the same as homosexuals. I have been called a
> fag.   (M)

> A lot of verbal abuse, usually they call me a faggot or queer. I have
> also been threatened with being beat up. . . . There was one time I
> had to quit a job because of verbal abuse.   (M)

Ironically, just as some homosexual men saw bisexual men as "really
homosexual," heterosexual men also did and subjected them to homophobic reactions.

Relations with Heterosexual Women

Three-fifths of the bisexuals had also experienced problems with heterosexual women. The same antihomosexual attitudes were said to be common, though the examples suggested a less hostile demeanor.

> I get the same homophobic treatment I'd get if I told them I was
> lesbian.   (F)

> I have dated a straight woman since January. I have not seen her for
> six weeks. When I told her I might be working on this bisexual
> research project, there was dead silence. We have not talked to each
> other since.   (M)

The bisexual women said that heterosexual women were concerned
about the possibility of sexual come-ons.

> They tend to be afraid of my friendship turning sexual. They are
> uncomfortable when I talk about my female relationships.   (F)

I think there is a bit of uncomfortableness with physical contact, that if and when we touch each other that means we are going to get it on with each other.   (F)

Heterosexual women were reportedly less belligerent to bisexual men than to bisexual women but there was still concern. In some cases, it centered on heterosexual women's fear about getting romantically involved with a bisexual man.

A little afraid I will leave them for a pretty man.   (M)

Some didn't want to date me because they feared I would not be sexually faithful to them. They were too frightened of me being attractive to men.   (M)

These reactions from specific groups show the difficulties bisexuals can face in the wider community. But these conflicts can be handled more easily by concealment than those in intimate relationships, where sexual preference is difficult to hide. In both cases, however, persons can be faced with the dilemma of risking rejection by disclosing their sexual preference or being discreet and hiding an important part of themselves. We next examine who our bisexuals told about their sexual preference and the price they paid.

Who Knows?

All of those we interviewed had disclosed their bisexuality—or "come out"—to someone (Table 11.2). The mean age at which they did this was 29.2 for the men and 26.7 for the women. The women generally disclosed their identity sooner after becoming aware of their bisexual feelings than did the men. The mean number of years between such awareness and disclosure was 8 years for the men and 6.5 years for the women. The interesting thing is the relatively late age of "coming out" for bisexuals compared with homosexuals.[1] This is because there is a less clear identity with which to "come out."[2]

Who bisexuals told first also seemed to be a function of their lateness in "coming out." Bisexuals usually told a partner first. This is because they did not come out until their late twenties, a time at which they were likely to have a partner to confide in. Among the men it was most likely to be a female partner and for half as many their spouse; among the women it was most likely to be a male heterosexual friend, a male partner, or a female partner, all about equally. Overall, the men were most likely to tell a woman first and the women were most likely to tell a man. This, of course, could be because of the usual partnership situation found among our respondents, namely, most had heterosexual primary relationships.

It is most difficult, but probably also most important for identity resolution, to come out to family members. It is our families that shape our identities in general. For a bisexual, family acceptance can provide support against negative community standards. Family rejection, on the other hand, reinforces a sense of alienation and isolation.

We asked the bisexuals to tell us who in their families, and then among others, knew or suspected they were bisexual. Among family members, the respondents were most likely to say that a sister knew (nearly two-thirds) followed by a brother, their mothers, and then their fathers (about a third). Overall this is similar to homosexuals' patterns of disclosure in that mothers are told before fathers, probably because, according to one investigator, children have a closer relationship with their mother than their father.[3] Most other relatives were thought to be least likely to know. Adding those who felt that a relative "suspected" as well, the figures increased to a range of nearly three-quarters (for a sister) to about a half (for fathers and most other relatives). The only difference by gender was that the women were more than twice as likely as the men to indicate that a brother knew.

Outside the family, over 90 percent said that their best homosexual friend knew. Somewhat fewer—close to 80 percent—said their best heterosexual friend knew (nearly all the women and about three-quarters of the men). They were followed by a large proportion of other homosexual friends (around 70 percent) and fewer of their other heterosexual friends (60 percent). Again, adding those who they think suspect changes the range to virtually all of the best homosexual friends to just over three-quarters of most other heterosexual friends. Generally our respondents were quite open about their sexual preference.

When it came to relations with a boss, around 40 percent had disclosed their bisexuality. About half mentioned they had come out to others at work. Disclosure at work then was more common than disclosure to parents. Respondents were least likely of all to tell neighbors, presumably because sexual preference is irrelevant to such relationships.

Part of this is related to their living in San Francisco and being members of the Bisexual Center. Here, pride in being bisexual was emphasized and social support given to those in the process of coming out. Many persons came to the center confused about their sexuality. At the center they were helped to "discover" their bisexual identity—part of which involved being encouraged to come out to others.

Reactions

As one might expect, the people to whom bisexuals were most likely to reveal their sexual identity were also the most accepting in their reactions. Among family members, sisters' reactions were mostly benign (acceptance, understanding, or tolerance), followed by those of brothers, mothers, and fathers (Table 11.3). The number of benign reactions ranged from over 80 percent for sisters to about half for both fathers and most other relatives. Four times as many women as men characterized their brothers as "accepting." Of the benign responses, "tolerance" rather than "acceptance" or "understanding" was most commonly attributed to family members, with the exception of sisters, who were typically seen as "accepting."

Since we choose our friends, unlike our family, we assume that friendship is based to some extent on acceptance and understanding. Thus virtually all of the respondents classified the reaction of their best heterosexual friend and approximately 90 percent of their other heterosexual friends as benign. Similar figures were reported with regard to best homosexual friend and most other homosexual friends.

Benign responses were less likely for categories of people other than friends. They were more comparable to the responses attributed to family members. About three-quarters of the bisexuals said that their employer and most of their work associates did not react in a negative way to the disclosure. Less than half said the same about most of their neighbors. A somewhat higher percentage mentioned a benign response from heterosexuals in general. About four-fifths answered likewise for homosexuals in general. These latter responses seem at variance with the more negative remarks they made that we report elsewhere in the book.

Discretion versus Disclosure

About half of the bisexuals said that there was someone they wanted to tell about their being bisexual who did not know, but they were afraid to do so (Table 11.4). Usually this was someone close to them, often their parents or children, but especially their mothers. Many felt uncomfortable about having to hide this aspect of their lives and also felt that being able to talk about it would make for a more honest and open relationship—one with less lying and concealment. Still, they were reluctant to come out, because they expected a negative reaction. Consider, for example, this heartfelt response:

> I wanted to tell my parents so bad that I had to go sit and talk to somebody about these feelings; it was killing me. And I had decided that I was going to tell them. It was a big part of my life then. But my mother says, "Hi, how are you? What have you been doing?" And, well, I can't say, "Well, Mom, I'm leading this women's rap group at the Bisexual Center." I can't tell her these things that are an important part of my life. And that hurts a lot that I can't do that. And then I went back East and found that I had a good time with them, with or without their knowing. It's like I want to tell them, and if they can't handle it that's their problem. At the same time, it's all this guilt. . . . They're coming out here in June and I have in my spare bedroom . . . all of these outrageous cards of women lifting

weights and of people with whipmarks and all kinds of stuff. They're probably going to see my place, and I'm not going to hide these things. And my children, my daughter who came to stay with me for a month, but ended up staying six months. I covered them up because she couldn't deal with it. Because she didn't want her boyfriend to see that her mother was weird. . . . But it's like that's the part that hurts me most, that I can't tell them. And I can't say, "I can't tell them." I guess I've chosen not to tell them.   (F)

On the other hand, about half as many said that there was someone who knew about their bisexuality who they wish did not. This also was often a parent, although just as many mentioned an employer or work associate. Frequently, the result of such persons knowing was estrangement.

Regarding parents:

Upon finding out [from an unknown third source], his parents made it clear that they were not proud of him, they were ashamed, they did not respect him, and they were afraid of what the neighbors would think if they found out. They had said they were afraid of being ostracized by neighbors about their son's bisexuality.   (Fieldnote)

As for one man's children:

He said he had chosen the wrong time to "come out." All his children were either in, or soon to enter, high school. And when word spread regarding their father's gayness, they received a lot of shit from their peers.   (Fieldnote)

Or in the words of one man who had lost friends:

I once belonged to a circle of heterosexual friends. I've been ousted and it doesn't feel too good. There's a general feeling of rejection by people I've known for years. Things that you think are stable aren't.   (M)

When we pressed the question of what it was like for them to have to hide their sexual preference, most of the bisexuals said they were angry about having to do so and that this society does not allow for the free expression of sexual differences.

With most people, if they are rejecting, I'll tell them to go to hell; who needs them?   (F)

Other respondents were not so much angry as frustrated and depressed at having to keep the secret.

I feel tense hiding from my parents, Mary's husband, and the world. (F)

I can't attach a word to it. I'm tired. I just don't like to hide myself. (M)

The next most common response was that hiding was the wisest thing to do given the costs of disclosure.

I wish I did not have to hide my bisexuality, but I feel there is nothing to be gained and only lost if it were known. (M)

I am a little more prudent now. I wanted to change the world. I experienced negative reactions and lost my naïveté. You can't proselytize all the time. (M)

Many were very selective when it came to disclosure. They reported a variety of thoughts and feelings.

I don't tell those people who I feel won't be receptive—my boss, people I barely know, people I don't trust. (F)

I don't like having to hide it. With my parents it's the only sensible thing to do. My mother is very Catholic and would start praying for me. My father would be very rejecting. He thinks it's abnormal and sick. (F)

Finally, some of our respondents felt that disclosure was a moral or political issue and that they could enlighten others by letting them know.

I wish it did not matter in this culture, but I feel it's important to be out now as a bisexual so that others can learn to understand us and know that we exist. (M)

It's real important for me not to hide it. Every time I come out I'm making a political statement—that I exist and bisexuality is a viable alternative and lifestyle. (F)

Receiving Social Support

Coming out was easier for those who had the support of others. Even outside the Bisexual Center, about three-quarters of the respondents said that they had found such support. Most said it came from friends:

Friends provide a lot of emotional support, and do not tell me I need to feel negative about my homosexual feelings. (F)

Many of the bisexuals said they had received support from other organizations in San Francisco such as the Institute for the Advanced Study of Human Sexuality, the San Francisco Sex Information service, and the Pacific Center. Others had been to more general support environments ranging from encounter groups to feminist networks.

Advice on Disclosure

The last question we asked about disclosure was: "Based on your own experience, what advice would you give another bisexual?" Most frequently people suggested finding a support group first, especially with other like-minded people (the Bi Center was often mentioned) or searching out close friends and relatives who would be accepting.

Find a support group first. Make a good friend in whom you can confide. (F)

To have a support group first, and keep the coming out message simple . . . that I'm bisexual and I still love you. (M)

The second most common piece of advice was to be careful about who you come out to and to weigh the risks. This was especially said in relation to work.

Basically I believe in coming out as long as you don't jeopardize your life or job. (F)

Come out as much as you need to with your closest friends. . . . It's nobody else's business. (M)

Test the water by finding out what the person's attitudes are about bi's in general. Make sure you can trust the person. (M)

Only come out if the position you are in is safe. Or if you are willing to fight to change and educate people about it. (F)

Finally, a common tip was to be sure of oneself—to have inner strength and confidence—before disclosing your bisexuality.

The more confident and okay you feel about bisexuality then the more accepting other people will be of it. (F)

Get really confident with yourself. But don't say bisexuality is better than other choices; that will alienate people. (M)

(1) Do it!; (2) Wait until *you* are ready; (3) You are more ready than you think; (4) If you feel good about the way you are, most other people will too. If you have doubts, so will they. (M)

A great deal of other advice was offered too. To paraphrase the remarks of those we interviewed: do not listen to negative judgments; do not expect everyone to be negative; keep a journal to chart your progress; the more you do it, the easier it gets; do not advertise or use it as a sexual come on; get counseling; do not feel you have to come out to others; read books and other literature on bisexuality; coming out is politically important. In general, our respondents had a great deal of advice for the closeted bisexual, most of it gained through their own experiences.

Managing the Bisexual Identity

As noted throughout this chapter, managing the public aspects of bisexuality was made easier by being a member of the Bisexual Center. Attending the center made it easier for bisexuals to define their identities, organize their intimate relationships, and deal with the outside world.

Minority groups often respond to negative social reactions by organizing and providing each other with social support. Part of this support can be a justifying ideology that counters negative attitudes and gives positive meaning to the minority condition. Attendance at the Bisexual Center exposed respondents to a set of understandings about the "true" nature of bisexuality captured in five basic ideas: (1) everyone is basically bisexual; (2) bisexuals have more options and avenues for sex; (3) being bisexual means sexual freedom and flexibility; (4) bisexual people are open to personal growth; and (5) sexual relations with both sexes means wholeness and completeness.

As we mentioned earlier, we met many people who were wavering over and contemplating whether or not they were "bisexual." By emphasizing positive aspects of bisexuality, the group ideology often made a critical difference in adopting the identity. The nature of this clarifying ideology emerged in our interviews and observations and was especially apparent in the Bisexual Center newsletter—*The Bi-Monthly* (hereafter abbreviated TBM).

First, the capacity for being sexual with both sexes was seen as the most natural state of sexual being. The most common ideas were that everyone "is born" or has the basic "potential" to be bisexual.

Perhaps we are all born with the potential to be bisexual, i.e., experiencing sexual attraction for both men and women. (TBM)

Whenever I hear bisexuality being discussed . . . it is usually in the context of . . . "Well, isn't everyone potentially bisexual?" (TBM)

Thus, some respondents referred to bisexuality as an "innate" sexual condition, as the most "natural" sexual state possible, even drawing on the world of infants and animals as evidence that all living beings start out that way.

Bisexuality is natural. It occurs in nature. It exists among other species. We spend too much time trying to define ourselves and we limit ourselves this way.   (F)

The higher primates include the monkeys, the apes, and ourselves. All mammals display bisexual behavior, and the primates are no exception. Monkeys and apes often engage in mutual masturbation, mounting, and other kinds of sex play with members of both sexes. (TBM)

The same message was relayed when the community uncovered cases in the outside world of people who were actively bisexual but refused to admit it. (Sometimes examples were made of those who identified themselves as straight or gay.) Usually people singled out friends, or friends of friends, or even well-known community leaders, and made apparent their bisexual tendencies. For example, one man told a story about two men friends of his who decided to live together. One of the men identified himself as bisexual, the other as gay. Though the latter man ardently insisted that he was gay, there were reportedly times when even he would come home and tell his bisexual partner about a sexual attraction to, or experience he had just had with, a woman.

I see my friend . . . on one level as being prototypically gay, i.e., he has done the Buena Vista Park scene, has had numerous homosexual experiences, and is worried about the AIDS problem. But at the same time, he has also been very strongly attracted to women. When he starts talking about himself as exclusively gay, I just sit back in wonderment, and know the next time I see the guy, he might be talking about settling down with a woman.   (M)

Probably the most remarkable case of "uncovering" we heard concerned a man and woman who were the leaders of a gay synagogue in the Bay Area. Though the person who told the story stated he was not sure that it was true, the man and woman were said to be secret lovers. What is crucial is not whether the account was true but the fact that such stories are told in the first place and thus have a "social reality." Thus, even among the most committed of gays, evidence of bisexuality was said to abound.

A second aspect of the ideology was that being bisexual carried with it added opportunities for sexual expression. While gays, lesbians, and straights claimed to focus exclusively on one sex, bisexual people prided themselves on having more available "options." Slogans such as "Being bisexual is twice as much fun," "Bisexuals have the best of both worlds," and even "Bisexuality doubles your sexual potential" were commonplace.

It's very exciting. When I do want to be turned on sexually, it always works, and I can go either way, for men or women. . . . It gives me a lot of options. Sometimes you get bogged down and in a rut, and nothing works sexually. When this is the case, you can always turn to someone else, of either sex, for help. If things aren't going well sexually with a male, I can always experience a female, or vice versa. (F)

Closely related to the notion of increased sexual opportunity was the belief that greater sexual freedom and flexibility went hand in hand with being bisexual. Usually this meant feeling less inhibited, less shameful, and more open to a range of sexual experiences—and thus a greater happiness.

Because I'm bisexual, I feel sexually like a free person. I don't feel ashamed of my body any more. I can just let go. When I have a climax, they are real intense, because I don't hold back any more. Being bisexual has helped me to realize that it's okay to be a real sexual person, which I am. (F)

Implicit in the notion of freedom and flexibility was a related theme of "personal growth." Often this came down to "growth" in terms of sexual expression. The bisexuals continually emphasized how they were looking to "expand" their sexual boundaries, to "broaden" their horizons, to stretch their sexual "limits," to try things they had never tried before—to enrich their sexual life. To them, this was what being bisexual was all about.

[My] focus [sexually] is upon discovering what my limits are, not ones that are impossible to reach, striving for that limit, and then once it is reached, reestablishing new limits. (M)

Coupled with this was the belief that sexual growth was more likely for bisexuals—that they were more open-minded.

I don't try to convert straight men to change their sexual behavior. It is their own responsibility to decide to remain straight or to open their minds to possible alternatives. For myself, I have given up the fear and ignorance around the [homosexual and heterosexual] options. (TBM)

The bisexual men and the women typically adopted the belief that because of their dual sexual preference they were blessed with special insight into sexual relations and were more sexually competent than the average homosexual or heterosexual.

I have found in general that bisexuals are better sexual partners than heterosexuals or gays. And I have been told by a female I had sex with first when I was fairly straight, and later after I had relations with men, that I was a better lover because of the latter. (M)

It gives you more of an idea of the way both men and women work, rather than just men and women if you related exclusively to one or the other. Physically, the way they think, their emotions, the way they express themselves.   (F)

Both sexes also tended to see themselves as humanitarians who were more tolerant and accepting of differences among people in the broader social world.

Empathy, not just with bisexuals, but with all minority groups who had to deal with social condemnation, excommunication, etc. Most bisexuals I've met are much more sensitive and aware of social injustices and of the feelings of others.   (F)

Finally, according to the ideology it was possible to achieve sexual wholeness, completeness, or a unity of sexual being through bisexuality. The *Bi-Monthly* newsletter was the most visible source of accounts of this nature, which stated that beneath the duality of sex there was oneness.

I feel that as bisexuals, exploring the middle ground between homosexuality and heterosexuality, we are in a position to make a unique contribution. . . . We are involved in the universal necessity to blend polarities, to integrate and express wholeness. (TBM)

The imagery of wholeness was further sustained by the Bisexual Center's organizational symbol, a display of two half circles enclosing a completed circle.

Regrets?

Issues of sexual identity can be a lasting social problem. Being bisexual can create conflict in relationships with family and friends and with both heterosexuals and homosexuals in general. Opposed to this, however, for the people we interviewed, was the social support and justifying ideology provided by the Bisexual Center. Interested in whether the "negatives" outweighed the "positives," we asked them whether they regretted being bisexual (Table 11.5). About three-quarters said they had no regrets whatsoever. Most of those who harbored regrets said that they felt this way because their lives were made more complicated by society's failure to accept them.

My life would be simpler if I was not attracted to females. I would not have to deal with not conforming to the heterosexual majority.   (F)

I don't regret being bisexual, rather I regret the social problems that result from the society I'm trying to live in—the denial of nonconventional sexuality, the fear and suspicion people have toward others who are sexually different. (M)

Self-acceptance usually occurred despite a lack of general social acceptance. What were once internal self-doubts about being bisexual appeared to evolve into external social hardships. It seemed better or easier to be what one was, a "bisexual," and to learn to handle the social penalties that might accompany this, than to deny an important part of oneself. Confronting these difficulties often served as a basis for further commitment to the bisexual identity. If, in 1983, these were unwelcome tasks, they were to pale compared with what was to come later as the AIDS epidemic began to grow.

II

BISEXUALITY, HETEROSEXUALITY, AND HOMOSEXUALITY

12

Surveying the Sexual
Underground

While our interviews provided us with a rich look at the experience and reality of being bisexual, they could not show how bisexuals lived their sociosexual lives compared with homosexuals or heterosexuals. The bisexuals themselves raised a number of questions of this nature. Were bisexuals more or less likely to feel confused about their sexual identity? Were there as many types of heterosexuals or homosexuals as bisexuals? How did their sexual behaviors and overall number of partners compare with the other groups? Were bisexuals more or less likely to have multiple relationships or to have trouble sustaining long-term involvements? How did the stigma of being bisexual compare with that of being homosexual or being an unconventional heterosexual?

We constructed an eight-page questionnaire to address these questions. Most of the questions we asked were drawn from our original interviews with bisexuals. A number of open questions were rephrased as closed items, and a few new questions were added. In the end, the questionnaire represented a condensed closed version of the interview schedule covering mostly the same topics.

Our main goal was to make comparisons between people who defined themselves as gay/lesbian, bisexual, and heterosexual. As with the interview study, we also wanted to make comparisons between men and women. Finally, although we needed a substantial number of cases for each sexual preference category in order to make statistically meaningful comparisons, it was important that each of the subsequent groups be comparable. Thus, we wanted organizations of people who had some familiarity with the sexually progressive philosophy of the sexual underground.

We therefore surveyed the populations of four San Francisco Bay Area organizations whose primary mission was to promote sexual freedom as well as to provide support, education, and information about sexual matters to anyone who sought it. In addition to the Bisexual Center, the Pacific Center served gays and lesbians; the Institute for Advanced Study of Human Sexuality provided professional degrees in the study of sex and promoted the same philosophies as other organizations; and the San Francisco Sex Information service (SFSI) was a telephone hotline that dispensed sex information to anonymous callers. (See Chapter 2 for a more complete description of these organizations.)

We mailed an initial version of the questionnaire in June 1983 to people on the SFSI mailing list. Some additions were made and the questionnaires were then sent to everyone on the mailing list of the other organizations between April 1984 and April 1985. We also completed a second mailing to the Bisexual Center during the same period. The mailing list for each organization contained the names of people who were paid members, or in the case of the Institute, who were students, faculty, or people who attended workshops. In addition to the mail distribution, the questionnaire was made available to people who visited the Bisexual Center, SFSI, and the Pacific Center over this same period of time. Copies were set in a stack on the reception desk where those who were interested could pick them up.

The front page of our questionnaire, with the address of the recipient, was removable so that it could be returned anonymously. A short letter attached to the questionnaire identified the nature of the research, asked the recipient to participate, stated that participation was voluntary and anonymous, and mentioned the approximate time it would take to complete the questionnaire (a half hour). We also included instructions about how to do the return mailing, and the letter was signed by the three principal investigators.[1]

The overall response rate for the study could not be calculated. Because packs of questionnaires were available to be taken at will at the four organizations, we do not know how many people opted not to take them. In addition, three of the organizations would not allow us to work directly with their mailing lists, since it was their policy that the names on their mailing lists remain anonymous. They conducted the mailing themselves after the printed questionnaires and required postage had been delivered to them. We do not know how many were undeliverable. Also, some people were on the mailing lists of two, three, or even all four organizations. (We made a point of asking many people at the Bisexual Center if they belonged to any of the organizations and many said yes.) This was the nature of the sexual underground, with a great deal of participation across different settings. Thus the number of overlapping cases could not be counted. To guard against repeat participation, we placed a note in the front of the questionnaire that asked people not to complete it more than once. In sum, we don't know how many people received the questionnaire, which makes it impossible to estimate a response rate.

A total of 702 people completed the mailed questionnaire. The breakdown by sex was 299 women and 392 men. By sexual preference, 192 identified themselves as heterosexual, 284 as gay or lesbian, and 217 as bisexual. There were also 4 respondents who indicated being transsexual, 2 respondents who selected multiple labels, and another 14 with no response on one of the two variables. We thus excluded twenty cases from much of the analysis and used 682 of the questionnaires. Combining their sex and sexual preference group, the number of cases used in the analysis were as follows: 105 heterosexual women, 85 heterosexual men, 94 homosexual women, 186 homosexual men, 96 bisexual women and 116 bisexual men.

In comparing the basic demographic profiles of respondents in these six groups (Table 12.1), two differences were apparent. One was in age. While the homosexual men were evenly dispersed in their ages (with the heaviest concentration between 30 and 39 years), the bisexual and heterosexual men were nearly all age 30 or beyond, with many in the 50 or greater age bracket. The homosexual and bisexual women fell primarily between the ages 25 to 44, but the vast majority of the heterosexual women were in the 30 or greater age category, with nearly half between 30 and 39 years. Thus, overall, the bisexual men and heterosexual men and women were older than the other three groups.

The second difference was in income. The men had higher incomes than the women, with over twice as many men reporting incomes of $30,000 or greater within all three preference categories. Also, heterosexuals—both men and women—had higher incomes compared with their bisexual and homosexual counterparts. From high to low, the distribution of people with incomes of $30,000 or more was: heterosexual men (63 percent), bisexual men (41 percent), homosexual men (38 percent), heterosexual women (32 percent), and bisexual and homosexual women (both 16 percent). All findings were checked for the effects of age and income. No significant differences changed when controlling for these variables.

Aside from age and income, the six groups appeared strikingly similar in many respects. The vast majority in all of the groups said they were college educated, and many had graduate and professional degrees. Nearly everyone was white. The most common religious denomination was Protestant, with somewhat smaller but relatively even numbers of Catholic, Jewish, or "other" respondents. Almost everyone said that they were either not at all or only a little religious. Around two-thirds of the men and three-fourths of the women said that they had moderate to strong feminist values.[2]

Our goal was to examine how bisexuals compare with those who adopt the major labels of sexual identity in our society—heterosexuals and homosexuals—even though the act of comparison itself supports the assumption that these groups are essentially different.[3] In particular we were interested in how sexual lifestyles and labels mix, the social implications of the choice of a label, and whether sexual preference was as fluid for others as it was for bisexuals. We explore these issues in the following pages.

Throughout these chapters we make two sets of comparisons. First, we look at men and women separately in our comparisons of sexual preference groups. We ask, for each sex, to what extent bisexuals, homosexuals, and heterosexuals are similar or different. Second, we compare men and women within each of the three preference groups: bisexual women versus men, homosexual women versus men, and heterosexual women versus men. The emphasis in our discussion is on significant differences where they appear. Also, in some of our analyses of social stigma, we isolate a subgroup we call "unconventional heterosexuals"— heterosexuals who were actively involved in swinging, group sex, sado-masochism, or cross-dressing. Further details of the analysis and presentation of the data for the mailed questionnaire study are found in Appendix B.

13

The Development of
Sexual Preference

In our earlier discussion of the emergence of sexual identity (Chapter 3), we outlined a pattern of sexual development among bisexuals. We saw an initial confusion over feelings, attractions, and behaviors, and we saw how bisexuals resolved this confusion by finding and applying a label for their sexual identity. Not all bisexuals settled into an identity, however, and many felt continuing uncertainty about their sexual preference. In order to understand how a person comes to identify as bisexual instead of heterosexual or homosexual, we need to make comparisons between the groups. Do the early sexual feelings and experiences of bisexuals differ from those of heterosexuals and homosexuals? Is continuing uncertainty about sexual preference more common among bisexuals, or is it found among other groups as well? How stable is the bisexual identity when compared with the homosexual and heterosexual identities? Examining sexual development among heterosexuals and homosexuals as well as bisexuals provides a broader idea about what goes into the development of sexual preference.

Pathways to Sexual Preference

Men

The age at which the men in our survey were first *attracted* to same-sex and opposite-sex persons was clearly related to future sexual identification (Table 13.1). Self-defined heterosexual men reported experiencing their first sexual attraction to *opposite-sex* persons at the youngest mean age (10.2 years) and the homosexual men (the 28 percent reporting this) the latest (14.5), with the bisexual men in the middle (12.8). For *same-sex* attractions, the homosexual men were attracted earliest (11.5), and the heterosexual men (the 30 percent reporting this) latest (21.9), with the bisexual men again in the middle (17.1). Bisexual men were like heterosexual men in that their attraction to women occurred earlier than their attraction to men. The reverse was found for the homosexual men.

A similar pattern emerged with first sexual *experiences*. Here, with *opposite-sex* partners, the bisexual men and the heterosexual men had the earliest initial heterosexual experience (15.9 and 16.7) and the homosexual men (the approximately 60 percent reporting this) the latest (17.7). For *same-sex* partners, the homosexual men reported the earliest sexual experience (14.7) and the bisexual men (17.2) and heterosexual men (the 48 percent reporting this) the latest (17.7). On the average, then, both bisexual and heterosexual men had their first sexual experience with a woman earlier than they did with a man. The reverse was found for the homosexual men.

In short, homosexual men were attracted earlier to *same-sex* persons and had *same-sex* sexual experiences earlier than did the bisexual men. This seemed to affect identity, as the homosexual men defined themselves as homosexual earlier (21.1) than the bisexual men defined themselves as bisexual (29.0). Moreover, the homosexual men "came out," i.e., accepted a public identity based on their sexual preference, earlier (23.6) than did the bisexual men (29.2).

Women

Similar patterns appeared for the women. Heterosexual and bisexual women reported experiencing their first sexual attractions to *opposite-sex* persons at the youngest age (10.4 and 10.9 years) and the homosexual women (the approximately 60 percent reporting this) the latest (14.3). For *same-sex* persons the homosexual women were attracted earliest (16.4) and the heterosexual women (the 51 percent who experienced this) the latest (23.6), with the bisexual women again in the middle (18.5). Both the bisexual and the heterosexual women were attracted to *opposite-sex* persons earlier than they were to *same-sex* persons. The reverse was not found for the homosexual women, as it had been for the men.

Age at the time of the first sexual experience also differed by sexual preference among the women. For *opposite-sex* partners, the heterosexual women had the earliest sexual experience (14.9) and the homosexual women (the 78 percent who experienced this) the latest (16.4), with the bisexual women most similar to the heterosexual women (15.1). For *same-sex* partners, the homosexual women reported the earliest experience (20.5) and the bisexual women (23.5) and the heterosexual women (the 51 percent who experienced this), who were similar, the latest

(23.0). Both the bisexual and the heterosexual women had their first sexual experience with an *opposite-sex* partner earlier than they did with a *same-sex* partner. This was also true for the homosexual women (heterosexual experience came first), which is the reverse of what we found for the homosexual men.

In summary, the homosexual women were attracted earlier to people of the *same sex* and had *same-sex* experiences earlier than did the bisexual women. This probably contributed to the difference in age at the time of their self-definitions, as the homosexual women defined themselves as homosexual earlier (22.5) than the bisexual women defined themselves as bisexual (27.1). Likewise, the homosexual women came out publicly earlier (23.0) than did the bisexual women (26.7).

Differences Between Men and Women

The men were likely to have had their first *same-sex* sexual experience earlier than did the women (Table 13.2).[1] This is probably because homosexual experimentation is more likely to be part of the adolescent male subculture than the adolescent female subculture.[2] On the other hand, the women, regardless of their sexual preference, had their first *opposite-sex* sexual experience earlier than did the men. The bisexual women were also first attracted to the opposite sex earlier than the bisexual men. The homosexual men were more likely to have had their first sexual experience with a same-sex partner, while the homosexual women were more likely to have had it with an opposite-sex partner. Finally, the bisexual women came out publicly earlier than did the bisexual men. This is probably due to the greater threat felt by men in moving away from exclusive heterosexuality in contrast to the support the women may receive from sectors of the women's movement.

The data provide a broad picture of the development of sexual preference. Among the homosexuals and heterosexuals, initial attractions and experiences seemed to mark the direction sexual preference would follow. The bisexuals, however, seemed more open to the effects of subsequent experiences that were opposite to the initial ones. Again, we see homosexuality as often an "addition" to an already developed heterosexuality among bisexuals. Even though many heterosexuals and homosexuals had subsequent experiences contrary to their initial ones, these did not seem to have a lasting impact as they did for the bisexuals.

One exception to the above pattern was the homosexual women, whose heterosexual experiences preceded their homosexual ones. We attribute this to the greater difficulty women have in avoiding heterosexual experiences. Women often suffer from what has been called "compulsory heterosexuality," being forced, cajoled, etc., into sex with men whether they are sexually attracted to them or not.[3] Our data do show that more homosexual women engaged in heterosexual sex than were sexually attracted to men but also suggest that "compulsory heterosexuality" is not confined to women. More homosexual men also engaged in heterosexual sex than reported sexual attraction to women.

A gap between attraction and behavior was also found for heterosexual men, many of whom probably had homosexual sex without being sexually attracted to men. We suspect that the motives for these experiences are curiosity, money, an orgasm and the like, rather than a sexual attraction. No such discrepancy was found for the heterosexual women, who had more of a balance between their sexual attractions and behaviors.

Confusion about Sexual Identity

A period of initial confusion about their sexuality is common among bisexuals, as described in Chapter 3. To find out how this compared with heterosexuals' and homosexuals' experiences, we asked two questions: "Have you ever been confused about your sexual identity? And, if yes: In what way(s) have you felt confused?"

The proportion of bisexuals reporting confusion about their sexual identity was not significantly greater than the proportion of homosexuals (Table 13.3). About two-thirds of each group reported having experienced such confusion. Perhaps the more interesting finding is that a substantial proportion (28 percent) of those who defined themselves as "heterosexual" had felt confused about their sexual identity at some time. We next examine the nature of this confusion among all three groups.

Heterosexuals

It is often assumed that people who call themselves bisexual are confused about their sexual preference. To our knowledge, few researchers have suggested that people who define themselves as heterosexual may also experience confusion about *their* sexual preference. This should not be surprising, however, because even the label "heterosexual" has to be applied to a combination of sexual feelings, sexual behaviors, and romantic feelings. Over a quarter of the heterosexuals mentioned feeling confused at some time about their sexual preference.

Some heterosexuals—mostly men—were confused about their relations with the opposite sex. Typical answers, in respondents' own words, were: I didn't instinctively know how to initiate sex with the opposite sex; I was scared of the opposite sex; I was unsuccessful with the opposite sex; I wondered if I feared the opposite sex more than being attracted to them.

Others were concerned about homosexual leanings. Here there were equal numbers of males and females. Typical responses were: I felt more comfortable with gays; I had homosexual fantasies; I wanted my emotional needs met by other women; I enjoyed the company and affection of lesbians; I had my first homosexual experience during group sex and liked it.

More than a few heterosexual women said that they had consciously considered becoming "bisexual." In respondents' own words: I paid attention to same-sex feelings as a result of the women's movement—so I tried to be bi; I'm affectionately bi—I think it would be more politically correct to identify as bi; I wondered if I would be happier as bi; I intellectually feel committed to bisexuality but have had limited same-sex contact; I want to be bi but I find the same sex threatening and confusing.

Homosexuals

Almost as many homosexuals as bisexuals said they suffered from confusion about their sexual preference. When homosexuals were asked how they felt confused, the most common reply was whether their sexual attractions really fit the boundaries of the label they had adopted. In respondents' own words: I had heterosexual leanings; occasionally I had heterosexual fantasies; I got turned on by reading about heterosexual sex; I thought I was bi for a brief period; I felt I might become heterosexual over time; I felt I hadn't given heterosexuality a chance.

In contrast, others were confused because of disappointments with the opposite sex. Heterosexual experiences did not fit well into their lives. In respondents' own words: I had homosexual fantasies and feelings during heterosexual sex in my marriage; heterosexual relationships weren't working emotionally or sexually; I never felt satisfied with the opposite sex; I lost interest in the opposite sex after two failed marriages.

More men than women actually said they tried or thought they might become "heterosexual." As they told us: I tried to conform to expected heterosexual roles, I tried to be a good heterosexual; I thought through strength that one could become heterosexual; I wondered if I should try heterosexuality again.

Some homosexuals mentioned a fundamental ignorance about homosexuality. They said: I had no awareness that homosexuality was a possible lifestyle; I thought I was the only gay in the world; I thought I was the only freak with homosexual interests.

Denial of homosexual feelings often created confusion too. As our respondents told us: It took me years to realize I had underlying homosexual interests; I didn't see the relevance of early homosexual experiences; I wanted to get rid of homosexual feelings; I hoped homosexual feelings would go away; I thought I was going through a homosexual stage; I thought I would naturally change and become heterosexual.

The absence of role models was a source of confusion, too, as our respondents said: I didn't realize I was a lesbian because no one said that I might be; there were no guidelines on how to act or express my homosexual attractions; I wasn't sure how being gay would fit into the scheme of society; I was confused about appropriate homosexual emotional and sexual roles.

Not surprisingly, reluctance to accept a homosexual identity was commonly linked to homophobia. This was reported more by men, who said: I felt it was unnatural to have homosexual attractions; sex with the same sex was unthinkable; I got turned off to the same sex through fear; I became frigid, abused alcohol and drugs, and rebelled against society; I wanted to feel normal about being homosexual but didn't; I was very angry about my homosexuality.

Even if not homophobic themselves, many had difficulty accepting the label homosexual in their lives because of its negative image. Most of the people who said this were men. They said: I was confused about the homosexual label and what it connotes; I was confused because I was exposed to negative stereotypes of homosexuals—queers, fags, swishy, sick, queens—I didn't want to think these things about myself; I was unsure if the stereotypes were true.

In this regard, the fear of stigmatization and other negative social reactions magnified their confusion; they said: I feared rejection by others on account of my homosexuality; I felt that being homosexual would doom me to life as an outcast; I worried about what people thought; conservative society represses and persecutes homosexuals; I was reared in a masculine-macho culture; I wouldn't fit in with old friends; I feared my ex-husband and child would find out I'm gay; It's hard to reject the negative messages from people I cared about; My parents caught me in a homosexual relationship as a teenager and sent me to a psychiatrist; I was discharged from the Navy for being homosexual.

And finally, there was regret over the loss of benefits that accompanied a heterosexual lifestyle. In our respondents' words: Missing the experience of having a family; would it be reasonable to develop a primary and domestic heterosexual relationship since I find the opposite sex more emotionally fulfilling? I have all the feelings for the opposite sex except sexual attraction; Sometimes I think a traditional heterosexual relationship would be much easier.

Bisexuals

The sources of confusion for bisexuals were often quite similar to those of heterosexuals and homosexuals. But being bisexual did add unique twists. First of all, as we saw in Chapter 3, a common source of confusion for the bisexuals was simply having sexual feelings for both sexes and not knowing what to do about them. They did not see themselves as either heterosexual or homosexual and were often unaware of the category "bisexual," which would have offered them another option to make sense of their sexuality. In our respondents' words: I suspected I was the only one who had such feelings; I wasn't aware there was such a thing as a bisexual lifestyle; I wasn't aware that being bisexual was an option till I heard the word "bisexual".

Even when they became aware of the category "bisexual," many were confused because they did not see themselves as balanced in their heterosexual versus homosexual feelings or behaviors. The label seemed to imply a need for similar degrees of both. They said: What does it mean to be bi—do I always have to have male and female partners? I had same-sex fantasies and casual sex experiences, but was affectionately attracted to the opposite sex—was I a true bi? Are my feelings strong enough to be described as bisexual?

Just the realization that they were not completely heterosexual or homosexual was enough to produce confusion and attempts or thoughts to view themselves as simply one or the other. As they said: I denied same-sex attractions; I devalued same-sex contacts and overvalued opposite-sex contacts; I doubted the genuineness of my heterosexual interests; I wondered why I couldn't be just heterosexual or homosexual; I really wanted to identify as gay and receive support from the gay community. With the men, this crossed over into homophobia or biphobia. They said: I felt panic when I realized I had same-sex feelings; I thought having feelings for both sexes was evil and this scared me; I responded with homophobia; I was biphobic—I wasn't normal.

Adding to this confusion was vacillation between their feelings and attractions for same and opposite-sex persons. In their words, I was confused by the fluid nature of my sexuality; I experienced swings in my primary sexual interests between a Kinsey 6 to a 2 approximately every five years for the past fifteen years; I experienced changes in my sexual preferences from day to day, even moment to moment.

Not only were doubts internally generated, but reactions from both homosexuals and heterosexuals contributed to confusion. Our respondents said: I felt pressured by others to be either homosexual or heterosexual; I was told by others that bisexuality does not exist; I was pressured by my psychiatrist, who told me bisexuality was unhealthy.

Finally, the absence of role models as living examples of how to "be bisexual" lent to a general feeling of confusion over how to translate feelings into action. They said: I had no knowledge of how to manage a bisexual lifestyle; I did not know how to act on my desires for both heterosexual and homosexual relationships; I didn't know how to function sexually with both sexes.

As noted earlier, the proportion of bisexuals reporting confusion about their sexual identity is not significantly higher than the proportion of homosexuals. Yet the reasons cited for this confusion suggest some interesting differences. Homosexuals were more likely than bisexuals to cite external pressures, such as the fear of stigmatization. Of course bisexuals experience confusion connected with these external pressures too, but the fact that they mention them less often than homosexuals could suggest that bisexuals' confusion is more closely connected with the struggle to define what their dual attractions mean and with trying to find the appropriate label for them. Perhaps there is more of a struggle to understand which behaviors fit the label and which label fits the behavior, since the label is less readily available in our society. Such an explanation would fit in with our finding that bisexuals define themselves as such later than homosexuals.

In Transition?

One belief about bisexuals we considered earlier is that they are in transition. That is, they are really homosexuals who have yet to acknowledge their true sexual preference. We found little evidence to support this view. We asked our questionnaire respondents whether they currently felt in transition in terms of their sexual feelings, the sex of their sexual partner (sexual behavior), their romantic feelings, and their identity (Table 13.4).

Among those who defined themselves as heterosexual or homosexual, almost all said that they were *not* in transition toward either becoming homosexual (for the heterosexuals) or heterosexual (for the homosexuals) on any of the four dimensions.

Among the men and women who defined themselves as bisexual, fewer claimed such stability. Still, three-quarters said they were *not* in transition toward an exclusive sexual preference on any of the four dimensions. At the same time, the bisexuals were the least stable in sexual preference in these ways.

We also asked respondents if they had ever knowingly tried to change their sexual preference (Table 13.5). A quarter of the bisexual men and about a third of the women said yes. A third of the homosexual men and about 40 percent of the women said the same. Attempts to change one's sexual preference were rare among heterosexuals.

The direction of the attempted change differed significantly between men and women. For the bisexuals, about 60 percent of the men and 40 percent of the women had attempted a change in the heterosexual direction, about 40 percent of the men versus approximately a quarter of the women in a bisexual direction (for greater balances in their homosexuality/heterosexuality) and a few of the men but a third of the women in a homosexual direction.

The homosexuals showed a similar and even greater difference. Three-quarters of the men had tried to change in the heterosexual direction compared with just under a third of the women. Conversely, approximately 60 percent of the homosexual women had wanted to move further in the homosexual direction compared with only about 10 percent of the men. Around 10 percent of both sexes wanted to move in a bisexual direction. Some of these people said that such attempts did lead to an actual change in their sexual preference. Close to half of the bisexuals said that they did change. Most actual change was reported by the homosexual women (about 60 percent of them) and least by the homosexual men (about 20 percent of them).

Despite this change, the vast majority of the bisexuals were stable in their sexual preference, which we argue has something to do with the fact that all have accepted the definition of themselves as "bisexual." As we suggested earlier, the acceptance of the identity may coexist with a continuing sense of uncertainty about how that label fits them. This differs from the heterosexuals and homosexuals, whose sexual identities carry within them a much clearer notion of how self-labeling and sexual behavior go together.

14

Dimensions of Sexual Preference

Whether people label themselves as bisexual, gay, lesbian, or simply think of themselves as heterosexual, the acquisition of a sexual identity is obviously a pivotal aspect of sexual preference. There is, however, more to sexual preference than self-definition. What exactly underlies the labels people adopt for themselves, and what implications does this have for an understanding of sexual preference? In the interview study, we addressed this question by examining the ways in which people who defined themselves as bisexual located themselves on our three dimensions of sexual preference along the Kinsey scale: sexual feelings, sexual behaviors, and romantic feelings. The answers indicated a range of sexual profiles among bisexuals. Once again, we focus here on the Kinsey scale, but this time we extend our analysis to examine whether there is a single sexual profile that characterizes people who define themselves as either gay, lesbian, or heterosexual, or whether people with different types of sexual profiles identify themselves under the same label.

We rejoin the people in our study at a point where they have defined themselves as bisexual, gay, lesbian, or straight. First, we report how the men and women in each group are dispersed along the Kinsey scale in their sexual feelings, sexual behaviors, and romantic feelings. Next we explore how the three scale measures combine into profiles to explore whether there are different heterosexual and homosexual types as well as the bisexual types outlined earlier.

Our goal is to shed light on the extent to which people become anchored in their sexual feelings, sexual behaviors, and/or romantic feelings. The major anchors are at either end of the Kinsey scale, in complete sexual exclusiveness for one sex or the other. Any increment on the Kinsey scale away from exclusiveness represents a divergence from these traditional bases of sexual identity. We would expect that complete anchoring should be most common among those who define themselves as homosexual and heterosexual and least common among those who define themselves as bisexual. But does this mean that sexual profiles do not vary within each group, or that there are no clear boundaries between the groups?

Independent Kinsey Scale Distributions

When someone claims to be heterosexual or homosexual, we usually think of him or her as completely heterosexual or homosexual. For those who identify themselves as bisexual, the common perception is that they have equal desires for and activities with both sexes (although we saw in Chapter 4 that this is not necessarily the case). We examine these conceptualizations among our respondents who defined themselves as "heterosexual," "homosexual," and "bisexual" (Table 14.1).

Heterosexual Men

Around two-thirds of the heterosexual men who filled out the questionnaire stated that their sexual feelings were focused exclusively on the opposite sex. About a third indicated having a small degree of sexual feelings for the same sex, that is, they scored themselves as a 1 on the Kinsey scale. A small fraction, roughly 3 percent, said that they were a 2 on this dimension; none indicated any greater degree of homosexual feelings.

As a group, the heterosexual men were predominantly exclusive in their sexual behavior. Ninety-one percent said they were sexual only with women. There were, however, almost 10 percent who indicated a small amount of homosexual activity, scoring themselves as a 1, but nothing beyond.

In terms of romantic feelings, over 80 percent of the heterosexual men said that they had an exclusive preference for the opposite sex on this dimension. Nearly 14 percent categorized themselves as 1's and about 3 percent as 2's on the Kinsey scale.

Heterosexual Women

Fewer of the heterosexual women were focused exclusively on one sex in their sexual feelings than were the heterosexual men. Almost half of the heterosexual women identified themselves as a 0 on the Kinsey scale, but over 40 percent considered themselves a 1, and the remainder fell between a 2 and a 4.

Regarding sexual behavior, the heterosexual women were distributed almost identically to the heterosexual men in the amount of same-sex versus opposite-sex behavior. Eighty-eight percent of the heterosexually identified women said that they only had sex with the opposite sex. All of the remaining respondents placed themselves at the next scale point of 1 on this dimension.

The heterosexual women were less exclusive than the men, however, on the dimension of romantic feelings. Just over three-quarters of the heterosexual women scored themselves as a 0, and one-fifth as a 1.

Thus, although most people who defined themselves as heterosexuals rated themselves as a 0 (exclusively heterosexual) along each dimension, the size of this majority varies. For both men and women the same pattern appears: sexual behavior is most likely to be exclusively heterosexual, followed by romantic feelings, and then sexual feelings. Very few in the "heterosexual" group went beyond a 2 (mainly heterosexual with a significant degree of homosexuality) on any dimension.

Even those who defined their sexual preferences as "heterosexual," then, were not necessarily exclusively so. A considerable number admitted to having sexual feelings toward the same sex, although few translated these feelings into sexual behaviors.

The "heterosexual" profile was more similar than different for men and women. The key exception was that heterosexual women were less exclusively heterosexual in their sexual feelings than were heterosexual men. We believe that in San Francisco, this is because feminists encourage women to develop sexual feelings toward other women.

Homosexual Men

Homosexual identified men were similar in their profile to the heterosexual men in their sexual feelings. Two-thirds ranked their sexual feelings as exclusive—in this case, as exclusively homosexual. About 30 percent indicated a small degree of sexual feelings for the opposite sex by scoring themselves as a 5. The remainder had sexual feelings they classified as a 4, significantly heterosexual.

Again, homosexual men were much like the heterosexual men in the profile of their sexual behavior. Over 90 percent were exclusive with the same sex. Nine percent were somewhat heterosexual in their sexual behaviors, scoring 5 with no one scoring themselves further than this point.

Compared with sexual behaviors, the homosexual men, however, were less entrenched at the exclusive end of the romantic feelings scale, just as their heterosexual counterparts were. Over 80 percent scored a 6, about 14 percent a 5, and the remainder a 4.

Homosexual Women

Homosexual women were like heterosexual women in that fewer of them were focused exclusively on one sex in their sexual feelings than were their male counterparts. About half indicated that they had sexual feelings only for the same sex. Forty-four percent had some sexual feelings toward the opposite sex, scoring as 5's, and three percent were 4's.

The same percentage of homosexual and heterosexual women were exclusive in their sexual behavior. Nearly 90 percent of the homosexual women only related sexually to other women. This also was virtually the same percentage as for the homosexual men.

A similar proportion of the homosexual and heterosexual women were also exclusive in their romantic feelings. About three-quarters of the homosexual women considered their romantic feelings to be exclusive, just over 20 percent scored themselves a 5—slightly heterosexual—and the remainder scored between a 4 and a 3.

As with heterosexuality, then, the sexual preference "homosexuality" is similar for men and women. The common exception (as among the heterosexuals) is that women are less exclusive in their sexual feelings. Perhaps homosexual women feel this way due to their experience of "compulsory heterosexuality," which involves aspects of "learning" that can result in sexual feelings toward the opposite sex.

Like those who defined themselves as heterosexuals, most of those who defined themselves as homosexuals were exclusive in their sexual preference (6's) along all dimensions. Like the heterosexuals too, the homosexuals were most likely to be exclusive in their sexual behaviors, followed by romantic feelings and then sexual feelings. However, a considerable number of homosexuals are not exclusively so on all the dimensions of sexual preference. Many admit to having sexual and romantic feelings toward the opposite sex, but as with the heterosexuals, these often did not result in sexual behavior.

Since sexual behavior is the most direct and potentially public source of one's sexual identity, this incongruity should not be a surprise. While people strive for consistency between behavior and identity (we found large inconsistencies between the dimensions to be rare), they also face social sanctions in moving away behaviorally from either exclusive heterosexuality or exclusive homosexuality. On the other hand, sexual feelings seem the most variable. We feel less control over our feelings (desires, dreams, etc.). In addition, they are less subject than behaviors to the controls of daily life. They can more assuredly remain private. And by definition fantasies are creations of the imagination and can breach any moral, social, or sexual boundaries at personal whim. Romantic feelings are the most complex of the dimensions. "Falling in love" seems to be a feeling that goes beyond sexual feelings and beyond one's control. Further, if we want to actualize a romantic relationship we must risk rejection, disappointment, etc. Romance is similar in some ways to pursuing sexual behaviors because both have public aspects.

Bisexual Men

In contrast to the heterosexual and homosexual men, the bisexual men were widely dispersed in their sexual feelings. Some 17 percent scored a 1, 23 percent a 2, 20 percent a 3, 28 percent a 4, and 9 percent a 5. Just 3 percent considered their sexual feelings to be either exclusively heterosexual or homosexual.

The bisexual men were even more varied in their sexual behaviors than their sexual feelings. Eleven percent were exclusively with the opposite sex, 8 percent with the same sex. Another 45 percent scored themselves as a 1 or a 2, compared with 23 percent in the categories of 4 or 5. Thirteen percent said they were sexual with both sexes equally.

The bisexual men were, in general, more heterosexual than homosexual in their romantic feelings even though the responses ranged from 0 to 6. Almost 60 percent classified themselves between a 0 and a 2— from exclusively heterosexual to mainly heterosexual but significantly homosexual—with just under a quarter in the 0 category. Another 20 percent felt their romantic feelings were equal for both sexes. Close to a quarter were more homosexual, with 5 percent scoring as a 6.

Bisexual Women

The bisexual women were also more varied in their sexual feelings than the heterosexual and homosexual women. Nearly half were either a 1 or a 2 in their sexual feelings, while some 22 percent had stronger same-sex leanings, scoring either a 4 or a 5. A large group, nearly one-third, were situated in the middle, as 3's. One reported being exclusively heterosexual and none as exclusively homosexual on this dimension.

With regard to sexual behavior, the bisexual women fell more in the direction of the opposite sex than did the bisexual men. About 20 percent ranked their sexual behaviors as a 0 and a few as a 6. In addition, more than half said they were a 1 or a 2. Eleven percent were equally balanced in their sexual behavior, and 14 percent scored a 4, 5, or 6.

In their romantic feelings, the bisexual women were also widely dispersed and again more in the heterosexual direction. Nearly 60 percent ranked themselves from 0 to 2, with 15 percent exclusively heterosexual. Around 20 percent considered themselves equally balanced in both directions. Seventeen percent were a 4 or a 5, and a few were exclusively homosexual. There was no significant difference between these scores and those of the bisexual men.

As one would expect, unlike those who defined themselves as heterosexuals and homosexuals, most of those who defined themselves as bisexuals were not exclusive in their sexual preference on any of the three Kinsey scale dimensions. Bisexuals were similar to those two groups, however, in that they were least exclusive in their sexual feelings, while their sexual behaviors and romantic feelings were more likely to be only with one sex or the other.

The general findings about bisexuals' position on the Kinsey scales were similar in the questionnaire study to those we discussed in Chapter 4. Contrary to the belief that bisexuals are somehow in the middle (between heterosexuals and homosexuals), we found few self-defined bisexuals locating themselves exactly in the middle on any of the dimensions. Rather, there was a strong tendency for them to locate themselves at the heterosexual side of the scales. That is, both bisexual men and bisexual women were more likely to be heterosexual than homosexual on all three dimensions.

As a group, the men scored more in the homosexual direction in their sexual behavior than did the women. The bisexual women were the most homosexual in their sexual feelings, but the bisexual men were more likely to translate those feelings into same-sex behavior than were the bisexual women. As noted previously, we believe this finding reflects differential socialization and opportunities that are organized along gender lines.

Scale Types

Examining the distribution of 0's, 1's, 2's, etc. for a group along the Kinsey scale does not tell us much about the specific profiles of individuals. Thus, we analyze how sexual feelings, sexual behaviors, and romantic feelings combine for those who label themselves in different ways. To accomplish this, we separated the respondents by sex and self-label and then calculated how each group scored on all three aspects of sexual preference simultaneously, just as we did for the bisexuals in the 1983 interview study (Tables 14.2 and 14.3). The result is a three-number profile that corresponds to sexual feelings, sexual behaviors, and romantic feelings respectively. (For example, 101 indicates a person exclusively heterosexual in behavior and somewhat homosexual in sexual and romantic feelings.) The data reveal two basic heterosexual and homosexual types along with the same bisexual types we discovered earlier. The obvious implication is that sexual preference has no necessary or essential borders.

The Pure Heterosexual and Homosexual Types

The heterosexual and homosexual respondents we classified as pure types said that their sexual feelings, sexual behaviors, and romantic feelings were all exclusively for one sex or the other. In other words, they were 000 or 666 on the three dimensions. This pattern held for about two-thirds of the heterosexual men and 58 percent of the homosexual men. In comparison, it was evident for just under half of the heterosexual and homosexual women. Hence, slightly more men than women showed a completely exclusive sexual preference. Overall, about a third to a half of the self-identified heterosexuals and homosexuals in our study were less than completely anchored in an exclusive sexual preference.

The Somewhat Mixed Heterosexual and Homosexual Types

The somewhat mixed designation among heterosexuals and homosexuals was quite common. Those who defined themselves as heterosexual or homosexual but who diverged from being exclusive with only one sex on any of the dimensions were included here. Just under 20 percent of the heterosexual and homosexual men ranked themselves as a 1 or a 5 on one scale measure and a 0 or a 6 on the other two. This same pattern accounted for about a quarter of the heterosexual and the homosexual women. In the vast majority of these cases, the one-point variation occurred in sexual feelings. Thus, the profile was 100 or 566. Although the overall difference is obviously small, it does show that for more than a few, the category "heterosexual" or "homosexual" is not completely closed.

Others displayed greater variations from the exclusive profiles. Among all four groups—heterosexual and homosexual men and women— over a quarter diverged one scale value or more on two or three of the preference dimensions, or they diverged at least two scale values on one of the preference dimensions. Only a small percentage, most of them homosexuals, diverged 3 or more scale points on any one dimension. For the homosexual men, the discrepancy was usually on romantic feelings, while for the homosexual women, it involved their sexual feelings.

The Pure and Mid Bisexual Types

The pure bisexual type was distinguished by complete balance and equality, meaning that they were a 3 on the three dimensions of sexual preference. Compared with the pure heterosexual and homosexual types, this profile was uncommon among the bisexuals. Only 7 percent of the men and a little under 4 percent of the women fit this pattern. Unlike being heterosexual or homosexual, there is no predominant sexual profile that describes most of those who adopt the label "bisexual."

The mid bisexual type was similar to the pure type but showed more variation around the center of the Kinsey scale. These were mainly people who scored a 3 on at least one dimension and between a 2 and a 4 on the other two.[1] Thus their profile was 343, 423, and the like. This pattern described about one-fifth of the bisexual men and women. Adding the pure and mid types together, only around a quarter of the bisexual men and women were clearly situated in the center of the Kinsey scale.

The Heterosexual-Leaning Bisexual Type

This type consisted primarily of those who had a numerical profile that was between 0 and 2 for their sexual feelings, sexual behaviors, and romantic feelings across the board.[2] In nearly every instance of the heterosexual-leaning type, sexual feelings were either the strongest homosexual component in the profile or respondents were as homosexual on sexual feelings as on one or both of the other elements. This was far and away the most common of the five bisexual types. It was evident among about 40 percent of the men and about half of the women.

There is a remarkable overlap between the heterosexual-leaning *bisexual* and the somewhat mixed *heterosexual* profiles.[3] In fact, nearly 90 percent of the males and females who made up the somewhat mixed *heterosexual* group, and about 60 percent of the males and about a third of the females from the heterosexual-leaning *bisexual* group *shared the same combinations of scale dimensions.* Among the men, the overlapping combinations were 100, 010, 110, 210, 101, and 111. For the women, they were 100, 110, 101, 111, and 112. Thus the only apparent difference in sexual preference between these specific persons is what they called themselves.

By looking at these profiles we can see how the boundaries of sexual preference can overlap. Maybe some of the self-defined heterosexuals who fell into the somewhat mixed category will at some time take on a self-identification as "bisexual." The opposite could also be true—some self-identified bisexuals in the heterosexual-leaning category might at some time label themselves as "heterosexual." People who fit the two profiles outlined could represent interesting prospects for researchers who want to investigate the question of transition in sexual identity as it is occurring. We will explore the issue of stability and change in sexual identity in greater depth in later chapters.

The Homosexual-Leaning Bisexual Type

At the other extreme from the heterosexual-leaning bisexual was the homosexual-leaning bisexual type. This profile, however, was not common. It consisted mostly of those self-identified bisexuals who ranked themselves from 4 to 6 on the three measures of sexual feelings, sexual behaviors, and romantic feelings.[4] Eighteen percent of the men and 13 percent of the women who defined themselves as bisexual displayed the predominantly homosexual scale combination.

There was also a great deal of similarity between the homosexual-leaning *bisexual* and the somewhat mixed *homosexual* types. We found that about 80 percent of the males and females in the somewhat mixed *homosexual* group and about 60 percent of the males and around 55 percent of the females from the homosexual-leaning *bisexual* group *shared identical scale profiles.* Among the men, the common overlapping patterns were 566, 656, 565, 555, 554, 464, 454, and 444. For the women, they were 566, 466, 565, and 555. As with the heterosexuals and heterosexual-leaning bisexuals, these could be the people who are most likely to change their sexual identity. Or perhaps some self-identified homosexuals have never been and will never be exclusive in their sexuality, and some self-identified bisexuals simply feel more comfortable living a predominantly homosexual existence.

The Varied Bisexual Type

The last bisexual type that we isolated was the varied bisexual. This profile was distinguished by at least one large discrepancy between two of the dimensions of sexual preference as measured on the Kinsey scale. In other words, there was at least a three-point scale difference separating sexual feelings, sexual behaviors, and/or romantic feelings in varied combinations. People in this group had a range of numerical scale combinations. Some were more heterosexual overall (e.g., 103, 203, 302), others appeared more homosexual (e.g., 366, 563, 463), and still others seemed to be a blend of both (e.g., 524, 414, 522). Eleven percent of the self-defined bisexual men and 13 percent of the women were in this varied group.

Only one bisexual showed a discrepancy of 4 scale points or greater across the three dimensions. It was produced by having more homosexual sexual and romantic feelings but no homosexual sexual activity. (The profile was 404.) Such wide variations were similarly rare among the self-identified heterosexuals and homosexuals. Also, as noted earlier, only a handful of people in the heterosexual and homosexual groups varied even 3 scale points on the different dimensions. These patterns show that large differences among the levels of the three dimensions are fairly uncommon, regardless of the way people label themselves. In general, the three dimensions tend to hang together in individual profiles even if they are not identical.

In conclusion, then, when we examine sexual feelings, sexual behaviors, and romantic feelings simultaneously, there is a great deal of consistency between the three dimensions for most people regardless of their sex or self-label. And where there was not consistency, very few people had a discrepancy that was very large. This was only evident among those who were classified as varied bisexuals. The most common profile among both the bisexual men and women was the heterosexual-leaning type, followed by the mid type. While most of our heterosexuals and homosexuals fell into the pure heterosexual and pure homosexual types respectively, approximately one out of six men and one out of four women were what we call somewhat mixed heterosexual or somewhat mixed homosexual types.

Persons of these types could be in transition from labeling themselves "heterosexual" to labeling themselves "bisexual" or vice versa; or from labeling themselves "homosexual" to labeling themselves "bisexual" or vice versa. In any case, these types clearly show how persons with similar or the same sexual profile may label themselves differently. These findings raise the question of whether people's sexual preferences can be in flux regardless of the labels they adopt. It is to this question that we turn in the next chapter.

15

The Instability of
Sexual Preference

Whether an explanation for sexual preference is biological, psychological, or sociological, the basic presumption usually is that sexual preference is set early and changes little through life. In this view, any changes
in sexual feelings, sexual behaviors, or romantic feelings are especially
unlikely after adulthood. Further, it would be sexual feelings that would
be least likely to change because they are assumed to be the most basic
element of sexual preference.

Despite these presumptions, evidence of substantial change in an
individual's sexual preference does exist. For example, one recent study
noted the effect of the feminist movement on women's sexuality, teach-
ing some women that relating both emotionally and sexually to other
women is an option.[1] Another study focused on a group of women who
were all heterosexual in behavior and identity before participating in
swinging.[2] As a response to their husbands' wishes and their observations of other women, they became involved in sex with other women,
and all of them eventually identified themselves as bisexual. Their bisexuality, moreover, was an addition to their previous heterosexual interest,
and they still preferred heterosexual sex.

Here we consider the possibility of sexual preference changing in the
course of a person's life. How common are changes in the direction of a
person's sexual feelings, sexual behaviors, or romantic feelings? When
are they most likely to occur? Are such changes more common among
bisexuals than heterosexuals or homosexuals?

The Kinsey scale can be helpful in measuring changes in sexual preference.[3] We asked participants in our survey two questions to examine
the nature and direction of this change. First, in order to measure short-
term change, we asked how they would have ranked their sexual feelings,
sexual behaviors, and romantic feelings along the Kinsey scale three years

earlier. Second, we asked about the greatest change in sexual feelings they had ever experienced, its direction, and the age at which it occurred in order to gain some measure of long-term change and perhaps some insight into the most important changes in their lives; assuming, that is, that sexual feelings are the most basic element of sexual preference. (We also compare ratings on the 1983 scales with the 1988 ratings of the reinterview respondents as a measure of change—see Chapter 23.)

Short-Term Change

Here we explore two aspects of short-term change: *prevalence*—the percentage of respondents who changed on any dimension; and the *magnitude* of their change, that is, the distance that those who changed traveled along the Kinsey scale (Table 15.1).

Heterosexual Men

The men who defined themselves as heterosexual showed a great deal of stability in their sexual preference along all three Kinsey Scale dimensions. Over 80 percent reported no change in the direction of their sexual feelings over the three-year period. Of the remainder who changed, in general, the range of change did not exceed one point on the scale (most often between 0 and 1 or 1 and 2). Almost 90 percent of the heterosexual men reported no change in the direction of their sexual behavior or romantic feelings over the three years. No one moved more than one point on the scale (most often between 0 and 1) on either of these dimensions.

Overall only 17 percent of heterosexual men reported change in either their sexual feelings, sexual behaviors, or romantic feelings over a period of three years. The most common change was small—usually between 0 and 1.

Homosexual Men

Homosexual men showed about as much short-run stability in their sexual preference as did the heterosexual men. Just over 80 percent reported no change in their sexual feelings, and few of those who changed moved more than one point (the most common pattern was between 6 and 5 or 4 and 5). Eighty-five percent of the homosexual men reported no change in the direction of their sexual behavior. Ten percent moved one point up or down the scale (usually from 5 to 6 or 4 to 5). When a greater change in behavior was reported, more than half moved across the scale (from 0 to 6 or 2 to 6). Regarding their romantic feelings, over 80 percent reported no change, and no one reported moving more than one point (usually from 4 to 5 or between 5 and 6).

Among men who defined themselves as homosexual, then, short-term change in sexual preference was not common—just under a fifth reported change on any dimension—and the change was usually small (most frequently between 5 and 6).

Bisexual Men

Bisexual men experienced the greatest change on all the dimensions. Only about half reported no change in their sexual feelings over the three years—much less than for the heterosexual and homosexual men. A third moved one scale point in either a heterosexual or homosexual direction, though more often in the latter. (The most common changes were 0 to 1, 1 to 2, etc.) The remainder had moved more than one point along the scale. These were almost always two-point moves toward the middle of the continuum from either direction. In terms of sexual feelings, men who defined themselves as bisexual occupied a number of scale points not always thought of as bisexual (i.e., a 1 or a 5). When they did change, the magnitude of change was not extensive for most of them.

Only about 40 percent of the bisexual men reported no change in the direction of their *sexual behavior*—this reflects the largest proportion to change and represented the greatest degree of change on any dimension among the male groups. Close to a third moved one scale point either way, and almost the same proportion moved at least two points. The one-point changes routinely occurred at the edges of the scale continuum (e.g., 1 to 2 or 6 to 5). Larger changes were often 2 points (e.g., 2 to 0 or 3 to 5). Sometimes the changes were very large ones (e.g., 0 to 6 or 4 to 1).

Almost 60 percent of the bisexual men reported no change in their romantic feelings, much less than for heterosexual and homosexual men. About 20 percent changed one point, and another 20 percent two or more points. There was no clear pattern to the larger changes, which occurred in both directions and ranged from two to five points. This pattern of change parallels that for sexual behavior—the magnitude of change was greater for those at the extreme categories.

There appears to be a different pattern of change in romantic feelings compared with sexual feelings for bisexual men. Change in romantic feelings seemed to be both more volatile *and* more stable depending on where the respondent was on the Kinsey scale three years earlier. The bisexual men who defined their romantic affections as exclusively heterosexual (0) did not move far from this; an element of stability is evident. But for those in one of the other categories, a change in romantic feelings was likely to be much *greater* than for sexual feelings.

Sexual feelings are often more diffuse and impersonal, romantic feelings more singular and focused. Thus, entering into a new personal relationship over the three-year period could radically affect the heterosexual-homosexual direction of one's romantic feelings. Further, forming a romantic attachment to another man (moving away from a 0) is a big step for the bisexual man.

In summary, short-term change was much more common for bisexual than for heterosexual or homosexual men. Among bisexuals, the largest changes occurred in sexual behavior. More reported changing, and changing more, in their sexual behavior than on any other dimension. Many also changed in their sexual feelings but not to any great degree. Fewer reported changes in their romantic feelings but among those who did the change was much greater than for sexual feelings, possibly indicating a change in the primary relationship from one sex to the other.

Heterosexual Women

Like the heterosexual men, the women who defined themselves as heterosexual showed stability in their sexual preference along all three dimensions. Eighty percent of the heterosexual women reported no change in the direction of their sexual feelings over the three-year period, and the change among the remainder was small (usually between 0 and 1). Similarly, just over 80 percent reported no change on the sexual behavior dimension, with about 15 percent moving one point (almost always from 1 to 0). In their romantic feelings, about 90 percent reported no change at all, with the remainder moving one scale point (almost always between 0 and 1).

Overall only 20 percent or less of the heterosexual women reported short-term change on any dimension. The magnitude of change was relatively small, most commonly between 1 and 0.

Homosexual Women

Women who defined themselves as homosexual also showed stability in their sexual preference. Three-quarters did not change in their sexual feelings over the three-year period. Fifteen percent changed one scale point in either direction (practically all between 5 and 6), while ten percent changed at least two scale points (over half of these cases involved wide swings in the homosexual direction—1 to 5, 3 to 6, etc.). Homosexual women were stable in the direction of their sexual behavior. Over two-thirds did not change (showing somewhat less stability than the homosexual men), another 20 percent had moved up or down one unit (between 5 and 6), and nearly 10 percent moved two or more points (usually from 0 to 6 or 1 to 6). About as many homosexual women, approximately three-quarters, stayed the same in their romantic feelings too. Seventeen percent moved up or down one scale point and just under 10 percent reported a more marked change.

Like the homosexual men, therefore, the homosexual women showed a great deal of stability in their sexual preference. At the same time, almost a third of the homosexual women did report change on at least one dimension. The homosexual women also reported greater short-term changes than did the homosexual men.

Bisexual Women

The women who defined themselves as bisexual showed the greatest change of the female groups. While just over half of the bisexual women reported no change in their sexual feelings, approximately a third moved one scale point up or down, and 17 percent noted a change of at least two points. Bisexual women who changed the direction of their sexual feelings from the extreme ends of the scale traveled further than did the bisexual men.

We also found substantial change in the sexual behavior of the bisexual women. Although 40 percent reported no change over the three-year period (almost identical with the men), widespread dispersion was evident for those who did change: about 40 percent moved up or down one point and about 20 percent moved two points either way. (Nearly all the small-increment changes were between 0 and 1 and 2 and 1. Over half of the large changes were again sweeping—e.g., from 1 to 5 or 6 to 0.)

Only about 40 percent of the bisexual women reported no change in their romantic feelings over the three year period (showing less stability than the bisexual men). Some one-third of the bisexual women moved one level up or down and slightly over a quarter moved two or more. Most of the small changes were on the heterosexual side of the spectrum (e.g., between 0 and 1 and 1 and 2). The large changes were wide ranging, with about one third–being at least halfway across the Kinsey scale. Change in romantic feelings was more common among bisexual women than among bisexual men.

In summary, short-term change was much more common for bisexual women than for heterosexual or homosexual women. In this they were similar to the bisexual men. The magnitude of change was greatest in their romantic feelings and sexual behavior. This was true for the men especially with regard to sexual behavior. For both bisexual men and women, substantial changes in sexual feelings also occurred, suggesting that sexual feelings are not particularly basic or fixed.

Major Long-Term Change

Rather than limiting our analysis strictly to short-term change of varying magnitudes, we asked "Has there ever been a *major change* in your *sexual feelings* as measured by the Kinsey scale over the years?" The prevalence of such change, from greatest to least, was as follows: homosexual women—just over two-thirds; bisexual women—just under two-thirds; bisexual men—more than a half; homosexual men—almost a third; heterosexual women—about 20 percent; and heterosexual men—15 percent (Tables 15.2, 15.3). Such change was least common among the heterosexual men and women. Overall, the women were more likely to indicate having changed in this respect than the men, especially the homosexual women.

We also asked about the direction of this change.[4] The percentage who experienced a change of at least two points (first percent) and at least three points (second percent) is as follows: homosexual women—55 and 50 percent; bisexual women—52 and 35 percent; bisexual men—40 and 18 percent; homosexual men—23 and 16 percent; heterosexual men—7 and 5 percent; and heterosexual women—5 and 4 percent. *For all six groups,* the vast majority of the major changes were in the homosexual direction.

For the heterosexual women, the change usually occurred after age twenty and was spread out from there across the age spectrum. Among the homosexual women, there was a major concentration. For 60 percent, the major change in sexual feelings occurred between the ages of twenty-one and twenty-nine. Cumulatively, 17 percent said the change came before age twenty-one, 78 percent before age thirty, and 100 percent before age forty (i.e., none of the women over forty reported a major change for age forty and beyond). The bisexual women were somewhat more spread out on the age variable. Cumulatively, 14 percent said the change came before age twenty-one, 54 percent before age thirty, 91 percent before age forty, and the additional 9 percent at age forty or beyond.

The age pattern for the few heterosexual men who reported a major change was widely dispersed. Among the homosexual men, cumulatively, 24 percent indicated that the major change occurred before age twenty-one, 56 percent before age thirty, and 85 percent before age forty. In comparison, only 3 percent of the bisexual men experienced this major change before age twenty-one, 30 percent before age thirty, and 68 percent before age forty. The remaining 32 percent experienced the major change at age forty or beyond. Thus, major changes in the direction of sexual feelings were more likely to occur *later* for the bisexual than for the homosexual men.

In sum, major changes in sexual feelings across the long term were likely to be reported more by homosexual women and bisexual men and women than by the other groups. Like the short-term changes, these long-term changes were mainly in the homosexual direction and occurred later for the bisexuals than the homosexuals (and later for the men than the women among both bisexuals and homosexuals).

Clearly, then, sexual preference is not always a fixed phenomenon. Some members of all of our groups changed along all of the dimensions of sexual preference. While heterosexuals and homosexuals showed the least change, we should not be blind to the fact that almost one in five of them reported change along one of the dimensions of sexual preference in the past three years. How can we explain the differences between the groups?

We propose that heterosexuals and homosexuals showed the least change because "heterosexual" and "homosexual" are labels with definable boundaries that can restrict the ability or desire to explore change— they act as "anchors" of sexual identity. Bisexuals showed the greatest change because the label "bisexual" is less clear and allows for a range of feelings and behavior—i.e., they have wider or less clear boundaries within which to explore their sexuality.

The magnitude and type of change shows further that people who accept the socially available sexual identities of "heterosexual" and "homosexual" are indeed more strongly anchored in their sexual preference than others. This is because changes from the edges of the Kinsey scale (0,6) are much less likely (and smaller) than changes that start away from the edges (1,5). The distances between various increments on the Kinsey scale, therefore, can be seen as far from uniform in their *meaning*. Moving from a 0 to a 1 or a 6 to a 5 may be a bigger step than moving from a 2 to a 4 or perhaps a 5 to a 2, as slipping the anchor from an exclusive sexual preference has more profound implications for one's sexual identity.

Moving from an exclusively anchored preference seems to make further change easier and more fluid. This would explain why some people, especially bisexuals, can move frequently and in a variety of ways on the Kinsey scale once they become unanchored, and why there are a greater variety of sexual profiles subsumed under the label "bisexual" than "heterosexual" or "homosexual."

As we have shown, sexual preference is different for men and women. So too is the case of change in sexual preference. The homosexual women stand out among the heterosexual and homosexual groups. They were less fixed on any dimension in the preceding three years, and they also were more likely to experience major long-term changes in their sexual feelings in the homosexual direction. This probably has to do with their involvement with San Francisco lesbian and feminist groups, to which many of them belonged. These groups promoted and supported radical changes in sexual lifestyles for women.

Differences between men and women were also found among the bisexuals. The women showed greater change in their romantic feelings than did the men. (This is consistent with our findings from the 1983 interview study that the bisexual women were more flexible in the direction of their romantic feelings.)

Shifts in sexual feelings across the life course also were much broader and more dramatic than those over a three-year period. In particular, the homosexual women reported the most extensive changes and at the earliest ages. Amazingly, one out of two moved at least halfway across the Kinsey Scale. Such sweeping fluidity was nearly as common among the bisexual women, but more spread out across their lives. The bisexual and homosexual men, while showing less sweeping changes, were likewise noticeably open to movement. Major changes in sexual feelings usually occurred in the homosexual direction. In the case of bisexuals, the homosexual changes were not enough to erase their heterosexual interests, and emergent same-sex feelings seemed to be gradually incorporated as an addition to their sexuality, possibly making sexual change less of an issue of identity for them than for the homosexuals. With the homosexuals, the shifts were more away from bisexual feelings and more toward exclusively the same sex. Thus bisexuals seemed to make a journey to bisexuality, and homosexuals seemed to make a journey through it. We think that bisexuals show more of a flexibility in their sexual choices, rather than a transition from heterosexuality to homosexuality.

16

Sexual Profiles

Earlier we described patterns of bisexual activity, notably the mix of opposite-sex and same-sex partners, opposite and same-sex sexual behaviors, and the sexual problems that can ensue. Here we compare these patterns with those of heterosexuals and homosexuals. Thus we ask such questions as: Are there differences in the number of sexual partners between the three groups and in the ways sexual activities become organized? Does the combination of same-sex and opposite-sex sexual behavior mean that bisexuals have more sex than heterosexuals and homosexuals? Or are bisexuals similar to the other two groups in their sexual frequency? Does the lifestyle of bisexuals create more sexual problems for them? Even more, who is most likely to contract a sexually transmitted disease? Finally, we examine differences in sexuality between men and women, exploring whether gender differences are similar across all three preference groups.

Sexual Partners

Comparisons with Heterosexuals

There was no significant difference between the bisexuals and the heterosexuals in their number of opposite-sex partners in the *twelve months* preceding the study. The overall median was two partners for both the bisexual and the heterosexual women and heterosexual men; one, for the bisexual men (Table 16.1). There was also no significant difference in lifetime opposite-sex partners. The median was twenty for men in each preference group; twenty-five and thirty, for the heterosexual and bisexual women.

In the twelve months before the study, though, more bisexuals than heterosexuals reported they participated in sexual threesomes (nearly a half versus less than 20 percent) and sex parties (approximately a third versus less than 10 percent). At the same time, both bisexuals and heterosexuals were about equally likely to have had an opposite-sex casual partner (from 48 to 59 percent) or an opposite-sex anonymous partner (between 20 and 30 percent) in the last twelve months. The patterns were similar for both sexes.

Comparisons with Homosexuals

Bisexual men reported significantly fewer *same sex* partners in the preceding twelve months than their homosexual male counterparts (median of three for the first group versus six for the homosexual men) and in their lifetime (median of thirty for the bisexual men versus a hundred for the homosexual men). There were no significant differences in numbers of partners between bisexual and homosexual women over either time period (lifetime medians were seven and ten partners, respectively; both reported a median of just one partner in the last twelve months).

Among the women, though, more bisexuals than homosexuals had participated in sexual threesomes (nearly a half versus under 10 percent). This was true too for participation in sex parties (a third versus 10 percent). The homosexual men were more likely than the bisexual men to have casual partners, anonymous partners, and partners whom they had paid for sex over the last twelve months. They were also more likely to participate in sex parties (just over a third versus about a fifth).

Total Number of Partners

When we added the number of opposite-sex and same-sex partners the men in our study had during the *preceding twelve months,* the homosexual men had the highest number (median of six), followed by the bisexual men (median of five), and the heterosexual men (median of two). The bisexual men had more same-sex than opposite-sex partners during those twelve months.

When we combine the number of opposite-sex and same-sex partners the men had over their *lifetime,* the homosexual men estimated the highest total (median of one hundred two), followed by the bisexual men (median of sixty-five), and then the heterosexual men (median of twenty). More of the bisexual men's lifetime partners were men. We also found that two-thirds of the homosexual men had at least one opposite-sex partner, and about half of the heterosexual men had had at least one same-sex partner in their lifetime.

Turning to the women and their total number of partners in the twelve months leading up to the study, the bisexual women had a median of three, the heterosexual women a median of two, and the homosexual women a median of one. Bisexual women had also had more men than women as partners in the preceding twelve months.

Combining lifetime opposite-sex and same-sex partners for the women, the bisexual women had the highest total (median of forty-four), followed by the heterosexual and homosexual women (medians of twenty-six and twenty-two, respectively). More of the bisexual women's lifetime partners were men. Homosexual women also had had opposite-sex partners. Indeed, more of their lifetime partners had been men. Also, more than 50 percent of the heterosexual women reported they had had a same-sex partner at least once in the past, though few mentioned more than one.

Differences Between Men and Women

The bisexual men reported more *same-sex* partners and a larger *total* number of such partners than the bisexual women over the last twelve months and in their lifetime. The women had a greater number of lifetime *opposite-sex* partners.

Among the homosexuals, the men had far more *same-sex* partners in the last twelve months and in their lifetime than the homosexual women did. The homosexual women had had many more *opposite-sex* partners over their lifetime. This difference, however, was not great enough to offset the large difference in lifetime same-sex partners, which gave the homosexual men a larger number of total lifetime partners.

The bisexual men had more anonymous and casual same-sex partners during the past twelve months than the bisexual women. Among the homosexuals, the men had more casual sex partners and partners in threesomes or at sex parties than their female counterparts. The homosexual men also had a much greater number of anonymous same-sex partners in those twelve months than did the homosexual women.

Thus do bisexuals, because they engage in sex with both men and women, have fewer *opposite-sex* partners in their lifetimes than heterosexuals? The answer, overall, is no. (We do find that they have *more* of some types of opposite-sex partners, such as those who participate in sexual threesomes and sex parties.) The same question can be asked with regard to their *same-sex* partners compared with homosexuals. Here the answer, overall, is that the bisexual men have fewer same-sex partners in their lifetimes and the bisexual women do not.

Does adding the number of both opposite-sex and same-sex partners make bisexuals higher in their total number of sexual partners than heterosexuals and homosexuals? Among the women, the answer is yes. The bisexual women had the greatest number of partners, followed by the heterosexual and homosexual women. For the men the answer is no.

The homosexual men had the greatest number of partners, followed by the bisexual men and then the heterosexual men.

Several factors underlie these differences. One factor is that male partners are easier to obtain—thus the highest total numbers for the homosexual men (who only, or mostly, seek other men), then bisexual men (who spread their energies between men and women), then hetero- sexual men (who only, or mostly, seek women). The lowest total of lifetime sexual partners is for the homosexual women (who rarely seek men as partners at all). A second factor is, as we found earlier, that the search for male partners begins at an earlier age for the homosexual men than for the bisexual men. This gives them a longer period in which to experience same-sex behaviors. Third, homosexual men are, in general, more involved in the homosexual subculture—where there are many available sexual partners—than are bisexual men, who are more tied to the heterosexual world. Finally, at the time of the study, in the bisexual world, as compared to the heterosexual world, there was stronger sup- port for nonmonogamy and experimentation (for example, engaging in sex with multiple partners—threesomes, group sex parties); thus the higher total number of partners for the bisexual women compared with the heterosexual women. Bisexual women who are open to experimenta- tion find many willing partners (especially men).

Sexual Activity

Comparisons with Heterosexuals

Overall the bisexual men and women showed no difference from their heterosexual counterparts in the incidence and frequency of most types of sexual behaviors in the preceding twelve months (Table 16.2). How- ever, they did report a higher incidence of involvement in sadomaso- chism with opposite-sex partners compared to heterosexuals. Overall, however, there were few other significant differences for other forms of unconventional sex (cross-dressing, spanking, anal and vaginal fisting, urination, and enema and feces play with opposite-sex partners). The one exception was that bisexual women reported a higher incidence of anal fisting than heterosexual women (10 percent versus 1 percent). Few bisex- uals or heterosexuals reported engaging in these more unconventional sexual activities with opposite-sex partners over the prior twelve months.

Comparisons with Homosexuals

More differences appeared when the bisexual men and women were compared with the homosexual men and women. The bisexual women masturbated somewhat more than the homosexual women (three- quarters versus about three-fifths reporting a frequency of at least once a week). Over four-fifths of the men in both preference groups said the same. But the bisexual men and women had a lower frequency than the homosexuals did in a number of same-sex sexual activities: masturbating and being masturbated by a same-sex partner (median values of one to three times a month for homosexual men and women, less than once a month for the bisexuals), performing and receiving oral sex with a same-sex partner (the same medians as for masturbating each other), and some anal sex activities (in most cases frequencies of less than once a month).

One difference that stands out for the men, but not the women, is that the bisexual men were less likely than the homosexual men in the previous twelve months ever to have engaged, or if they had, to have engaged less frequently, in nongenital activities involving kissing, hugging, cuddling, stroking, and massaging with same-sex partners. (Three-quarters of the homosexual men versus half of the bisexual men reported this a great deal with same-sex partners, over four-fifths of the women in each group said the same.)

In terms of unconventional activities, the bisexual men reported a lower incidence than the homosexual men for urination play. There were no other significant differences. Few bisexuals or homosexuals reported engaging in these more unconventional sexual activities with same-sex partners over the past twelve months.

In sum, bisexuals' sexual activities with the opposite sex were similar to those of heterosexuals, but their same-sex behaviors showed a number of significant differences from those of homosexuals. Overall, a significantly larger proportion of bisexuals than homosexuals had not engaged in certain same-sex activities during the past twelve months. This was especially true for bisexual men not having anal sex, and for bisexual women not engaging in same-sex masturbation or oral sex during the twelve-month period.

Total Sexual Frequency with a Partner

Did the bisexuals have a greater *total* sexual frequency than the heterosexuals and homosexuals? To address this question we asked how often they had engaged in sexual activity with another person—male or female—in the past year. For each sex, the homosexuals showed the lowest overall frequency of sexual activity. Among the men, the bisexuals had sex most often, but not that much more than the heterosexuals. This difference was even narrower among the women, where the bisexuals and the heterosexuals had very similar total frequencies.

The proportion who had sex with a partner once a week or more was, from highest to lowest: just under two-thirds of the bisexual men, just over 60 percent of the heterosexual women, just under 60 percent of the bisexual women, over half of the heterosexual men, about 45 percent of the homosexual women, and around 40 percent of the homosexual men.

Differences Between Men and Women

What stands out in the male-female comparisons is that the bisexual men showed a greater frequency of a variety of homosexual acts than the bisexual women, especially the following: masturbating a partner, masturbating in front of a partner, receiving oral sex, and finger-anal sex. The same was found among the homosexuals. In addition to the above behaviors, the homosexual men reported higher frequencies of self-masturbation, performing and receiving oral-anal sex, plus anal fisting and enema play. On the other hand, the bisexual and homosexual women reported substantially higher levels of same-sex kissing and cuddling, etc. than their male counterparts. Thus typical gender differences in sexual expression seem to appear regardless of sexual preference.

In conclusion, do bisexuals, presumably because they have sex with both men and women, have less heterosexual sex than heterosexuals and less homosexual sex than homosexuals? No and yes. While overall the frequency of heterosexual activity is not significantly different between bisexuals and heterosexuals, we did find that the bisexuals had less frequent homosexual activity than homosexuals. This supports our earlier suggestion that our bisexuals had a heterosexual bias and that their homosexual behavior was an addition to the heterosexual behavior.

Bisexuals not only had less homosexual genital sex than did the homosexuals, but also, among the men, fewer nongenital behaviors such as same-sex hugging and kissing. This reflects the bisexuals' lesser acculturation to what are often considered to be "advanced" homosexual practices.[1]

Adding the frequencies together (regardless of whether the partner was of the same or opposite sex), bisexuals were more sexually active than heterosexuals. The bisexuals and heterosexuals in turn were more sexually active than were the homosexuals.[2] While the homosexual men had more partners than the other groups, every other group, especially the bisexuals, seemed to have sex more frequently.

In the early 1980's, then, bisexuality did not seem to involve simply substituting one sex for another in the search for sexual satisfaction. Rather, same-sex and opposite-sex partners were combined, with the result that bisexuals had a greater frequency of sex.

Sexual Problems

Our final question was whether the bisexuals we surveyed had more sexual problems than the heterosexuals and the homosexuals, and whether the sexual problems they reported differed in character. We also asked them to rate the severity of reported sexual problems on a scale from "a little" to "a great deal."[3]

Comparisons with Heterosexuals

One might hypothesize that bisexuals have more sexual problems with the opposite sex than exclusive heterosexuals, thus the reason for their involvement in homosexual sex. Among both men and women we did not find a great deal of difference between the bisexuals and heterosexuals with regard to any of a range of problems with *opposite-sex* partners that we asked about (Table 16.3). The most common problems reported by men of both preference groups were difficulty finding suitable female partners and difficulty telling women their sexual needs. In the case of the women, the most common problems were difficulty staying sexually aroused and difficulty telling their sexual needs to partners.

Comparisons with Homosexuals

It could also be hypothesized that bisexuals have more sexual problems in their homosexual life because they are less acculturated to homosexuality. However, even though the differences were small, we found that the homosexual men cited somewhat more problems with *same-sex* partners than the bisexual men (Table 16.3). The homosexual men rated their communication with male sexual partners as more problematic than did the bisexual men; they cited more problems in feeling sexually adequate and also noted more difficulty in feeling good about their sexual experiences with men. Among the women, the only difference was that the bisexuals were more likely than the homosexuals to report problems in finding a suitable same-sex partner.

Differences Between Men and Women

The overall trend that appeared was that, among *all* three groups, typical sex problems that characterize men and women appeared. Women were more likely to have problems with sexual arousal, reaching orgasm, and reaching orgasm too slowly. Men were more likely to reach orgasm too quickly. Also, respondents saw their male and female partners in these ways. However, few rated these problems as severe for themselves or for their partners.

Sexual preference did make a difference though. Here, the outstanding group was the bisexual women. In their *same-sex* activity, compared to bisexual men, the bisexual women were more likely to report orgasm problems, staying sexually aroused, and being able to tell their partners about their sexual needs. They also were less likely to feel sexually adequate and to feel good about their sexual experiences. Gender differences among heterosexuals and homosexuals were less evident.

In sum, did the bisexuals differ from the heterosexuals and homosexuals in the types of sexual problems they reported? They did, although these problems were not rated as severe and the nature of this difference was not the same for bisexual men and women. Except for having had more difficulty than the homosexual women in finding suitable same-sex partners, bisexual women had a profile similar to both heterosexual and homosexual women. On the other hand, in general the bisexual men seemed to do better compared to both heterosexual and homosexual men, especially in feeling sexually adequate (regardless of the sex of their partner).

We theorize that bisexual men may have fewer problems than heterosexual men because having both male and female partners and a greater number of total partners leads to greater feelings of sexual confidence. In addition, bisexual men do not seem to be operating from unhappy heterosexual relationships, so additional partners are more likely to be a supplement to and not a substitute for their involved relationships. Another possible reason that bisexual men have fewer problems than homosexual men might be because they expect and look for less in same-sex relationships. Same-sex activities can function as mainly an additional sexual outlet and are not as often predicated on an emotional relationship. So there is less to get upset about. It might also be that bisexual men take any sexual problems that do arise with a same-sex partner less seriously than homosexual men. For the latter, problems with homosexuality may be enveloped in larger questions about being gay. For the former, such problems are unlikely to have similar meaning because of more marginal involvement in the homosexual subculture. In contrast, with bisexual women, it could be they have more problems than homosexual women in finding suitable same-sex partners because of their lesser integration into the lesbian world and the hostility that they feel from that world. This is also reflected in their having more sexual problems in their same-sex activities than bisexual men do.

Sexually Transmitted Diseases

Given the emerging climate of AIDS in late 1984 and the widespread stereotype of bisexuals as disease spreaders, we examined how the bisexuals fared with respondents who were heterosexual and homosexual in terms of the incidence of sexually transmitted diseases (STD's).

Comparisons with Heterosexuals

Because the bisexuals had more anonymous sex partners than did the heterosexuals, one might expect them to have more STD's (Table 16.4). Over 60 percent of the bisexual men and women had contracted an STD in their lifetime, compared with slightly less than half of the heterosexual men and women. Even more, among those who indicated having done so, the bisexuals were much more likely to do so repeatedly. About a quarter of the bisexual men and women had contracted STD's four times or more compared with only 2 percent of the heterosexual men and 8 percent of the heterosexual women.

At the time of the questionnaire study, bisexuals were beginning to become aware of and concerned with AIDS. Herpes was also a big worry among both bisexuals and heterosexuals. Among the bisexuals, 34 percent of the men and over 40 percent of the women said they had cut back their heterosexual activities at least "somewhat" because of concern about STD's. For the heterosexuals, the corresponding figures were 19 percent for the men and 39 percent for the women. The differences between bisexuals and heterosexuals were not significant.

Comparisons with Homosexuals

The bisexual men had fewer anonymous sex partners than did the homosexual men, but their experience with STD's was not significantly different: 69 percent of the bisexual men and 73 percent of the homosexual men had ever contracted a STD, and close to a quarter of each had contracted STD's four times or more.

The bisexual women had more anonymous sex partners and more male sex partners (who had additional numbers of anonymous partners) than did the homosexual women. Thus, as expected, they did show a significantly greater prevalence of STD's. Just over 60 percent of the bisexual women but less than 40 percent of the homosexual women had ever contracted a STD. Over a quarter of the bisexual women had contracted STD's four times or more, compared with only 8 percent of the homosexual women.

Did a concern with STD's lead to a cutback in homosexual activities? Among the bisexuals, almost three-quarters of the men versus 20 percent of the women said they had cut back their homosexual activities either "somewhat" or "a great deal" compared with about 70 percent of the homosexual men and 7 percent of the homosexual women. The difference between bisexuals and homosexuals was significant for the women but not the men.

Differences Between Men and Women

The major differences in contracting STD's between men and women were found among the homosexuals. Homosexual women reported the lowest rate of incidence of all groups; homosexual men, the highest. Accordingly, more homosexual men than homosexual women had cut back on their sexual activities. Although there was no significant difference in the incidence and frequency of STD's among the bisexual men and women, the men were more likely to have made a substantial cutback in their same-sex behaviors.

Also, although heterosexual men were no more likely than heterosexual women to report STD's, more heterosexual women than men had made cutbacks in their sexual activity. No significant difference was found among the bisexual men and women in the proportion who had substantially cut back their sexual behaviors with the opposite-sex.

As the 1980's progressed then, we found that our respondents were very active sexually. An ominous sign of this activity was their experience with sexually transmitted diseases. But we were seeing the last days of the sexual revolution. People were beginning to cut back on their sexual activities. Neither we nor they realized at the time just how big the changes would be!

17

Intimate Relationships

Sexual preference cannot be separated from involved relationships, because it is often through these relationships that sexual feelings and behaviors are realized and sexual identity maintained. The importance of *significant* others—in distinction to more casual sex partners—is often ignored in studies of sexual preference; sexual preference is considered a property of individuals, measured solely by the Kinsey scale. We go beyond this narrower view by trying to understand the individual in a web of many types of social relationships without which his or her sexuality would have no meaning or sustenance.

This concern led us to ask about the number of significant involvements people had with each sex, whether or not they were monogamous, their opportunities for meeting partners, and the longest relationship they had ever maintained. We wanted to know which patterns of relationships were most common among women and men who adopted a particular sexual identity, how stable such arrangements were, and what role sexual preference played in a person's ability to form and sustain them. Given that many bisexuals see multiple relationships as inherent to their sexual preference, are they less able to sustain significant relationships than heterosexuals or homosexuals?

The Number of Involved Relationships

We asked all three preference groups to report the number of men and women with whom they were significantly involved in "an affectional and sexual way" (Tables 17.1, 17.2, and 17.3). The term "significant involvement" thus implied some degree of personal investment and commitment and was distinct from the number of sexual partners. Studying the number of involved partners allowed us to assess how often people who defined their sexuality in different ways had multiple relationships with each sex. We could also examine the common belief that bisexuals are incapable of being involved and committed, and that if they are, they cannot sustain their interest in one partner or one sex.

Opposite-Sex Relationships

The heterosexual men had more opposite-sex relationships than did the bisexual men. Fourteen percent of the heterosexual men reported no significant involvement with a female partner, about a third had one significant relationship, and about a half had two or more. About a quarter of the bisexual men had no involved female relationships whatsoever, a half had one, and some 20 percent had two or more. (Just under 7 percent of the men who identified as *homosexual* said that they currently had at least one significantly involved sexual-affectional relationship with someone of the opposite sex.)

The women presented the same basic picture, i.e., the heterosexual women were more likely to have involved relationships with men than were the bisexual women. Among the heterosexual women, 13 percent reported no involved relationships with the opposite sex, almost two-thirds one involvement, and some 20 percent two or more. Among the bisexual women, about a quarter had no involved relationship with a man, about a half had one, and the remaining one-fourth had two or more. (A few *homosexual* women, about 8 percent, told us they currently had at least one sexual-affectional relationship with a man.)

Same-Sex Relationships

The homosexual men had more same-sex relationships than did the bisexual men. About a third of the homosexual men said that they had no such relationships with males, a third had one, and a third had two or more. Over half the men in the bisexual group reported no same-sex involvement, another quarter indicated one, while about 20 percent had two or more male relationships. (Just under 4 percent of the *heterosexual* men indicated currently having an involved same-sex sexual-affectional relationship.) Thus, the homosexual men were much more likely to have one or more same-sex involvements. Many bisexual men had no same-sex relationships whatsoever.

The homosexual women stood out as unique when it came to their involved relationships with females. Very few—only 7 percent—indicated being in multiple same-sex relationships. The vast majority, about 70 percent, said that they had one involved female relationship.

The bisexual women, in contrast, were much less likely to have an involved relationship with a female. Over half indicated no such relationship. Of those who were in same-sex relationships, single involvements were about as common as multiple ones. (Three percent of the *heterosexual* women also said they currently had at least one involved relationship with a person of the same sex.)

Total Involved Relationships

When we look at involved relationships with both the same and the opposite sex together, a more detailed relationship picture emerges, especially among the bisexuals. About half of the bisexual men and the bisexual women were involved in multiple relationships (i.e., two or more involved partners). This was well beyond the percentages for the homosexual or heterosexual men or women. Only 16 percent of the bisexual men and women were uncoupled. The remainder (around 40 percent) were in one involved relationship.

In these patterns of relationships, we again see a strong heterosexual leaning in the case of bisexuals. For example, the most common arrangement was one involved relationship with the opposite sex (just under a third of the men and women indicated this pattern). Only about 8 percent of the men and women indicated one involved relationship with the same sex. Obviously, the bisexuals had a variety of mixed-sex relationship combinations as well. About 12 percent indicated having an involved relationship with one man and one woman, while a third had multiple ongoing relationships with other combinations.

Regardless of sexual preference, most of the people in our study were in coupled relationships of one type or another. But the bisexuals stood out. They were more likely to have multiple involved relationships rather than the single involvements that characterized the heterosexual men and women and the homosexual women. Their involved partners were usually of the opposite sex, which corresponds with the "add-on" nature of bisexuality for our respondents. Along this line, more had an involved relationship *only* with the opposite sex, and fewer had an involved relationship *only* with the same sex. Those who mentioned a same-sex relationship usually had one or more involvements with opposite-sex partners too.

Monogamous or Nonmonogamous Relationships

As we have noted, at the Bisexual Center people often spoke of having open sexual relationships. They thought it was unreasonable to limit their sexual desires or activities in any way. But was this merely ideology, or was it really put into practice among bisexuals? Others could often be heard debunking as myth the perception that most bisexuals were "promiscuous." Were the bisexuals different from heterosexuals and homosexuals when it came to having sex outside of their relationships?

Overall (Table 17.4) the bisexual men displayed a greater propensity to engage in sex outside of their involved relationships than their hetero-sexual or homosexual male counterparts.[1] Among the men involved in just one significant relationship, close to half of the bisexuals had outside sex, compared with just over a third of the homosexuals and 10 percent of the heterosexuals. Men with two or more relationships, no matter what the preference label, were much more likely to be nonmonogamous outside of those involvements than men with one relationship. That is, the greater the number of significant partners, the greater the likelihood that the men also had additional "outside" partners.

For the women, the differences were even greater. Bisexual women with one significant involvement were much more likely to indicate being nonmonogamous than the heterosexual or homosexual women. (The breakdown was close to half compared with only about 15 percent for the heterosexual and homosexual women.) As with the men, the incidence of sex outside of their relationships was higher among the women who had two or more relationships across all three sexual prefer-ences. Thus, the bisexual respondents were substantially more likely to be nonmonogamous outside of their involved relationships than were either the homosexuals or heterosexuals.[2]

Opportunities for Meeting Involved Partners

What opportunities did people with different sexual preferences have for meeting partners? Did more opportunities arise inside or outside the sexual underground? If inside the sexual underground, what types of social settings—the bars, the baths, swing houses, or group sex parties—were most conducive to meeting prospective partners? Or were the vari-ous support groups where many went to talk about their sexual identity and lifestyle more likely to be a source of partners? How about settings more removed from the sexual underground? How important were or-ganized mainstream activities such as work, school, and religious or recreational groups? Or contacts established through same- or opposite-sex friendships and conventional parties? In a sense, how far did people have to look to meet their involved partners? To what extent did those we surveyed rely on the sexual underground as their main setting for meeting partners?

Opposite-Sex Partners

Both the heterosexual and the bisexual men most often had met an involved partner of the opposite sex through people or places that were *not* explicitly part of the sexual underground, usually through friends, conventional parties, or mainstream activities such as work, school, or recreational groups (Table 17.5). Less than 10 percent of the heterosexual men mentioned either bars and swing and group-sex parties or underground support groups as meeting places. Such settings were more popular among the bisexual men—nearly a quarter mentioned underground support groups.

The heterosexual and bisexual women also typically met their male partners in settings outside the sexual underground. Fewer of the heterosexual women than the bisexual women mentioned underground support groups.

Same-Sex Partners

The sexual underground was much more important among the men when it came to same-sex partners. About a third of the homosexual men said that they met their current significant male partner(s) through underground venues such as bars, the baths, and sex parties. (Few mentioned underground support groups.) This was about the same percentage who mentioned friends and conventional parties or mainstream activities as venues. The bisexual men relied on underground support groups more, but the underground sex scene the same, as places to meet significant same-sex partners. Nor was there much difference in meeting a significant same-sex partner through mainstream activities, friends, or conventional get-togethers.

About 40 percent of the homosexual women met their significant same-sex partners through friends, parties, and other mainstream activities. Fewer mentioned underground support groups and very few said the underground sex scene. The bisexual women, as a group, relied much more on underground places. For example, nearly a quarter said that they met a significant same-sex partner through an underground sex setting.

These results show the importance for bisexuals of the "sexual underground" for opportunities to meet significant partners (especially same-sex partners). They relied on this way of meeting partners more than did homosexuals and heterosexuals.

The Longest Relationship

In our fieldwork at the Bisexual Center and throughout the sexual underground, we heard time and time again from gays, lesbians, and heterosexuals alike that bisexuals were incapable of sustaining long-term involvements and of forming lasting commitments. Along with this was the belief that they could not remain satisfied relating to only one sex. Given a new sexual opportunity with a member of the other sex, bisexuals were seen as likely to "jump ship." Thus they were considered basically untrustworthy. We therefore decided to make some comparisons to test these assumptions. We asked the heterosexuals, homosexuals, and bisexuals the length of their longest sexual-affectional relationship (including marriage), the sex of the partner involved, and if the relationship was still intact (Tables 17.6, 17.7, and 17.8).

The Length of the Longest Relationship

The heterosexual women reported having had the longest relationships, followed by the bisexual women, and the homosexual women the shortest ones. On the one hand, nearly half of the heterosexual women reported a relationship that had lasted for ten or more years, compared with about a third of the bisexual women, and only 12 percent of the homosexual women. On the other hand, just over a quarter of the heterosexual women, about a third of the bisexual women, and approximately half of the homosexual women indicated that the longest relationship in their lives had lasted four years or less.

Just over half of the heterosexual men, nearly 40 percent of the bisexual men, and approximately 20 percent of the homosexual men said that they had been involved in a significant relationship that lasted ten or more years. About a quarter of the heterosexual men said that their longest relationship had been four years or less, compared with just over a third of the bisexual men, and approximately half of the homosexual men.

The heterosexuals seemed to have experienced the longest relationships, and the homosexuals the shortest ones, with the bisexuals falling somewhere in the middle. Thus, contrary to popular belief, a substantial number of bisexuals seemed to be capable of long-term involvement.

The Sex and Status of the Longest Relationship

As one would expect, the heterosexual women almost always reported the longest relationship ever for them had been with the opposite sex. About half said it was still intact. Among the homosexual women, 70 percent said that their longest involvement had been with the same sex, and 30 percent said with the opposite sex. Around one-third were currently in the relationship. The bisexual women were again in the middle but much closer to the heterosexual women, with 90 percent reporting the longest relationship that they ever experienced had been with the opposite sex. Only about a third, however, were still in that relationship.

Virtually the same patterns again held for the men we surveyed. The heterosexual men's partners were almost always of the opposite sex and about one-third of the men said that they were presently involved with the person. Among the homosexual men, the breakdown was about 80 percent same-sex and about 20 percent opposite-sex for the longest relationship, with almost a third saying that they were presently involved with this person. By comparison, over 80 percent of the bisexual men reported their longest relationship had been with the opposite sex, with about 40 percent answering that the relationship was continuing. A strong heterosexual leaning was evident for this latter group, as it was with the bisexual women.[3]

Effect of Sexual Preference on Expected Duration

Given the pervasiveness of the belief that bisexuals are incapable of forming lasting and committed relationships, we took the argument one step further by examining whether people with different sexual preferences had different outlooks about relationships (Tables 17.9 and 17.10). First, we asked those who were in a significantly involved relationship to estimate how long they expected that relationship to last. (For those in multiple relationships, we specified that the answer should be framed in terms of the relationship they felt would last the longest.) Second, we asked a more general question of everyone: to what extent did they think the direction of their sexual preference (or unconventional sexual practices in the case of heterosexuals) hurt their ability to sustain a long-term relationship?

There were no substantial differences among the men across the three preference groups on either question. In particular, most men who were in significantly involved relationships were quite optimistic about how long they would last. Three out of five of the heterosexual men and over half of the homosexual men answered "for more than ten years." (This was the longest response category that could be selected.) Virtually the same proportion of the bisexual men also said this. Further, the majority of the men surveyed (69 percent, 54 percent, and 54 percent, respectively) felt that their sexual preference or practices had no negative impact whatsoever on whether they could remain in a committed involvement.

For the heterosexual, homosexual, and bisexual women, the majority from each group (71 percent, 57 percent, and 59 percent, respectively) expected their current relationship to last for over ten years. As such, few seemed to hold a pessimistic view about the integrity of their significant relationships. Also, while most of the women felt that their sexual preference or practices had no negative effect on their ability to sustain a long involvement, this view was more common among the bisexual and heterosexual women (about 80 percent for both) than among the homosexual women (about 60 percent). The bisexual women, too, were less likely than the men to indicate that sexual preference affected their ability in this regard. Optimism, then, is not the province of one sexual preference group more than another. It should be remembered, however, that in general our groups comprise people who are not at the beginning of their sexual careers and who are strongly committed to their sexual identities.

Making Sense of Long-Term Commitment

To delve deeper into this issue we asked the people in our study to explain *why* their sexual preference (or unconventional sexual practices in the case of heterosexuals) either did or did not affect their ability to sustain long-term relationships. In formulating this question, we were thinking about the special circumstances of relationships between people in the sexual underground—for example, being a same-sex couple, being nonmonogamous, participating in swinging or group sex, practicing sadomasochism, or simply being out of the ordinary. We wanted to know to what extent these diverse ways of being sexual made a difference in long-term commitment.

Ways Heterosexual Preference Affects Commitment

Among the heterosexuals, the most common factor that was said to affect the duration of an involved relationship *negatively* was an active personal desire for or interest in nonmonogamy. For some nonmonogamy carried with it a lot of internal stress and strain. Others had problems simply because their partners did not approve. In general, nonmonogamy was mentioned with about the same frequency by both sexes.

Two additional factors were said to have a negative overall effect as well. A few people in the study, especially the women, pointed out that they lacked the sexual desire or sexual facility necessary to keep a relationship going. In their words: I believe I'm too restrained in sexual expression to satisfy my wife completely (M); I need to be freer (less in the head, more attuned to the body) (F).

In contrast, others, primarily men, felt that their current partners did not have what it might take to keep them happy sexually. As they said: I expect more variety, intensity than my partners (M); the lady is sexually very conservative (M).

Still other heterosexuals listed factors that had nothing to do with their sexuality per se (which was the question) as shaping the fate of relationships. This included factors such as honesty, communication, friendship, having partners with patience and tolerance, doing fun things together, a good mental or emotional attitude, and one person's words, "just plain love." Rather different things seemed to matter to different people.

Ways Homosexual Preference Affects Commitment

The homosexual men and women suggested a much different set of factors that made it difficult for them to sustain long-term relationships. Compared with the heterosexuals—who focused on their interest in nonmonogamy and feelings of sexual dissatisfaction—the homosexuals emphasized a range of broader social experiences. Many of both sexes noted that there was a great deal of stigma and adverse pressure that accompanied homosexual coupled relationships.

A few of the men, but more often the women, pointed to the homosexual community itself as the basic problem. They mentioned that there were no committed homosexual couples to set an example for involved relationships, and that other homosexual people offered no social support. As they said: There are no role models for gay males (M); the women's community does not promote long-term relationships (F).

For the homosexual men, the question of social support carried over into the widespread belief that the gay subculture was a culture of promiscuity and easy access to sex. Nonmonogamy seemed to be viewed as inevitable. In our respondents' words: The easy availability of potential sexual partners and the "aura" of promiscuity in the gay community tend to always threaten the stability of most long-term relationships. At least mine (M); living in San Francisco, "available" men are always "available." And they expect you to be single (M).

Others suggested the reason was that in homosexual relationships, there were few if any of the external factors like those that cemented and bonded heterosexual couples together such as a traditional state-sanctioned marriage and children. This type of answer was equally common among the women and the men.

Most homosexual respondents, however, stated that being homosexual had no effect *in and of itself* on their ability to sustain lasting relationships. When asked why not, the men, in particular, along with a smaller contingent of women, said that what really mattered was a range of nonsexual factors. Some emphasized that relationships that stood over time depended largely on the individual characteristics of the people involved—for example, one's personality type, degree of self-esteem, and overall maturity. Others talked about interpersonal conditions such as shared values, common interests, communication, emotional compatibility, mutual determination, trust, the desire to share, the amount of love, basic respect, finding time for each other, and the willingness to make adjustments. As one man so aptly put it, "Individuals learn to deal with one another regardless of [sexual] orientation."

A number of the homosexual respondents, especially the women, said that they were currently involved in a same-sex relationship that they felt would last indefinitely, and this indicated to them that long-term relationships were possible regardless of sexual preference.

But a substantial number of homosexual men stated that they simply wanted nothing to do with long-term involved relationships. They did not see this as an intrinsic aspect of their sexual preference, but rather as an individual or personal interest. As they said: Even in the past when I was in love with a man, I tired of him sexually after six to eighteen months; I would probably be active regardless of my sexual preference. I dislike monogamy; I don't think I have the energy or desire to nurture a relationship.

Ways Bisexual Preference Affects Commitment

The bisexuals, in contrast to the other two groups, often said that the nature of bisexuality had a negative effect on the stability of relationships over time. Some—both men and women—mentioned being attracted to both sexes or being unable to focus exclusively on one sex. They said: No one . . . sex seems to satisfy me (M); I go between one sex and the other sexually every two to three years and my sexual fantasies change frequently (F); being bisexual I am beginning to think it may make it difficult for me to settle for either one man or one woman (F).

Others said that being bisexual meant that they naturally leaned toward nonmonogamy, which was seen as an intrinsic aspect of their sexual preference. This predilection made long-term relationships especially difficult, again because potential or actual partners preferred exclusivity. In any case, it made things more complex. Answers such as these came more often from the men who made reference to the reactions of their partners or spouses. That is, it was not so much that their preference for both sexes necessarily made a relationship more tenuous. Rather it was that the other person, usually an opposite-sex partner, simply could not handle the situation. As our respondents said: My need to be with another gender partner is sometimes interpreted as rejection by the first gender partner (M); some women fear I'll leave with a man and vice versa (M); I have trouble finding someone to accept my flexibility. Let's face it, some spouses just aren't real open to this sort of thing (M).

Many bisexuals saw no connection between their sexual preference and their ability to handle or manage involved relationships. Like the heterosexuals and homosexuals, more than a few of the men and women just dismissed the issue point blank. They said: The question pisses me off (M); I'm totally comfortable with my sexual orientation and it has no bearing on the length of my relationships (M); it is simply not a question of sexual practices or orientation for me (F).

The most frequent and detailed set of replies used to dismiss the preference-relationship connection emphasized that lasting and successful relationships depended on many other factors besides one's sexual preference. This view was not unique to one sex but was shared by both. These were the same kinds of nonsexual factors that were mentioned by the heterosexual and homosexual respondents: I believe the ability to sustain long-term relationships is mostly dependent on communication, trust, self-esteem, commitment, caring, and sensitivity (F); the ability to sustain a long-term relationship . . . depends on a sensible match of partners, the communication skills and flexibility needed to negotiate . . . , a willingness to commit, the support of friends to share good times and bad (M).

None of our findings in this chapter give much support to the negative beliefs that bisexuals say people hold about them. There is evidence enough that bisexuals are capable of forming involved relationships and that they are capable of feeling committed to one sex (albeit over-whelmingly the opposite sex). They *are* much more likely to be non-monogamous, but this is not because they are untrustworthy but because open multiple relationships are an important part of their lifestyle.

People of all sexual preferences make choices about whether they want to be in relationships, and if so, the kinds of relationships they want to have. These choices, however, depend on their sexual preference and the cultural context in which they live. Put another way, there are differ-ences in the amount and types of opportunities for involved relation-ships. Theoretically, bisexuals have the greatest number of choices be-cause they desire sexual relations with both men and women. Being in, or having access to, the sexual underground also expands their oppor-tunities. But they also experience the greatest difficulties because of reac-tions from both the heterosexual and homosexual worlds, and moreover they must deal with some problems that seem intrinsic to being bisexual.

18

Managing Identities

Sociologist Erving Goffman describes a stigma as "an attribute that is deeply discrediting," a social status that marks those who carry it as being "of a less desirable kind," or an associated condition that transforms its owner "from a whole and usual person to a tainted, discounted one."[1] Certainly many of the people in our study had been treated as if they were tainted, less desirable, unimportant, or less than whole. People involved in the sexual underground are likely to be touched by stigma to varying degrees. Engaging in homosexual sex, practicing non-monogamy, sadomasochism, swinging, or group sex did and often does lead to social ostracism.

Here we examine how the persons in our study, who were all in some way affiliated with the sexual underground (if only as members of the organization through which they were contacted), dealt with their potentially discrediting social status. To what extent did others know about their sexual preference or unconventional behaviors? Did they tell others directly about their sexual preference or practices? How did people react once they found out? How many felt socially isolated because of their sexual preference or practices? With whom did they most often socialize? Was it harder to maintain social relations with some groups than with others? Finally, to what extent did the social reality of carrying a stigma make bisexuals, homosexuals, and unconventional heterosexuals regret their sexual preference or behavior?

The Experience of "Being Out"

Did bisexuals let others know about their sexual preference more or less frequently than the homosexuals and unconventional heterosexuals? Were bisexuals hiding behind "heterosexual privilege," presenting only their heterosexual side to the world while keeping their homosexual side secret?

It would seem to make sense that they might manage the stigma of being bisexual by passing as heterosexual, especially because most of the bisexuals we knew were predominantly heterosexual in their sexual feelings, sexual behaviors, and romantic feelings. Why should they risk disclosure of their homosexual behaviors and its possibly severe consequences when same-sex relationships were less central to their lives? Since they were less enmeshed in the homosexual world and more enmeshed in heterosexual social circles, it would be easy to present their heterosexual side to the public.

We asked the people we surveyed to estimate how many family and traditional friends on the one hand, and acquaintances and work associates on the other, knew or suspected that they were homosexual, bisexual, or an unconventional heterosexual. This division between family and friends versus acquaintances and work associates was a way of distinguishing between primary versus secondary associations, albeit a rough one. The heterosexuals were asked to answer only if they were actively involved in swinging, nonmonogamy, group sex, SM, or cross-dressing, all of which presumably were practices that could carry social stigma if publicly known. We also asked whom they had directly told about their sexual preference or practices, and how many of the people in each group were accepting or rejecting of them as a consequence (Tables 18.1 and 18.2).

"Being Out" with Family and Friends

Among the women, the homosexuals were the most likely to say family and friends knew about their sexual preference, followed by the bisexual women and the unconventional heterosexual women. In particular, about 90, 60, and 45 percent of the women in the three groups, respectively, said that many to all of these people knew of their sexual preference or practices. Similarly high numbers had disclosed their sexual preference directly. That is, about 80 percent of the homosexual, 70 percent of the bisexual, and 55 percent of the unconventional heterosexual women claimed that they took the initiative to tell these people firsthand.

The same basic patterns were evident between the homosexual, bisexual, and unconventional heterosexual men, though a smaller proportion of the men in each group indicated being out compared with the women. Over three-quarters of the homosexual men indicated that most of their family and friends knew of their sexual preference or practices, followed by about half of the bisexual men and only about a third of the heterosexual men. When asked if they had taken the initiative to disclose this type information themselves, about two-thirds of both the homosexual and bisexual men said that they had done so with many to all of their family and traditional friends who knew. Less than half of the men in the unconventional heterosexual group had done so.

Thus with family and traditional friends, the homosexual women were the most likely to be out—and to be out by their own doing. This was followed by the homosexual men, the bisexual women, then the bisexual men, with the unconventional heterosexual men and women far behind.

"Being Out" with Acquaintances and Work Associates

When it came to secondary associations, the homosexual, bisexual, and unconventional heterosexual men were generally less likely to be out and to have made direct disclosures, compared with their primary associations. The same order between the groups held, however. Over two-thirds of the homosexual men reported being out with many to all of their acquaintances and workmates, versus about a third of the bisexual men, and just 14 percent for the unconventional heterosexual men. As far as direct disclosures were concerned, almost half of both the homosexual and bisexual and a third of the unconventional heterosexual men said that they had told many to all of the acquaintances and work associates who knew about their sexuality.

The distributions among the women were less clear, though the women, like the men, were less likely to be out and to make direct disclosures to acquaintances and work associates. Some 60 percent of the homosexual women indicated that many to all of these people knew they were homosexual. The same was true for half as many of both the bisexual women about their bisexuality and the heterosexual women about their unconventional practices. Roughly 40 to 50 percent of the women in the three preference groups said that they told many to all such people who knew. There were no marked differences between men and women in being "out" to acquaintances versus work associates. The homosexual men and women seemed to be "out" the most, followed by the bisexuals and then the unconventional heterosexuals, and no one group was much more likely than another to make direct disclosures. One reason disclosures to those closest to oneself were more common was probably that it was more important that intimate friends should know than others. On the other hand, perhaps sexual secrets are less easy to hide from family and friends.

Reactions of Family and Friends

How did family and friends react once learning about the person's sexual preference? The homosexuals reported the most positive responses, the

unconventional heterosexuals the least, with the bisexuals falling in between. Among the homosexual women, nearly two-thirds said that
many to all of their family and traditional friends either had responded or
would respond with acceptance. Less than 10 percent said that many or
all of their family and friends had displayed or would display rejection.
About 40 percent of the heterosexual women reported the same level of
positive reaction, and close to a third the same degree of negative reaction. For the bisexuals, the figures were well over a half for acceptance
and about 15 percent for rejection. The same patterns and distributions
held for the men, and there were no notable differences between the two
sexes.

Reactions of Acquaintances and Work Associates

As with family and traditional friends, the homosexual men perceived
more favorable reactions from secondary contacts (65 percent) followed
by the bisexual men (40 percent) and the unconventional heterosexual
men (35 percent). The homosexual men reported no overall differences in
actual or putative reactions compared with their primary relationships,
and the bisexual and unconventional heterosexual men experienced somewhat less overall acceptance.

Among the women, the order among the preference groups was the
same as with the men (comparatively, 50, 40, and 30 percent). Also,
across the board, compared with the situation with family and traditional
friends, the women perceived a smaller proportion of favorable and a
greater proportion of negative reactions.

Why were the homosexual men and women "out" more often and
why did they report greater levels of acceptance? The differences in the
three sexual preference groups may be related to the fact that homosexuals are the largest and most visible population among the three groups
in the Bay Area. There is a strongly supportive environment, and the
public is used to social contact with homosexuals.

The Nature of Social Relationships

Being out as a "bisexual" carries a unique type of social status, one that
could most aptly be termed "double marginality." Bisexuals always risk
being stigmatized from two directions: by heterosexuals for their homosexual inclinations and by homosexuals for their heterosexual inclinations. Certainly this was the feeling among the bisexuals who frequented
the Bisexual Center, where many spoke about "not fitting in with" the
gay or straight world.

Given the intense divisions and rifts that bisexuals perceived in their
relations with homosexuals and heterosexuals, where did the bisexuals
seek or find refuge? One obvious place was the Bisexual Center. According to its leaders, the Bisexual Center was formed in part so that bisex-

uals could meet other bisexuals, an otherwise difficult task in a social world dominated by heterosexuals and homosexuals. But aside from the center, there were few places where bisexuals could meet others who shared the same sexual preference and similar social concerns. This complaint was voiced frequently among the bisexuals in San Francisco. What, then, was the social landscape for these people really like? How did it compare with the situation faced by homosexuals or heterosexuals?

To help profile these social relations, we asked the people in our study how often they socialized with same-sex heterosexuals, same-sex homosexuals, and same-sex bisexuals. (We focused on social relationships with the *same sex* because this was the most common form of nonsexual socializing.) In addition to assessing these relationships, we asked "How uncomfortable do you feel—or think you would feel— in the company of gay men, lesbians, heterosexual men, heterosexual women, bisexual men, or bisexual women?" Immediately following was a second checklist that read: "How negative do you feel toward" the same six categories. Both sets of questions distinctly required respondents to assess the six categories of people specified.

Extent of Social Contact

Who socialized with whom? The homosexual men socialized the most with other homosexual men, to a much lesser (yet still substantial) degree with heterosexual men, and to the lowest degree with bisexual men (Table 18.3). The heterosexual men socialized primarily with other heterosexual men, often avoiding homosexual and bisexual men alike. The bisexual men, in contrast, said they socialized the most with heterosexual men, then homosexual men, and least of all with other bisexual men, though they tended more than those in the other groups to relate more with everyone equally. The breakdown was about 70, 60, and 50 percent of the bisexual men who said they socialized either somewhat or a great deal with men from the three preference groups respectively.

The same patterns held for the homosexual, heterosexual, and bisexual women. Homosexual women said that they socialized the most with other homosexual women; the percentages declined but were still high for contact with heterosexual women, but socializing with bisexual women was virtually non-existent. The heterosexual women socialized the most with other heterosexual women and much less with both homosexual and bisexual women. For the bisexual women, three-quarters reported that they socialized somewhat or a great deal with heterosexual women, just over a half said the same for homosexual women, and a similar number said the same for other bisexual women.

The data suggest that homosexual and heterosexual respondents form more distinct community enclaves while the bisexuals are less restrictive in the direction they look for social relationships. Of course, one reason that all of the groups socialized least with bisexuals is probably that bisexuals are more difficult to locate and identify. The fact that bisexuals socialized the most with heterosexuals reminds us again of the heterosexual aspect of bisexuality for this group.

Quality of Social Contact

Although the bisexuals often felt stigmatized and discriminated against by heterosexuals and homosexuals alike, they may not have been as disliked as they thought (Tables 18.4 and 18.5). The bisexual and homosexual men said that they felt most *uncomfortable around* and most *negative toward* heterosexual men and lesbians. For the heterosexual men, it was homosexual men and lesbians. None of the three groups had such feelings to the same degree toward bisexual men and women. The proportion of the bisexual and homosexual men who felt uncomfortable around heterosexual men and lesbians was in the 40 to 50 percent range, with somewhat smaller numbers, 30 to 45 percent, reporting negative feelings. The breakdown for the heterosexual men was close to half who felt uncomfortable around lesbians and homosexual men, and over a quarter who felt negative toward them.

The patterns varied slightly among the women. In general, the homosexual women were more likely to indicate feeling uncomfortable around and negative toward every group (with the exception of themselves) than was any other group. They especially stood out in their discomfort around and negative feelings toward heterosexual men, with over three-fourths reporting one type of feeling or the other. The bisexual women, like the bisexual men, most often mentioned unfavorable feelings about lesbians and heterosexual men, especially the latter. For the heterosexual women the most common answer was also lesbians, with about 40 percent feeling uncomfortable and close to a third negative in some way.

In summary, we found first that there was mutual dislike between lesbians and every other group, but the heterosexual men were disliked almost as much. Second, heterosexuals seemed to dislike homosexuals of the same gender. Finally, bisexuals did not figure predominantly as a group that was particularly disliked. Certainly part of the reason is that bisexuals are hard to identify. Bisexuals were not a group that needed to be reckoned with *per se*. However, this could also reflect a degree of misinterpretation on the part of the bisexuals. As a group, bisexuals might be less often the objects of negative feelings than they believe themselves to be.

Isolation and Sexual Preference

To tap the quality of social relationships in a different way, we asked: "How socially isolated do you feel at the present time?" For those who answered affirmatively to any degree, we asked if they felt that their isolation was a result of their sexual preference or practices, and if so, why (Table 18.6).

About three out of five of the homosexual, the bisexual, and the heterosexual men said that they felt socially isolated to some degree. More revealing, however, was that the bisexual men were the most likely to conclude that their social isolation *was a consequence* of their sexual preference or practices. (Sixty percent of those who felt isolated believed it was because of their bisexuality.) The homosexual men reported this effect less often, the unconventional heterosexual men least.

Women's answers were similar. About the same proportion of women in all three preference groups felt isolated. Over a half of the bisexual women felt that their social isolation was related to their sexual preference or practices, compared with only about 20 percent of both the homosexual and the unconventional heterosexual women. Thus while it was common for members of every group to feel socially isolated, the bisexual women and men were much more likely to link such feelings to their sexuality.

Reasons for Feeling Isolated

As most unconventional heterosexuals reported no connection between their sexual patterns and social isolation, there is not much to report. Among those who did see a link, however, the one common theme was that they attributed their feelings of isolation to a unique aspect of their sexuality that hindered the formation of involved relationships. As they told us: I'm more comfortable involved with a man who is married. I do not know anyone in San Francisco in this situation whom I find attractive (F); I am very confused about how to find a woman who will accept me as a transvestite and want an intimate relationship with me. I'm fearful of rejection. I feel awkward pretending to be normal on dates (M); I choose to associate strictly with S & M people both socially and sexually. That limits my desire to interact and thus my interactions with non S & M people (F).

The homosexual men in particular provided a number of revealing answers on the links between sexuality and social isolation. Many homosexual men spoke of the negative social reactions they had experienced in the broader social world (i.e., from heterosexuals). They said: Even though I'm liked quite a lot, I seem to be excluded from a lot of activities (M); relationships with heterosexuals are difficult due to stereotypical judgments on their part (M).

Other homosexuals said that they felt socially isolated because of the climate and makeup of the homosexual world itself. In particular, the men talked about being unable to connect with same-sex partners who wanted to relate on an emotional and caring level. As they told us: It's hard to connect with someone who is not only sexually compatible, but emotionally compatible also (M); I tend to have more romantic feelings as opposed to purely physical ones, which are not satisfied very often (M).

When it came to finding partners, the homosexual men also cited age as a complicating and thus isolating factor. Older men in particular said that it was not easy to meet other homosexual men in their age bracket. A few said the homosexual culture was not kind to older men, that youth and consequent physical attractiveness were critical to fit in. In their words: Gay men my age are very few in number, very hard to find, and rarely available if you find one (M); Available gay men in my age group are a small percent of the population and cannot be distinguished from other men. There are very few places to meet them (M).

Last but not least, homosexual men were becoming more aware of AIDS. At the time of our survey in 1984 and the first part of 1985, early on in the spread of the disease, AIDS was beginning to have a profound impact, isolating some gay men from both heterosexuals and homosexuals. The men described struggling with their fear of AIDS, with the possibility of death, with having to cut back on sexual activity, and with the effect AIDS had on gay relations. They said: I have withdrawn somewhat from "the scene" partly due to fear [of AIDS] and partly due to bitterness (M) AIDS . . . is isolating many gays from each other. It has also isolated me from straight friends who don't want to hear about it (M).

The bisexual men and women, like the homosexual men, often linked their feelings of isolation to a lack of social support, a lack of acceptance, and to the negative social reactions they experienced in relation to their sexual preference. Some referred to the social world at large as the source of conflict; others specifically to both homosexuals and heterosexuals. Descriptions by bisexuals about the social reactions they experienced often had an element of anger and contempt. They said: Out in the world it is not okay to be bisexual. People react with either sick humor or disgust (F); bisexuality is rarely accepted as a valid orientation by heterosexual or homosexual people for their own phobic reasons. Being out as bi is more difficult (M).

The bisexual women at times pointed the finger directly at other women, but especially lesbian women, whom they saw as flatly rejecting of them. A few of these bisexual women stated that they, in turn, tended to be rejecting toward lesbians. Whatever the pattern, it led to feelings of social isolation. They told us: My recent new relationship with a man has caused me to feel cut off somewhat by lesbian and women friends (F); about a year and a half ago I stopped being exclusively lesbian and ended a relationship with a woman in which I didn't feel fully sexually adequate and fulfilled. Anyway, becoming "straight" isolated me from the lesbian feminist social and political circles I was at least peripherally involved in. The feeling is mutual—I shy away from lesbian feminist activities and also am shunned by women who previously were my friends. I'm very disillusioned (F).

Many of the bisexual men and women did not feel socially isolated from being openly stigmatized and rejected *per se;* rather they did not feel comfortable in the social world of homosexual or heterosexual people. They felt trapped between two cultures. They said: In San Francisco, lesbians tend to be separatist, straight women sexually demanding, and gay men seem either to want lovers or tricks with little room for Mr. In-Between (M); sometimes I have difficulty towards totally straight or gay people. I feel like I have very little in common with them (F); I have mostly straight friends with whom I can't really discuss my activities or desires because it would embarrass them. I tend to be bored in gay circles (M).

The bisexual men and women also frequently said that their social isolation was the result of restricted opportunities for sexual partners. The bisexual women in particular pointed to difficulties they had in locating sexual same-sex partners. This complaint, it should be noted, was more common among the bisexual women than the homosexual women. As they said: I would prefer to have more female sexual partners in my life now. Because I'm a suburban housewife, the opportunities to make that happen are frequently not available to me (F); I wish I had a way to meet affectionate bi ladies. I would enjoy being much more active.

Some of the bisexual men and women reported feeling isolated because they lived outside the San Francisco Bay area. Some couldn't find same-sex partners or other bisexual partners. They said: I am bisexual, into S & M, and live outside the Bay Area! I feel cut off from contacts and support (M); I live in a small town and find it difficult to relate on an honest level with very conservative people. . . . So true friends are harder to find (F).

Time and time again the bisexuals who attended the Bisexual Center talked about how they shared a variety of issues with gay and lesbian people because of their common participation in homosexual activity. As far as the bisexual men were concerned, at the time of the questionnaire study, AIDS did not seem to be one of these issues. Only a couple of the bisexual men made brief mention of the disease as a source of isolation. In fact, the bisexual women referred to AIDS (or STD's in general) with greater frequency than did the bisexual men. Both the bisexual men and women mentioned only a general fear and a concern about the effect the disease had on their sociosexual lives.

Regrets about Sexual Preference or Practices

So far we have touched on a variety of social consequences that accompanied being bisexual, homosexual, or an unconventional heterosexual: the willingness to be out, the reactions of others, limits to socializing with people from other preference groups, feelings of uncomfortableness and negativity toward people of the same or a different sex and preference group, and the experience of social isolation. How did these factors all add up? What toll did they ultimately take on the self-acceptance of those who lived and experienced them? We asked how much, in general, the people in our survey regretted being the kind of sexual persons they described themselves to be (Table 18.7).

The vast majority in each of the six groups said that they had no regrets. This was the case for 70 percent of the homosexual, 76 percent of the bisexual, and 85 percent of the heterosexual men. It also held for 79 percent of the homosexual, 76 percent of the bisexual, and 86 percent of the heterosexual women. Regardless of the sexual preference/practices category, most of those who indicated that they had some degree of regret answered only "a little." Thus while there were often marked social ramifications for people who were bisexual, homosexual, or un-conventional heterosexuals, very few had any noticeable regrets. They seemed able to compartmentalize or separate the social from the personal aspects of their sexual identity. Being in San Francisco certainly protected them from many social pressures from the outside world. And knowing others who shared a similar outlook obviously made it easier to feel good about whatever sexual identity they espoused.

III

AFTER AIDS

19

The Emergence of AIDS

Quarantine and its moral equivalent, stigmatization, function to maintain a boundary we find essential to our psychological well-being. We feel the need to separate ourselves not only from sickness and death, but from the perils of what has been called the "spoiled identity"—the fall from grace, the contagion of dirt. . . . AIDS patients attract fear and dislike precisely because of these forces. Our fright is enhanced by the marginality of their social position. . . . They possess a profoundly ambiguous and inarticulate status—they have, to many minds, been left out of the patterning of society, and they are to blame.[1]

If I had contracted the AIDS virus I would not have gotten "half-AIDS" because I was bisexual! (M)

As we were concluding our initial interview study in the spring of 1983, a phenomenon was emerging that was to rapidly change the social context of sexual preference—the AIDS epidemic. Although AIDS did crop up as a concern to a few of our 1983 respondents (for whom herpes was the disease they most feared), it had not yet received the immense attention it ultimaely did. Indeed, the first official attention paid to AIDS in San Francisco was the proclamation by Mayor Diane Feinstein on Monday, May 2, 1983, that the first week of May was to be "AIDS Awareness Week." The proclamation contained the following language:

San Francisco is encountering an epidemic of a new life-threatening health problem called AIDS. . . . AIDS has affected hundreds of Bay Area citizens since 1980 . . . the incidence of AIDS is doubling every six months and . . . AIDS is a health problem of urgent national concern.[2]

It was not that we and others were unaware that something was going on that was important. Already by then we had seen a TV talk show in San Francisco in which an interview with a guest—a man with AIDS—was interrupted by a studio technician who refused to put a microphone around the guest's neck because of a fear of infection. The interview had to be continued with the guest speaking to the interviewer through an off-camera telephone. But even with this attention, we and others were still unaware of the magnitude of the effect that AIDS would have on the country as well as our own study. We left San Francisco thinking that the sexuality of the bisexuals we interviewed would remain pretty much the same during the time that we were writing up our results. This was not to be the case.

As most of the world now knows, AIDS (Acquired Immune Deficiency Syndrome) is caused by a virus (HIV—Human Immune Deficiency Virus, discovered in 1983) that infects the cells that defend the body against infection. This leaves the person infected with HIV susceptible to illnesses that a healthy immune system would repel. A person may have the HIV virus for many years before the symptoms of AIDS appear. At present, the incubation period is thought to be up to twelve years. AIDS is the terminal phase of the HIV infection, and the two most common opportunistic illnesses that cause death are Kaposi's sarcoma (KS), a form of cancer, and *Pneumocystis carinii* pneumonia (PCP), a lung infection. (Other diseases associated with AIDS are candidiasis, a severe yeast infection, persistent cytomegalovirus (CMV), meningitis, pervasive herpes, and toxoplasmosis.) The virus is spread primarily through the transmission of infected blood and semen.

Today the two main ways of contracting the virus are by sharing drug needles and syringes with an infected drug user or by having sex with someone who is infected. By 1989, it was estimated that 1 to 1.4 million persons in the United States were infected with HIV. Fifty-three thousand had died. By 1993 there were 253,448 AIDS cases in the United States and 169,623 persons had died.[3] At present there is neither a cure nor a vaccine for AIDS.

Approximately three-quarters of those who have contracted AIDS have been homosexual or bisexual men. Often, however, these two groups are not distinguished. The emphasis is placed on *same-sex behavior,* so that "homosexual" comes to stand for both groups. AIDS first emerged in the U.S. in the "homosexual" population, an ominous portent for this group, as casualty and cause of AIDS became conjoined in the minds of many people, both lay and professional. AIDS first emerged as a clinical entity in 1981 when physicians in New York, Los Angeles, and San Francisco began seeing young men with KS and PCP, diseases hitherto found only among Central Africans and older men of Mediterranean origin. This pointed to an underlying immune deficiency since these diseases do not usually affect adults with normal immune systems. Remarkably, these young men were all homosexual. Thus the editorial in the Centers for Disease Control (CDC) bulletin *Morbidity and Mortality Weekly Report* read:

> The occurrence of *Pneumocystis* in these five previously healthy individuals without a clinically apparent underlying immunodeficiency

is unusual. The fact that these patients were all homosexuals suggests an association between some aspect of a homosexual life-style or disease acquired through sexual contact and *Pneumocystis* pneumonia in this population.[4]

This link to homosexuality was reflected in the terms first applied to the phenomenon, GCS (Gay Compromise Syndrome) and GRID (Gay Related Immunodeficiency). The connection between sexual preference and AIDS was thus established early. The term AIDS (Acquired Immune Deficiency Syndrome) was adopted by the CDC only in 1982. And because early cases were urban gay men, the syndrome was presented to be a sexually transmitted disease. This had important implications, as AIDS could be seen as a direct result of homosexual practices (especially anal intercourse), the supposed homosexual "lifestyle" (e.g., large numbers of sexual partners), and the use of recreational drugs like "poppers"—amyl nitrate—which were said to affect the immune system. According to Seale:

> [O]nce doctors are convinced that a disease is transmitted through sexual contact, many assume that any patient must have acquired it by promiscuous sexual activity. Those afflicted are assumed to be guilty till they can prove their innocence.[5]

The social construction of AIDS was further reinforced by the CDC's emphasis on *who* got sick rather than on the disease itself. Thus the emergence of the "4H's"—homosexuals, heroin addicts, hemophiliacs, and Haitians as the major initial "risk groups" for AIDS. The latter three groups were established as separate risk groups in late 1982, despite efforts by epidemiologists to attribute homosexuality to them in an attempt to maintain that all cases of the disease were caused by sexual contact. When drug addicts, hemophiliacs, and Haitians were classified as separate, nonhomosexual risk groups, terms such as GCS and GRID were abandoned and replaced by the term AIDS. It was not until 1986 that the CDC changed its emphasis from *who* one was to *what* behaviors a person engaged in—from risky people to risky practices. Even with the appearance of AIDS among other groups, however, "sexual orientation persisted as the defining characteristic of the person with AIDS."[6]

Thus a clear conceptualization was not found for AIDS. It was more a social than a medical problem, a disease restricted to socially marginal persons, caused by some aspect of a deviant lifestyle. Homophobia thus found a new component. By 1982 AIDS was beginning to receive increasing attention in the media, which reinforced the ideas coming from the medical and scientific world. *Time,* for example, emphasized "hotbeds" of homosexual sex and "living playgrounds of homosexual promiscuity" in describing the homosexual lifestyle in one of its earliest articles.[7] The full media onslaught began in 1983 and helped create the "AIDS panic." Rumors flew that AIDS could be spread by casual contact such as sneezing, kissing, handshakes, etc.; restaurants with gay waiters began to lose business; a major airline considered barring passengers with AIDS; police and firemen in some cities were issued surgical masks; undertakers refused to bury those who had died from AIDS; people with AIDS were discriminated against with regard to schooling, jobs, insurance, dental treatment, and so forth. And who was to blame? Those engaging in homosexual behavior, who were believed to have caused the epidemic in the first place, were seen as being duly punished. Thus Patrick Buchanan's widely reported statement in 1983 that homosexuals "have declared war on nature and now nature is exacting an awful retribution." This fusion of sex, sin, and death with its theme of divine punishment was taken up by the religious right as conservative forces grew increasingly strong in mid-1980's America. Thus "the homosexual" became socially reconstituted in terms reminiscent of pre-1969. According to one commentator:

> Hence the incomparably strange reincarnation of the cultural figure of the male homosexual as a predatory, determined invert, wrapped in a Grand Guignol cloak of degeneracy theory, and casting his lascivious eyes—and hands—out from the pages of Victorian sex manuals and onto "our" children, and above all onto "our" sons.[8]

The stigma of the disease was also transferred to those who had nothing to do with homosexuality. Thus the case of Ryan White, a thirteen-year-old hemophiliac who had contracted the virus in the course of blood treatment and was barred from school by administrators in Kokomo, Indiana, in 1985. And later, in 1987, the Ray family in Arcadia, Florida, whose three HIV-positive hemophiliac children were readmitted to their school only by court order. The community boycotted the school and the family's home was burned to the ground.

It was becoming rapidly evident, however, that AIDS would not be confined to the socially marginal. The death of movie star Rock Hudson in 1985 was a cultural turning point. Despite his homosexuality, here was someone who was well known to the public. An article in *Newsweek* in 1985 on AIDS stated: "Once dismissed as the 'gay plague,' the disease has become the No. 1 public-health menace."[9] Beginning in late 1986 media coverage centered on the threat to the "heterosexual population." Otis Bowen, Secretary of Health and Human Services, warned in 1987 that the disease was "rapidly spreading" to the wider population and would ultimately make the Black Death that wiped out a third of Europe in the fourteenth century seem "pale by comparison."[10] There was now a new fear. Public health officials had done little to inform people about how the virus was transmitted. Not until 1986 did Surgeon General C. Everett Koop issue the first public report that used explicit language instead of euphemisms like "the exchange of bodily fluids." Seeing AIDS as a gay disease thus caused delays in education, funding, and an accurate perception of the disease.

The growing awareness by the public that AIDS was not a disease of homosexuals, however, did not necessarily improve the social status of those engaging in homosexual behavior. The very nature of the terms "wider population" or "general population" served to maintain a boundary to isolate and condemn the "original sources of the problem." Indeed, *U.S. News and World Report* in 1987, after describing three cases—a divorced mother of two, a soldier who had had sex with a prostitute, and a single female attorney—all "heterosexual" and decidedly nonmarginal, began its discussion by saying, "The disease of *them* suddenly is the disaster of *us*" (original italics).[11] This division was buttressed by the distinction between "innocent" and "noninnocent" victims. As early as 1985 in discussing Ryan White, *Newsweek* had stated, "ignorance about the way the disease is transmitted can make pariahs of even the most blameless of its victims." The distinction still continues. Dr. Anthony Fauci, Director of the National Institute of Allergy and Infectious Diseases, had this to say as late as 1988:

The severity of the epidemic is catastrophic in the male homosexual population and among intravenous drug users. But what is encouraging is that we are not seeing any significant spread into the general population.[12]

Homosexuals and IV drug users were seen as a "disposable constituency," certainly not part of the "general population," a vague and neutral term that "stands in opposition to the descriptive terms applied to most PWA's (persons with AIDS)—homosexuals, gays, junkies, IV users."[13] Moreover, the term functions to promote the image of the public as

virtuously going about its business, which is not pleasure seeking (as drugs and gay life are uniformly imagined to be), so AIDS hits *its* members as an assault from diseased hedonists upon hard-working innocents.[14]

The idea of AIDS as a gay disease still persists despite new knowledge. In addition, the reference to an undifferentiated general population makes it difficult for people to see *whose* behaviors are the most problematic (new marginal groups have been "discovered" as threats, however, e.g., prostitutes). Even new safe-sex guidelines have not been accepted as easily as expected. The media are still reluctant to permit safe-sex advertising, and none of the newsweeklies has yet published advertisements for condoms. President Reagan did not use the term AIDS until September, 1985, and his administration showed great reluctance to provide money for AIDS research and education. President Bush did little more. President Clinton, however, has promised to make AIDS policy a key issue. The homophobia at the national level was seen in the Helms Amendment, a 1987 response to a safe-sex campaign. Jesse Helms, a senator from North Carolina, introduced an amendment to a bill that would have allocated nearly a billion dollars for research and education about AIDS in fiscal 1988. The amendment, however, prohibited any funds that would provide information that might "promote, encourage or condone homosexual activities or the intravenous use of drugs." On the Senate floor he fulminated about "perverts" and suggested that civil rights be waived in favor of quarantining those with AIDS. The Senate agreed that safe-sex education for homosexuals was unnecessary and the amendment, slightly changed, was passed 94 to 2. The House also passed the amendment, 368 to 47. Thus, the group most affected by AIDS did not qualify for federal funding.

This should not have been too surprising to the homosexual community and its organizations, which since the beginning of the epidemic had come to realize that most help would have to be self-help. They developed organizations like the Gay Men's Health Crisis in 1982 in New York City to educate about AIDS, counsel and care for patients, and lecture to health-care professionals. And the Shanti Project in San Francisco, as early as 1981 began to focus on the psychological needs of persons with AIDS as well as their loved ones and family members through volunteer counselors and support groups. Gay groups provided the single most effective program dealing with AIDS, the information campaign promoting the necessity of "safe sex," detailing explicitly which sexual behaviors were the most risky in transmitting the AIDS virus, and emphasizing the use of condoms.

Sex, which had been at the forefront of the gay liberation movement with its celebration of sexual freedom in the 1960's and 1970's now came to mean something different. Instead of a symbol for freedom, unfettered sexuality became associated with death. In the space of a few years, areas like Castro Street in San Francisco, which had become the embodiment of a freewheeling lifestyle, was transformed by the AIDS epidemic.[15] By 1988 there had been over five hundred deaths attributed to AIDS in San Francisco, and the medical director of the AIDS office of the San Francisco Department of Health was quoted as saying, "Half the gay men in the city are infected and a majority of them will develop AIDS."[16] This was projected in a city where the gay population has been estimated to be between 40,000 to 70,000 individuals.

Direct action by those engaging in homosexual sex, however, has slowed the spread of the virus. The San Francisco AIDS Foundation found that the proportion of men with more than one sex partner declined from 49 percent in 1984 to 36 percent in 1985.[17] It also found a decline in oral and anal sex. The San Francisco Men's Health Study of 1,000 homosexual and bisexual men found a dramatic reduction in number of partners and active anal intercourse.[18] A study of 655 homosexual men conducted between 1982 and 1985 by the University of California Medical School found a big decline in anonymous sex (e.g., visiting gay baths and backroom bars). The average number of partners per month decreased from 5.9 in November 1982 to 2.5 in November 1984, and 80 percent of the sample said they followed safe-sex guidelines.[19]

Thus the epidemic is said to be over in San Francisco. Among one group of 350 men the rate of new infections per year was 21 percent in 1982 and one new infection in 1986 and 1987. This does not mean that there will not be more deaths. According to the *San Francisco Examiner*, there were more than 5,500 cases of AIDS in the city in early 1989 with a projected 17,000 within the next five years—and these projections are low.[20] But when we look at the rate of new infections, the spread of the virus has slowed markedly.

Because AIDS hit homosexuals first, because homosexuals have been so vocal about it, the disease came to be seen as a plague of the "homosexual." This characterization ignored many of the people who engaged in same-sex behavior but who did not identify themselves as "gay" or "homosexual," including bisexuals. While we did find the conjunction "gay and bisexual men" in official statistics, bisexuals had not received much public attention. After all, there was a well-defined public image of what a "homosexual" was, but no such clear image existed for the "bisexual." In addition, the "homosexual" in the AIDS crisis served the function of maintaining the boundary between the infected and the "general (heterosexual) population," whereas most people preferred to ignore bisexuals because they violated the boundary. Not until relatively late in the epidemic, when the "general population" could no longer maintain that it was not at risk, did journalistic references to bisexuals begin to appear as explanations for the "spread." Thus the emergence of the "epidemic's new *bête noire,* the bisexual . . . a creature of uncontrollable impulses . . . whose activities are invariably covert."[21] This person, horrific in his duplicity—a "homosexual *posing* as a heterosexual"[22]—brings the virus from the homosexual world to the heterosexual world. (Such a belief, of course, denies the reality of bisexuality as a distinctive sexual preference.)

The *Atlantic Monthly* in 1987 reported that "The potential role of bisexuals in heterosexual transmission of AIDS has been gravely underestimated," supported by statistics of people who thought their heterosexual partner may have had same-sex relationships.[23] New stereotypes were provided by an investigator of sexually transmitted diseases:

A certain kind of bisexual man is not immoral but amoral as regards sexual candor. He is less apt to feel the guilt or conflict that a gay man might going both ways.[24]

Later that year *Newsweek* featured "bisexuals" on its cover. The featured article suggested that bisexuals were becoming the "ultimate pariahs of the AIDS crisis," representing a "new dimension of the deadly crisis."[25] Thus the belief that bisexuals spread the deadly disease to the heterosexual world was publicly acknowledged. This led the public to fear bisexuals' duplicity—their "dangerous double life"—whereby they hide their same-sex behavior from unsuspecting heterosexual partners.

To make matters worse, bisexuals were considered to suffer from a certain sexual "compulsiveness," making them all the more dangerous. For example, the article claimed, "Many bisexuals seem reluctant or unable to deny themselves the exotic other side of sex." This was buttressed by a quote from a doctor from Virginia whose credentials are not mentioned: "These are tortured souls in a wretched situation." "Homosexuality" retains its aura of danger, but in the guise of the "bisexual" who begins to emerge in the public mind as another category to add to those who make up the major "risk groups" for the transmission of HIV. Sexual preference again is interpreted as having less to do with sexual desire than with a medical threat. The very fact that the "bisexual" is less socially recognized than the "homosexual" makes him a greater threat. "Amoral," "duplicitous," and "compulsive," the bisexual stands as a secret agent spreading a deadly disease to the unsuspecting public.

Given the widespread publicity about AIDS, the social reconstitution of same-sex behavior, and the firsthand experience of friends and lovers sick and dying around them, it is not surprising that homosexuals and bisexuals came to see themselves, their sexuality, and their moral status differently as a result of AIDS. It certainly became clear to us that we could not write a book on what it means to be "bisexual" based solely on our earlier data. So much had happened, the social context of sexual preference had so radically altered, that a return to San Francisco and a second wave of the interview study were essential.

After AIDS: The 1988 Follow-Up Study

We conducted the follow-up study in 1988, five years after our initial research. We had kept a record of the names, addresses, and phone numbers of everyone who had participated earlier. Despite the time lapse, some of the people were easy to find with a simple phone call. Others had moved or changed their telephone numbers with no forwarding information. We tried to solve this problem by calling everyone listed under a given name in the Bay Area telephone directories. And we asked others for information on those who could not be reached.

Two research assistants helped track down the bisexuals we had interviewed in 1983. Working in Indiana, they placed calls to San Francisco and combed phone directories for leads on missing cases. When they made contact, they described the follow-up research and scheduled meetings with those willing to participate. Only a few people refused, and most of them changed their minds after we called them back personally. (As it turned out, most of them had balked for personal reasons. One woman had put on a lot of weight, over fifty pounds, and was too embarrassed to meet us in person. We reassured her that we were her friends, after which she decided to participate.) We ultimately obtained sixty-one of the original one-hundred self-identified bisexuals. When we compared their 1983 answers to the answers for the total group in 1983, the similarity was remarkable.[26] Of course, demographically, they were five years older (Table 19.1); 76 percent were between thirty and forty-nine years of age. They had also increased some in terms of income, reflecting a pattern of income rising with age. Again, the repeat group was predominantly college-educated, most often Protestant but with a smaller yet relatively equal number of Catholic and Jewish respondents; nearly all were white and employed. There were twenty-nine men, twenty-seven women, and five transsexuals. We also interviewed another six transsexual bisexuals for the first time to augment the latter group.

In August 1988 we conducted face-to-face interviews with the sixty-one bisexuals and gave them a questionnaire to fill out on their own. The interview and the questionnaire were administered in a single sitting, and it generally took forty-five to seventy-five minutes to answer all the questions.

The closed and open questions we asked in the interview focused on changes the bisexuals had experienced over the last five years in the way they defined themselves, perceptions about ideal relationships, the structure of involved relationships, conflicts with homosexuals and heterosexuals, and the willingness to tell others about being bisexual. In each case we asked specifically if the changes were related to AIDS. Additionally we asked them how they currently felt about nonmonogamy, how their sex life changed in the face of AIDS, and what it had been like to be bisexual during the AIDS crisis.

The questionnaire included only closed questions. It contained items about one's location on the Kinsey Scale, number and kinds of sex partners, frequency of different types of sexual activities, and regrets over being bisexual that were worded exactly as the initial interview study had been. A few new items asked respondents to assess how much they had changed over the last five years in each of these areas and about the extent of their safe-sex practices. Another question ascertained their HIV status.[27] We coded and analyzed the data as we had in the previous studies.[28] Further details of the analysis and presentation of the data for the 1988 follow-up study are presented in Appendix C.

At the same time we returned to San Francisco's sexual underground to see what changes had occurred since 1983. One evident change was that the Bisexual Center was no longer in existence. But a new organization called the Bay Area Bisexual Network had been formed, with about five hundred members who gathered regularly at different meeting places for discussion, lectures, and the exchange of information. Some of the members were people who had been previously affiliated with the Bisexual Center. The Bisexual Center had closed down in part because the president of the organization and his wife, who together had provided much of the time, energy, and money needed to keep the place running, had divorced. No one had stepped in to fill the vacuum created by their absence. Most of the former leadership were now devoting themselves to AIDS-related work.

The gay bath houses in San Francisco had also closed down, though one house on the Berkeley side of the Bay was still operating and business was brisk. One popular underground scene from 1983 that was still operating in 1988 was the SM club, but the "piss and fist" club frequented by some of the bisexuals in 1983 had closed down. The San Francisco Sex Information service, the gay/lesbian-based Pacific Center, and the Institute for Advanced Study of Human Sexuality were still intact and going strong. The swing house that was described in the 1983 fieldwork was also still in operation. Finally, group-sex parties, which were commonplace in 1983, had been transformed into what were called "Jack- and Jill-off" parties—gatherings of men and women who participated together in masturbating themselves or others.

Some of the institutions of the sexual underground thus remained intact, but none had escaped the biggest change of all—the alteration of the meaning of sex in the face of the AIDS epidemic. Gone was the hedonistic flavor of the early 80's when sex was celebrated and explored. Now sex could hardly be discussed without reference to the fear of contracting the AIDS virus. San Francisco, with its concentration of homosexuals and bisexuals, had suffered from the onslaught of the disease and was a very different place.

20

Bisexuals Face AIDS

We have already seen in this study that sexual preference is not always fixed but can be extremely malleable for some people. The bisexuals we studied often changed their behavior and feelings in response to changing circumstances, a fact that led us to emphasize environmental over internal factors in the construction of a bisexual preference. Like other researchers who agree with this emphasis, we focused on the role played by face-to-face relationships in changing sexuality, especially intimate relationships and subcultural support (e.g., from the Bisexual Center). Soon after we returned to San Francisco we realized that our notions of the "environment" were too narrow.

On our return in 1988 San Francisco was in the grip of a plague, a sociohistorical event that threatened the stability of many of the city's institutions. Obvious effects could be seen on health organizations; less obvious effects were evident on the real estate market as gay partners who jointly owned houses were separated by death; or on political and economic leadership as young professionals in their prime died. No one has systematically examined how environmental changes of such magnitude affect sexuality.[1] But any future environmental theory of sexual preference must make allowance for such large-scale or unique historical events that fundamentally affect people's lives.

To prepare for what follows in the rest of the book, we would like to outline how such a large-scale event as AIDS affected the social environment that shaped sexual preference in San Francisco. Because such effects are most clearly mediated through the experiences of individuals, we rely on answers to one simple question to sketch the contours of the change: "What has it been like being bisexual during the AIDS crisis?"

The Pall of Death

The most common response to our question was that sex and death had become intertwined and that just as sex had been a pervasive part of their lives, death now occupied such a role. The theme of AIDS and death had come to dominate the lives of bisexuals:

> It is extremely abnormal to be surrounded by talented, witty, spiritual, beautiful young men, and some women, who die before they reach their prime. I can't go by one day without hearing about AIDS and disease and death.   (F)

> AIDS permeates my whole life, and that has nothing to do with my bisexuality. What affects my personal life, both sexual and emotional, is the toll of seeing my friends die.   (F)

For those living in San Francisco, AIDS became *personalized* through the loss of friends, lovers, and acquaintances to the disease. Above all, AIDS had a concrete personal meaning for a great many of the bisexuals. The reality of AIDS, through personal loss, was described in the following ways:

> Initially it was an intellectual sadness, being much more objective and removed from the situation. But it's moving towards a deep emotional sadness, being overwhelmed by grief and loss because the circle is closing in. It's becoming more personal. . . . It's possible that two of the most important people in my life won't be around in two or three years.   (F)

> Like a pebble dropping in a pond. The first ring was watching friends and acquaintances who were gay dealing with their anxieties and the sickness and death of their friends. The next ripple was bi men realizing that they're at risk—going through anxiety and dealing with sickness and death of their friends and themselves. Three or four close friends and previous lovers of mine who are bi males are dead or dying of AIDS. The third ripple is bi women who've had sexual contact with bi men or bi women who are partners with bi men. The fourth ripple is seeing the anxiety filtering to the heterosexuals who are aware that I'm bi and have had high-risk partners. The heterosexuals are now seeing themselves at risk. These rings hit a year or two apart. The pebble dropped in about '82.   (F)

> It has been a bitch—gut-wrenching, devastating. The first gay male nurse that I loved died and shattered me to pieces. Then another and

another and another, and by the time the last one I loved died, I was numb. (F)

The terms used to communicate their feelings—sadness, grief, devastation, despair, and helplessness—bear witness to the depth of the effect of AIDS. The world once taken for granted by these people had been radically altered. As one man said:

Life will never be the same again. Can you forget the ones who are no longer living? Can you speak their names aloud or are they also part of the "closet"? (M)

Since AIDS had forced confrontation with death (among a population that was far from old), most of the bisexuals we interviewed had come to reevaluate the role sex played in their lives. What does this mean for identities such as "bisexual," which are centered around a certain sexual proclivity? The answer, among homosexuals at least, was stressing the relational aspects of homosexuality rather than the purely physical. Because social institutions were less developed among bisexuals, they were slower to make this change. When they were confronted with this wide-scale life-threatening event, though, they began to find the pleasurable aspects of sexuality less globally important, and they began to emphasize relationships that could counter the threat, fear, and despair of death. Because of the impact of AIDS on the nature and perception of sexuality among bisexuals, a more encompassing view of sexual preference was obviously required.

The End of Sexual Freedom

If it had been their sexuality that defined and confirmed the identity of the people in our study as "bisexuals," then AIDS meant that they now had to reevaluate what once was identity-confirming. The most commonly expressed view was that the days of sexual freedom were over. The celebration of sexuality as a liberating force that was so apparent in our 1983 study now seemed almost completely absent.

AIDS had taken away a lot of freedom that existed for a while. The idea that "deviant" sex was criminal changed with the liberation that occurred in the 60's and 70's. Now this freedom has gone because of AIDS. (M)

There aren't as many people that I feel it's really safe to be open with about my sexuality as years ago in San Francisco when exploring sexuality was my primary focus. I miss the freedom to explore the horizons of sexuality that we had in the late 70's. (M)

To a large degree too, the pall of death overshadowed the joy and spontaneity that once accompanied sexual expression.

> I've had a sense of loss and mourning over sex. A sense of sexual freedom is lost that I experienced before. Sex is not an important part of my life now. It's not as much fun as it used to be.   (M)

> It [AIDS] has taken a great deal of joy out of life. Part of the wonder of life has been a spontaneous eruption of joy in a sexual encounter and we can't allow that any more.   (M)

Thus certain ways of behaving sexually died somewhat along with those afflicted with AIDS as those who were living became repressed and even celibate:

> I had periods when I've felt nonsexual—more so than I want to. In response to seeing a lot of illness. My close friend died Wednesday. It's hard not to internalize this sex-death equation.   (M)

> It's not so much being bisexual through the AIDS crisis that is significant, it's a matter of being sexual. I have seen the gay community go through a period of sexual repression where people have stifled their sexual impulses. I've seen a lot of people retreating from their own sexuality.   (M)

Not everyone felt this way. Some reevaluated their sexuality in positive ways. As one woman commented:

> Among many people, the focus on sexuality and multiple partners is "passé." For many of us, we realize we were searching for feeling attractive, accepted, okay with who we were, and that this is superficial. Many people have quit drinking or doing recreational drugs, and this has changed the sexual dynamic drastically. It is more than wearing thin on life in the fast lane. As times have become more severe, people are searching more for meaning in sex and less for simple recreation and diversion.   (F)

The major question faced by the bisexuals in our study was how to be sexual at all in the face of an epidemic. We found not only that they had greatly reduced their sexual frequency but also that these changes had far-reaching consequences for their sexual identities.

Because having many partners became defined as a health risk, a movement toward monogamy occurred. Monogamy no longer seemed to be inherently at odds with a bisexual identity. Now it seemed a responsible way to organize a relationship. Changing to monogamy was not easy, however. In the past, some of the bisexuals—the men especially—

had had many anonymous sexual contacts. Now they had to learn new ways of meeting and relating to people. Couples had to learn new ways of relating as they closed their open relationships. A partner might be expected to meet needs once satisfied outside the relationship. Such difficulties were acknowledged by the bisexuals:

> AIDS has made me more conscious of the responsibility we take on whenever we open ourselves up emotionally to another. It has imbued my relationship with a greater mindfulness, a greater sense of risk we take in human interaction and a respect for the possibilities, dangers, and rewards of doing this.   (F)

Clearly, then, large-scale changes in the environment can affect the nature of involved relationships and thus sexuality. Wars remove men from families and away from the controlling environment of community norms. Other disasters lead to a greater reliance on primary groups. It is more difficult to see how such large changes affect sexual preference, but we suggest that it had an impact among bisexuals in two ways: First, we have seen how the bisexual identity rests to a great degree on an open relationship. As AIDS made monogamy more attractive, there could have been a weakening of the sense that one was bisexual. In addition, as the culture of sexual experimentation in San Francisco receded, the ideological basis for bisexuality virtually disappeared. The most obvious symbol of this was the closing of the Bisexual Center, one reason being that its leaders felt that AIDS-related work was far more important than propounding the bisexual lifestyle.

AIDS also affected sexual preference in a second way: through changes in sexual opportunities. Unconventional sexual behaviors require a variety of institutions where like-minded people can meet. A change in the environment can markedly alter these institutions. The AIDS epidemic altered the structure of sexual opportunities in San Francisco. There were, for example, about twenty gay baths in operation before the AIDS epidemic. In 1988 there were none. Nor were any of the backroom bars (where sex occurred) still in operation.[2] A bisexual man who frequented these institutions summed up the problems:

> My sexual proclivities have decreased because there are not that many places to be active any more. I'm starting to look for quality not quantity of sex. Maybe it's a good talk with a friend I need, not fucking. Sex took the place of intimacy in the past and now I realize this.   (M)

Later we examine how the notion of ideal relationships changed, to what extent monogamy became a goal, and the implications of these changes for sexual preference and self-labeling.

The Fluidity of Sexual Preference

Given that AIDS has been called a "gay disease" and that bisexuals are widely thought of as carriers of the disease, could the disease change a bisexual preference? Was their dual attraction fixed, or could it be given up easily? If so, were they "really" bisexual? All of these questions reflect on the wider question of the adaptability of sexual preference to environmental change. What is changeable and what is not?

We found that the major change for the bisexuals was their avoidance of men—particularly bisexual men—as sexual partners. Women were especially likely to do this.

> I wouldn't sleep with bisexual men at this point and I would have in the past. [Why?] Because they could possibly be carrying the [HIV] virus. It seems risky to sleep with men who have been sleeping with other men.   (F)

> I'm not interested in bisexual male partners because I think they are dangerous. It seems that female partners are much safer right now.   (F)

> It's been comforting to be able just to relate to females and I feel that's an easy and valid option and a safe one too.   (F)

Not only did bisexual women reject men as sex partners, but to a lesser degree bisexual men did as well.

> I've stopped having sex with men. AIDS was a big reason. It was just not worth it. I was also afraid that women would not want to be involved with a bisexual man. My identity as a bisexual has diminished as I don't act on my bisexual feelings.   (M)

> Since I feel flexible in my sexuality and can choose between genders, I've made a conscious effort to choose women and avoid the AIDS problem.   (M)

Some women, then, moved in the homosexual direction because of their fear of bisexual men. Others confined their opposite-sex relationships to heterosexual men, who appeared safer. Some men, on the other hand, also rejected bisexual and gay male partners and confined themselves to heterosexual women because they saw bisexual women as less safe. An interesting twist here was that some men moved more in a *homosexual* direction as they responded to the fear women had of them and to the resulting decrease in heterosexual opportunities.

Thus the AIDS crisis forced many bisexuals to examine their sexual preference and to make painful choices. They were more aware of the flexibility of their choices, at least insofar as their sexual behavior was concerned. Later we will examine how these changes in sexual activity affected the way that they labeled themselves.

The Bisexual Label

Most of the people we interviewed recognized that bisexuality had now become more widely recognized than ever before. As one woman said, "Finally, 'bisexual' is a word that exists." But the majority regretted that bisexuality was known mainly in relation to AIDS.

> AIDS has made us more visible, but as potential infectors of the heterosexual community. It's a negative not positive visibility.   (F)

> It's been a hell of a time for being recognized as bisexual, that bisexuality is an option. It's been very hard suddenly seeing your lifestyle in a lot of publications as being the reason for the spread of AIDS. We wanted recognition. We got it, but suddenly it's like we're lepers.   (M)

This belief, that bisexuals are disease carriers, so prevalent in current depictions of bisexuality, has had a great impact on bisexuals. After their concern with death and dying, it was the most frequent concern the bisexuals in our study expressed in response to the question of what it's like to be bisexual during the AIDS crisis.

Reactions to this stereotype varied. Some sought to defend bisexuality or saw merit in its recognition.

> Most of the stuff coming out of the media about bisexuality is negative. It makes me feel it's more important than ever to come out about being "bi." A lot of people are using AIDS to force bi's back into the closet.   (F)

> Through the AIDS crisis I've been acknowledged for the first time as existing as a bisexual. I feel good that I've been acknowledged but bad that it's in terms of a disease. The stigma is better than not having a category from a political point of view.   (M)

Some suggested that it was important to educate people that bisexuals were not just disease carriers.

> AIDS is something I'm more cognizant of as it's affecting me and my friends and I need to set an example as a person who is no longer fearful of expressing myself sexually with responsibility. I educate others—I've been a facilitator in AIDS education workshops.   (M)

Some respondents, though, had taken these negative social messages to heart. They saw the end of a particular community and lifestyle and a return to secrecy.

As the closet closes in, I have tended not to continue the coming out process as extensively as in the past. I guess I take "heterosexual privilege" (as I live with a woman) to shield myself from being branded a sexual outlaw or one of the slimy bi's who are infecting the straight community.   (M)

In San Francisco, questions of sexual identity are often embroiled in sexual politics. As AIDS had hit both bisexuals and homosexuals the hardest, it affected their relationships with one another. On the whole, the bisexuals we talked to said that relationships with homosexuals had improved because of the common threat of AIDS.

There's been an incredible display of support and caring from the gay community to everyone afflicted with the virus.   (F)

It's meant knowing more and more different people who have been infected with HIV or have died. It's strengthened my ties to the gay and lesbian communities through all of this.   (M)

On the other hand, others reported worsened relationships because of lingering prejudice against bisexuals, a prejudice still attributed to lesbians:

There are a lot of lesbians who now just want to sleep with lesbians and don't want to be with you if you've slept with a bi woman.   (F)

Most of this prejudice was similar to what we had discovered in 1983 but was enhanced by the heightened fear of bisexuals as carriers of the HIV virus, especially bisexual women who might infect lesbians through involvement with bisexual men. What was new was disagreement between bisexuals and homosexuals about how to face the epidemic—whose epidemic was it?

I'm furious at the divisiveness I see. There are PWA's (persons with AIDS) period, not gay PWA's, bi PWA's, and so on. We're fighting the same virus, the same medical system, the same pain.   (F)

In contrast to relations with other sexual minorities, relationships with the dominant heterosexual world seemed to have worsened across the board. Since bisexuals had drawn the attention of the public only as disease carriers, some were angry toward society and withdrew from heterosexual contacts.

It's been awful. We are seen as carriers of disease to the heterosexual community. This has made me limit my social and sexual relationships more and more to gay ones.   (M)

I feel anger that people would consider me dangerous if they knew I was bi when I know I haven't been infected with HIV. I feel anger that I also can no longer donate blood.   (M)

Others thought that the increased fear of AIDS would be used to strengthen the homophobia that already existed in the heterosexual world:

There's less acceptance than in the 70's. The whole Falwell thing— brand of religion—cut its teeth on homophobia. This and feminism all came together in the 80s with a disease that was made for them to create a hatred of the homosexual man.   (M)

Societies under some real or imagined threat often seek a scapegoat for their ills, usually a powerless minority who comes to symbolize both the problem and its resolution. Bisexuals fit this scapegoat role by providing the imagined route for AIDS to enter the healthy heterosexual population. Being the scapegoat can affect sexual preference in a variety of ways. Most important, we argue, is that it hinders the "coming out" process. As we will show, bisexuals were less likely to make their sexual preference known in an environment that became increasingly homophobic. This, we believe, can retard the development of a bisexual identity. On the other hand, the severity of the situation can lead to coalition building and/or better organizational efforts among sexual minorities. An increasingly defensive community and an increasingly aggressive minority movement heat up the crucible of sexual politics so that identity choices can be more salient than they would be at other times.

We will develop some of these points more in the following chapters. At the moment we simply wish to reiterate the necessity of considering large scale sociohistorical events in any theory of sexual preference. All aspects of the bisexuals' sexual preference seemed to be touched by the emergence of AIDS: their frequency of sex; their number and balance of same sex/opposite sex partners; their view of sexual pleasure versus intimacy; their choice of some sex acts over others, and so on. And this has occurred through those factors in the social environment that we have described—involved relationships, group ideologies, group support, the sexual politics of minorities, and the wider community in which they became involved. AIDS has sharply increased the importance of these environmental factors.

21

Changes in Sexual Preference

How did the sexual preference of the bisexuals we interviewed change over five years? In our original study, about 40 percent of the bisexuals said a change in their self-definition was possible, but most of them thought it was unlikely (Chapter 3). Compared with heterosexuals and homosexuals, however, the bisexuals in our comparative study seemed more likely to have experienced major changes in their sexual preference (Chapter 15). Five years after we first interviewed the bisexuals in San Francisco, what changes had actually taken place? Did they change along any or all of the three Kinsey scale dimensions? Did they still define themselves as bisexual? Did they remain the same in their sexual identity yet change on the Kinsey dimensions, or vice versa? And what were the reasons for any changes? Was AIDS the main cause or were there other factors?

In this chapter, we look for answers to these questions in three ways: first, we compare how the bisexuals scored themselves on the Kinsey scale in 1983 and 1988. Second, we asked those who had changed why they thought they had done so and, over this five-year period, if they had called into question whether they were bisexual. Finally, we asked whether they still labeled themselves as bisexual.

Change as Measured by Kinsey Scale Scores

Earlier we reported that change in *sexual behavior* was more common than change in sexual feelings or romantic feelings, because behavior is most affected by environmental factors and by the immediate situations in which people find themselves. Change in *sexual feelings* may be more significant, however, because these feelings are the most basic element of sexual preference (Chapter 15). How had the people in our study changed on each dimension? Was AIDS a force powerful enough to change not only their behavior but also their sexual feelings, romantic feelings, and sexual identity? Without reminding them what they had said five years earlier, we asked them to score themselves on each of the three dimensions again, and then we compared the results (Tables 21.1 and 21.2).

About 40 percent of the women reported the same score on their sexual feelings as in 1983. Nearly a quarter scored more in the heterosexual and about a third more in the homosexual direction. Thus close to 60 percent of the women had changed on this dimension.

The results for the men were similar. About a third reported the same score as in 1983. About another third scored more in the heterosexual direction and the remaining third more in the homosexual direction. Thus, about two-thirds of the men changed on this dimension.

The pattern for romantic feelings was similar to that for sexual feelings. Over a third of the women reported the same score in their romantic feelings as in 1983. About a third scored more in the heterosexual direction and the remaining third more in the homosexual direction. Thus, about two-thirds of the women changed in the direction of their romantic feelings.

Again, differences between women and men were slight. About a third of the men reported the same score on their romantic feelings as in 1983. Nearly 40 percent scored more in the heterosexual direction and over a quarter more in the homosexual direction. Two-thirds of the men, then, changed on this dimension.

There were more changes in sexual behavior than in sexual or romantic feelings. Only 17 percent of the women reported the same score in their sexual behavior as in 1983; over 80 percent had changed. Change occurred equally in the heterosexual and homosexual directions, with just over 40 percent of the women scoring in each direction. Change was the rule and not the exception then for the women.

Only 14 percent of the men reported the same score in the direction of their sexual behavior as in 1983. As with the women, change occurred equally in the heterosexual and homosexual direction, with just over 40 percent of the men scoring in each direction. Like the women, close to 90 percent of the men reported a change in the direction of their sexual behavior.

Overall the results were the same for both men and women. Sexual feelings and romantic feelings were the most stable, but a large majority (almost two-thirds) of the bisexuals changed on these dimensions, with roughly equal percentages in the heterosexual and the homosexual direction. Sexual behavior changed more. Close to 85 percent indicated a

change in direction, again with roughly equal percentages in the heterosexual and the homosexual direction. Indeed, about 40 percent of the
men and 45 percent of the women changed to exclusive sexual behavior
with either male or female partners.

For those who changed on the Kinsey scales, however, the amount
of change was not very great. Some two-thirds changed only *one* Kinsey
scale score in either the heterosexual or homosexual direction, and most
of the remainder moved two Kinsey scale scores. Very few reported
more major changes.

Reasons for Changing

To understand what these changes meant and how the people we interviewed accounted for them, we asked our interviewees to describe how
they had changed and why. Their answers are reported below.

In the Heterosexual Direction

Most of the bisexuals who changed in the heterosexual direction attributed this to a new heterosexual relationship or the deepening of an
old one.

> I'm in a relationship with a man right now. The last time I was
> interviewed I was not interested in relating to men. I'm not really
> interested in men now except for this one man.   (F)

> Recently I've developed romantically with two females. I have yet to
> have a romantic relationship with a man. One of the relationships
> with a woman has been the deepest connection I've had with any-
> body.   (M)

> I value my relationship with my wife even more than I did five years
> ago. I feel she has given me more freedom, and I realize how unique
> she is, and I feel closer to her.   (M)

Others cited an increase in heterosexual social contact in general
since 1983, e.g., in their friendships, as a result of a change in jobs, or in
the social opportunities that had come their way.

In turn, some people attributed a move in the heterosexual direction
to a decline in homosexual social contact, e.g., a failure to find a same-
sex partner, or a general dissatisfaction with homosexual relationships. A
number of people said that the closing of the Bisexual Center had limited
their opportunities to meet same-sex partners. Others were unlucky at
developing homosexual relationships or were dissatisfied with what these
entailed:

I'm mostly attracted to women, but I haven't had much sex with women. It's not because I haven't put myself out there; I've become more active in the lesbian community socially and tried to pursue some women, but it hasn't worked out. (F)

I quit the Bi Center because I realized I wasn't the type of individual that bi men were interested in. I felt they were more into sex and guys with great physiques and not so much personality and caring. (M)

It's easier for a male and female to bond—men and women don't have to create a ritual, men and men have to create a bridge. Also what are our expectations—if it's not to have children, what is it? This just keeps it [homosexual relations] a sexual thing. (M)

One theme that was prominent among men who experienced a change in the heterosexual direction was a dissatisfaction with the impersonality, lack of intimacy, and instability of homosexual relationships. Some also mentioned the continuing stigma of homosexuality as a reason for change, especially the fear of public discovery of their homosexuality.

As important as the above for changes away from the homosexual direction was the fear of AIDS. AIDS was cited more frequently by men and more for a change in sexual behavior than on the other two dimensions.

The AIDS crisis affected my behavior. . . . I tested negative and I want to stay that way. I've heard a lot about safe sex but it's hard to be totally safe. (M)

Many men changed in order to protect their heterosexual partners, not just themselves. Other men felt that their success in achieving heterosexual relationships would be affected because of the risk of AIDS contracted from same-sex partners:

AIDS was a big reason. My anxiety about getting it was not worth engaging in sexual behavior with men. I was afraid that women would not want to be involved with a bisexual man. (M)

In addition, engaging in homosexual sex had become more difficult because of AIDS.

I've had a long-term heterosexual relationship in the past five years. I've found few men willing to commit themselves. Because of AIDS, too, men seem to be withdrawing and becoming asexual. And it becomes complicated to find out from a man how many partners they had, whether they've had the AIDS test, etc. I also became a single parent too and this makes me less willing to take risks. (M)

I was dissatisfied with the lack of wholeness in relating to men. Sex is easy to get, but in the age of AIDS, I fear getting AIDS or giving it to my partners. There's a big hype in homosexual personal ads, a big distinction over who's positive and who's negative. I tested positive four years ago. It certainly affects one's perception of relationships.   (M)

The last time I went to a bath house there was hardly anybody there. And that was my main gay life—bath houses. There's just a lack of availability of men who are willing to have sex because of the fear of AIDS.   (M)

As all of these cases show, the role of AIDS in changing sexual preference was sometimes direct and immediate if a person feared for his own or his partner's life. Other times it was more indirect in its effects, as in the case of the man who became a single parent. AIDS also interacted with other factors to facilitate change, as it did for the man dissatisfied with the emotional quality of same-sex relationships. And finally, the fear of AIDS seemed to be a catalyst for changes that might have occurred anyway.

I'm married now and my wife really doesn't approve of me expressing my bisexuality. Because of AIDS, I'm not going to be promiscuous as I once was. I personally believe there's nothing wrong with having multiple partners, but marriage should be a thing of trust. I don't like to be unfaithful too because of the disease problem.   (M)

In some cases AIDS probably was only a secondary reason that people had for changing their sexual preference:

Part of the change has to do with the AIDS scare. If not for this I'd be more experimental. A more important reason though was deciding to go for a long-term committed relationship, which culminated in my getting married. Doing this was precipitated by more important factors than AIDS. I still retain my bisexual feelings but I don't act on them sexually.   (M)

Other reasons people gave for changes were: deciding that the heterosexual label more accurately fit them; problems of self-acceptance; a result of undergoing therapy; a spiritual transformation; a desire for monogamy; wanting a traditional marriage; and having a baby. This last case is instructive as it shows how a change in sexual preference can be affected by typical life events, which are often underrated in academic theories of sexuality.

Fewer erotic dreams and fantasies about women were due, I'm sure, to the birth of my daughter. During my pregnancy, I fantasized almost exclusively about men and have felt more interested in men since that time. Being in the company of other tired mothers seems to have dampened my erotic interest in women.   (F)

In the Homosexual Direction

The reasons people gave for changing in the homosexual direction were not unlike those for changes in the heterosexual direction. Thus the most frequently mentioned were becoming involved in a new homosexual relationship or deepening an existing one:

> I've been in a monogamous relationship with a man. Previously I've been fifty-fifty in my relationships with men and women. I still have heterosexual desires, but I made the decision to be monogamous with this man. Monogamy is safe emotionally, but sometimes I wish for the good old days of three-ways and women.   (M)

> I've been in a committed relationship with a woman since 1983. If some guy came along who was attractive to me, I'd possibly go for him if I wasn't in a committed serious relationship.   (F)

A general increase in homosexual social contacts also played a role in changes in the homosexual direction. One woman said that she became more involved with the Bisexual Center during this period, which allowed her to explore her homosexual feelings. On the other hand, a decrease in heterosexual contacts sometimes enabled the change too.

A change in a homosexual direction for some resulted from heightened dissatisfaction with heterosexual relationships or a lack of opportunities and/or the interpersonal skills necessary to maintain them. Some men spoke of not meeting many women, of having difficulty attracting those they met because of being shy or not aggressive enough, or of not meeting the type of women they wanted.

Women spoke of similar reasons but were far more likely to complain about what they saw as typical male-female relationships:

> I had a relationship with a bi male which ended badly. I felt betrayed and that I didn't want to be with men again. Emotionally it's more satisfying to be with women—their ability to deal with and talk about their feelings, to give and receive comfort.   (F)

> Women turn me on more. Being with a woman, there's a lot of things they don't have to explain. They know, they are one, so it makes it easier to get emotionally close to them.   (F)

Noticeable among those dissatisfied with heterosexuality were some women who attributed their change to the women's movement:

I think feminism, the whole feminist movement, has a lot to do with these changes. I don't want to put up with the garbage necessary to be with men. I want more equality in relationships, which I feel with women. Most of the men I'm around don't take me seriously. They discount what I have to say, treat me as an inferior.   (F)

A considerable number of women mentioned AIDS as a reason for becoming more homosexual—that sex with men, especially bisexual men, was more risky than sex with women.

Most of my men friends are infected with the virus. And so I rely on women more for support. . . . I just have more close women friends I've turned to for support because of the AIDS situation.   (F)

AIDS just makes it easier not to deal with men. I don't want to deal with sperm. That's how I look at it.   (F)

The concerns these women voiced affected some bisexual men, who said the reason for their decrease in sex with women was a reflection of the difficulty they experienced in finding female partners because of AIDS:

One of the factors influencing the breakup of my last relationship was her fear of me as an AIDS risk. I know the AIDS crisis has made me less comfortable putting myself out there as available to women.   (M)

Some men, as we have seen, cited AIDS as making them more desirous of monogamy per se. Such monogamous relationships were sometimes homosexual ones.

I would have continued including women in my life, but I got scared. Even taking safe sex into consideration, it was wiser to be with one man.   (M)

Another theme that appeared among the men was that society's treatment of the AIDS crisis made them angry, and this increased their identification with homosexuals. They became more involved in the politics of what they referred to as "the AIDS movement"—the organized homosexual approach to helping AIDS victims and fostering AIDS education. As one man said: "I felt more compassion and love toward men as a result of AIDS."

Other reasons given for a change in the homosexual direction were similar to those of the people in our study who moved in a heterosexual direction: people changed as a result of therapy or spiritual reasons; because of problems of self-acceptance; or because they felt that the homosexual label was a more accurate one. Some reported a more complete acknowledgment of their homosexual feelings:

> I've acknowledged that I've been attracted to men all my life. I was in denial all my life before. I now have a greater awareness of my feelings. (M)

Questioning the Bisexual Label

All of the people we interviewed in 1988 defined themselves as "bisexual" in 1983, but as we have seen, a large proportion reported changes along the Kinsey scale since that time. How did such changes affect the way in which they labeled themselves? Over a third had experienced doubts about their self-identification, calling into question whether they were bisexual.

The major reason for this self-questioning was fluctuations in their attractions to men and women:

> I was making love to a woman and I thought, I really want a man's cock. I never thought this about a vagina when I was with a man. I got to think I was preferring men over women. (M)

> Despite having a sexual relationship with a woman, I did not really feel turned on to her as much as I had with some men. (F)

> There were times when my attractions to women were so strong that I didn't feel attraction for men and vice versa. (M)

> Every time I meet an obnoxious macho male, women turn me on more, and I forget temporarily about all the nice men there are. I feel more gay when in the presence of lesbians and I feel more heterosexual when in the presence of men. (F)

This concern about fluctuations seems to reflect the belief that bisexuality should involve more balanced feelings toward both the sexes. In other cases such fluctuations rarely occurred at all because the person was interested in or presently in an exclusive relationship.

> I always question my [bisexual] sexual identity because I'm in a committed primary gay relationship. (M)

Some questioned their bisexuality because they felt that they lacked the social skills or the opportunity to develop relationships with both sexes. They admitted they would be more "bisexual" if they could be more active with same-sex partners. For example:

I question it [my bisexuality] because opportunities for same-sex relationships have been few and far between. (M)

I have greater success with women. I seem to have difficulty in getting satisfactory homosexual relationships. (M)

Finally, another reason for calling into question the bisexual label was social pressure.

I sometimes fall into defining myself by other people's definitions. (F)

I've questioned it [being bisexual] all along. . . . The culture constantly forces you into either being "heterosexual" or "homosexual." (M)

All of these people experienced over the five-year period many of the confusions that those in the 1983 study faced earlier in their bisexuality. And the reasons were pretty much the same—mainly dealing with the fluctuating nature of their desires and commitments, and pressures to deal with these by adopting a particular label.

Changing the Bisexual Label

Of the people we interviewed who called their bisexuality into question, three-quarters changed how they labeled their sexual preference. Only one respondent changed the way she defined herself to heterosexual. This was a woman who said she experienced decreasing attraction to people of the same sex. About three-quarters of those who questioned their bisexuality now referred to themselves as "homosexual," "gay," or "lesbian." The women who defined themselves as lesbian said they changed because of increased social contact with homosexuals, or becoming involved in a new or deepened same-sex relationship.

My feelings for women were always there and I allowed myself to go with them more fully. I acted on my feelings with women only and actually was not engaged in sexual behavior with men for a long time. (F)

I've been in a primary relationship with a female for seven years and now I don't miss men at all. Five years ago I could relate to both men and women, but I'm happy where I am now. I know where I fit in. I'm happy with this lifestyle instead of constantly fighting my lesbian feelings. Sexually and emotionally I just relate to women better. (F)

I've been dealing with women for the last 5 years. I haven't slept with any men. Since I'm not actively sleeping with men I can't really call myself heterosexual, or bisexual, for that matter.   (F)

Part of their interpretation was the sense that they were really lesbian all along. One woman voiced the strong belief in the homosexual world that bisexuality was simply a transitional stage toward accepting a true homosexual identity.

It was a natural unfolding after I broke up [with a man]. Bisexuality was just a "stepping stone" to becoming lesbian.   (F)

Another important reason for labeling themselves "homosexual" or "lesbian" was a loss of interest in or a dislike for heterosexual relationships. As noted previously, this was not so much because they disliked heterosexual sex per se, but either because they had had negative experiences with men or felt that they could have a deeper and more equalitarian relationship with another woman.

I guess women turn me on more than they used to, emotionally and physically. They are who I want to be with. Being a woman makes it easier to get emotionally close to other women.   (F)

The men who adopted the label "homosexual" or "gay" did so for similar reasons—increasing social involvement with homosexuals and a change in a committed relationship:

I've acknowledged my homosexual side now. I participate more in the homosexual community and my friends are homosexual. At the Bi Center I was just coming out. I haven't had much contact with the bi community since. "Gay" is an easier term to use for me.   (M)

I have been in a committed relationship with a man for the last five years—"married" with a double ring ceremony and all. Thus I assume I am a homosexual.   (M)

In 1983 I was married and had a child. After the divorce I had choices to make. I met a man three years ago who also had a child. I'm an active member of the Gay Fathers Organization. My gay relationship is the basic cause of this change. I still have my bisexual feelings.   (M)

As in the case of the women who now labeled themselves "homosexual" or "lesbian," we were struck by the retrospective interpretation among the men that they were really "homosexual" or "gay" all along. Bisexuality was just something they had experimented with until they could acknowledge their "true" selves.

I've acknowledged I've been attracted to men all my life. I was in denial for a long time.   (M)

When I joined the Bi Center I was exploring what I was. I was basically homosexual but had heterosexual feelings. I was very curious, the feelings were there, and I thought they could be developed. But I never met a woman or had heterosexual sex.   (M)

Again, we heard the belief that bisexuality is a transitional phenomenon.

For the first thirty-six years of my life I was strictly heterosexual. My wife was stuck in her roles and I wanted to explore new horizons. I divorced my wife, then began dating persons of both sexes, then just men, and then fell in love with a man. I don't meet many women in my social life now. Bisexuality was a "staging ground," a middle or transitional period for me.   (M)

The remaining people in the study who changed now rejected labels altogether. One woman simply felt that labels were of no use to her.

Being around the Bi Center there was a lot of emphasis on labeling. Since leaving, these labels aren't important to me.   (F)

The remainder of those who changed their sexual identity complained about the difficulties in applying the label "bisexual" to their specific circumstances.

The sexual label was a box. It just seemed like there's a set of assumptions about how bi persons are supposed to be. It focuses in on who you have sex with and how often and not the more complex elements that are part of our sexuality and loving people, or romantic interactions with people.   (F)

I label myself as a human being. I don't like labels, though. I'm afraid too of placing myself in a category I can't maintain. I'm not bi because I don't have the time or energy to divide myself equally between the two sexes.   (M)

The difficulty of fitting one's fluctuating sexual feelings and behaviors to a general label like "bisexual" was something we had also found in 1983. This seems to be a situation endemic to bisexuality, the "continued uncertainty" we described in Chapter 3.

In the end, the people we interviewed changed in many of the ways they had predicted they would in 1983 (see Chapter 3). Then, many said change was possible—and our 1983–1988 comparisons show that this could happen (even though the change was not often radical).

Among those who felt they could change in 1983, a change in sexual behavior was considered more likely than in their sexual or romantic feelings—and the 1983–1988 comparisons show the greatest change was indeed in sexual behaviors. Bisexuals in 1983 predicted that the change was equally likely to be in the heterosexual or homosexual direction— and, again, this is what we found in the 1983–1988 comparisons. And, finally, the reasons people felt that they might change in 1983 centered mainly around entering an exclusive relationship—a major reason for the actual changes that occurred between 1983 and 1988. The respondents in 1983 did not, however, foresee the effect AIDS would have on them.

AIDS was not cited as a reason either for calling into question one's sexual identity or changing it. But AIDS *was* an important reason cited for changes along the Kinsey dimensions. Our fieldwork suggested that the people we studied saw AIDS as an external threat for which they had no responsibility. Therefore, they might have felt that such a threat could alter their behavior but could not affect their identity. They might have believed that if not for AIDS, their lives would not have changed much at all. Given this, they could retain the identity "bisexual" despite changes in sexual and romantic feelings and sexual behavior. Conceptualized as an important aspect of who they were, and achieved after some struggle, the bisexual identity was not something they could easily give up.

Those who changed their bisexual identity thus appear to have reacted more to this uncertainty about their "real" sexual preference and to have changed their sexual identity as a way to make sense of it. Why some should have changed their identity and not others, we cannot say. Of course, who can say how these people will see themselves in another five years—given that the label "bisexual" seems to have less of an "anchoring" effect than those denoting an exclusive sexual preference.

22

Change and the
Transsexual Bisexual

The transsexuals in our study not only had problems with their sexual identities (how they defined their sexual preference), but far deeper problems with their gender identities (their claims to being a man or woman). As we saw earlier, many of them relied on their sexual behaviors to establish their identities; to prove she was "really" a woman, the male-to-female transsexual had to demonstrate attraction to men. If the events of 1983–1988 led to changes for the nontranssexual bisexuals, then we might expect changes to be at least as extensive for the transsexual bisexual. After all, if the direction of their sexual behavior has important implications for their gender identity, and sexuality has changed because of the AIDS situation, then they may have even more difficulty in achieving a stable sense of gender identity. How the transsexual bisexual fared follows. We refer to those we surveyed by pseudonyms. Remember too that in all but one case, these are male-to-female transsexuals.

Change on the Kinsey Dimensions

When asked how they had changed along the three dimensions of sexual preference over the last five years, the transsexuals reported a great deal of stability, especially in their sexual and romantic feelings. The greatest change was in sexual behavior.

Some of the reasons for change are similar to those given by non-transsexual bisexuals. Beth, for example, moved from San Francisco and away from the bisexual community. The removal of opportunities for finding homosexual partners (i.e., women) meant that she had become exclusively heterosexual in her sexual behaviors (her feelings and affections had remained the same—both equally heterosexual and homosexual). Tomi had become more heterosexual in both her feelings and behaviors, although the change was only slight (she presently scored as a 5 on all three dimensions). This was because she had withdrawn from her lesbian relationships, relaxed her radical feminist position, and became less critical of and more open to experimentation with men.

On the other hand, Jolene had become more homosexual in her behavior (though still equally heterosexual and homosexual in her feelings) because of an increasing acceptance of feminism. Allied to this was the fear of AIDS, which also led to her rejecting men. AIDS was similarly the major reason given by Linda for becoming exclusively homosexual in her sexual behavior.

Other reasons for change among transsexual bisexuals were related to their gender concerns and were different from those of the other bisexuals. Leslie had become more heterosexual in feelings, affections, and behaviors because she had overcome her fear of men and wanted to identify increasingly with heterosexual women by dating more men and progressing to heterosexual sexual intercourse. She saw heterosexual sex as validation of her gender. A similar stance was taken by Brooke, who had also become more heterosexual on all three dimensions. Her feelings were more heterosexual because since receiving estrogen treatment, she said she had been treated more like a woman by men. Dating men and engaging in sex with them, she believed, would enhance her gender as a woman. Both these cases were interesting in that even though each had become more heterosexual in response to gender concerns, they still retained their bisexual sexual preference. These preferences were, however, quite different: Leslie had an overall Kinsey mean for all three dimensions of 1.7 and Brooke an overall Kinsey mean of 3.7.

Melanie said she had become more homosexual in her feelings. She feared that she might really be a "lesbian" though she wanted to be a heterosexual woman. Apparently she had accepted the medical definition of what a transsexual's sexual preference should be (heterosexual), yet this was at variance with both her strong homosexual feelings and affections. This confusion was manifest in her ambivalence over self-labeling (and also perhaps in the fact that she was sexually inactive).

Pat was also sexually inactive, due to a strong fear of AIDS. He is a *female-to-male* transsexual whose affections had changed in the homosexual direction, i.e., toward men. This was because he felt more comfortable relating to homosexual men. This respondent, though preoperative, liked to dress as a man and felt that this aspect of gender validation was more acceptable to homosexual or bisexual male partners. He was, however, very effeminate in appearance.

The final case among the changers, Sammi, illustrated how the vagaries of life can affect the direction of one's sexuality. She said that she had become much more heterosexual in behavior (though still equally heterosexual and homosexual in sexual and romantic feelings) because she began engaging in heterosexual prostitution in order to finance her transsexual surgery (transsexual prostitutes sometimes say this to justify their prostitution). She claimed that in the past her heterosexuality validated her gender as a woman, and she had also continued her heterosexual prostitution long since her surgery. Thus mixed motives seemed to be involved in the change in her sexual preference. She said she enjoyed the work as well as the financial remuneration it brought.

Changes in Self-Labeling

Some of the changes the transsexuals had experienced during the five-year period affected how they defined their sexual preference. Over this period, four of the eleven reported calling their bisexuality into question, i.e., thinking they might not "be" bisexual. Of these four, three reported they had changed in the way they presently labeled themselves. Another reported not experiencing confusion yet had changed her self-label.

Leslie, as we saw earlier, moved in the heterosexual direction on all three of the Kinsey dimensions and also reported thinking she was not bisexual during this period. But her self-labeling had changed from possibly "homosexual" to "bisexual." This appeared to be closely connected to her transsexuality. She originally moved in a heterosexual direction as she accepted her gender identity as a woman. In the course of this change, her experiences with both sexes were positive, so that her sexual preference did move from what she thought was "homosexual" to the point that she saw herself as "bisexual." As she said:

> Living as a woman opened my experience to both genders. And men began relating to me as a woman. I thought that I would continue to just have sexual feelings for women, but I gave men a chance.

This case again illustrates well the independence of gender and sexual preference.

Jolene had become more homosexual in her behavior because of her acceptance of feminism and her fear of AIDS. She had called her bisexual identity into question because she had come to enjoy women more. She did not want to rule men out, however. (She was still bisexual in her feelings and affections and reported that most of her sexual fantasies were heterosexual.) Nonetheless, she changed from a "bisexual" to a "lesbian" label because it seemed more in accord with her political stance. In this case, transsexualism and sexual preference seemed to be unconnected. Her self-labeling was affected by the same factors (i.e., AIDS, feminism) that affected the nontranssexual bisexuals.

Pat was the female-to-male transsexual whose romantic feelings had changed in the homosexual direction. He thought he was not "bisexual" because he was becoming increasingly attracted to men, especially homosexual men. His self-labeling had moved from "lesbian" to "bisexual" to a somewhat confused situation in which he then defined himself as a "gay-bisexual." Part of this confusion seemed due to his celibacy, in that he had no sexual partners to give him a sense of sexual preference. This celibacy was due to an overwhelming fear of AIDS.

Not every transition in self-labeling showed the consistency of the above three cases, i.e., thinking that they might *not* be bisexual followed by a change in label. Beth said that over the five-year period, "I've had increasing feelings that I'm a lesbian and I'm really trying not to be." (She reported a Kinsey score of 4 and 5 on sexual and romantic feelings, respectively.) She had *not* changed the way she self-labeled, however; she still considered herself "bisexual." In fact, she had originally accepted the bisexual label as a way to handle her confusion. Two friends had taken her to the Bisexual Center, and she had learned that there was a category of sexual preference other than "heterosexual" or "homosexual." She became "proud" to say she was "bisexual," but because she had been celibate for a long while (five years), she felt that this undermined the reality of the label for her. It does seem that the bisexual label was a convenience for her and that a resolution of her feelings might lead to an active sex life that could change her self-labeling. Because of her persistent feelings toward women and her fear of AIDS, we think she might adopt the label "lesbian" in the future.

The final case deserves separate mention from the four above. All of the four had called their bisexual label into question—three had changed the way they self-labeled (and the fourth case could too). The final case, Tomi, reported *not* experiencing doubts about being "bisexual," yet as a result of her strong feminist position she had changed her self-label from "bisexual" to "lesbian." This label was "more descriptive of the way I am, a female-oriented female, which fits my identity as the type of female I am."

This case shows that when male-to-female transsexuals complete their gender change, this in no way predicts the *type of woman* they will become (i.e., in their sexual preference or social and political attitudes). And the type of woman they become can determine the gender of the partners they will seek, hence the way they are likely to label themselves. Thus Melanie and Tomi illustrate a more general finding that appears for the bisexual group as a whole—that external factors can affect one's self-labeling. Though *both cases* shared strong homosexual feelings, Tomi, because of her association with feminists, felt secure in accepting the label "lesbian." Melanie denied these feelings and seemed to lack the social support or sexual experiences that would make her comfortable in accepting any label, so she appeared to label herself "bisexual" as a convenience.

Changes in Sexual Behavior

Actual sexual behavior, we have seen, played a role not only in the self-label the transsexual might adopt, but also in confirming gender. Thus the sexual patterns of transsexual bisexuals were very complex. Not only did they experience changes that were the same as the nontranssexual bisexuals but also changes that affected their gender identity. The interaction of gender identity and sexual behaviors produced a variety of effects. Two things unique to transsexual bisexuals that were especially likely to affect their sexual behavior were their surgical status (pre- or postoperative) and their experience with hormonal therapy.

*Pre*operative male-to-female transsexuals are "women" who have a penis. What does this mean when they have sex with women? Both of our preoperative male-to-female transsexuals reported having penile-vaginal sex in the last twelve months. One approach to the inconsistency involved a reinterpretation of the penis by both partners. Tomi had penile-vaginal intercourse with her female partner, yet because she was accepted by her partner as a woman, this was considered "lesbian sex" by both of them. Sandra, the partner of a preoperative male-to-female transsexual, also reinterpreted the genitalia:

> I'm in love with a woman. When I met her she was a man. She presents herself as a female now. But she really hasn't changed as a person—I'm still in love with the same person. . . . The only word that would work for this relationship is "bisexual" because of the gender question—she's got a woman's breasts and a man's penis. . . . [Do you still have intercourse?] Yes, we're very lucky the hormones have not prevented her penis from working, although it doesn't get as hard as it used to.

A more common solution for the male-to-female transsexual was to ignore or gloss over the penis altogether by engaging in acts such as providing oral sex, sadomasochism, etc., where the penis was not necessary. In any case, we did find it difficult to ascertain from the transsexuals exactly what role the penis played in their woman-to-woman relationships, i.e., preoperatively or before atrophy of penile tissue through hormonal therapy. (Another study found that eight of sixteen preoperative male-to-female transsexuals used their penises during sex with women. Their accounts centered around the fact that the penis gave them pleasure so why not use it. Also, some fantasized the penis was a vagina and they were women.)[1]

What about sex with men? The obvious problem for the preoperative transsexuals was that sexuality could not escalate to vaginal intercourse because there is no vagina. They seemed adept at concealing that they had a penis and that they were having substitute sexual behaviors with men, but this pattern lasted only for a short time. Sometimes they could no longer resist demands by their male partners for intercourse, so the relationship ended. In other cases, the transsexuals had male partners who knew their preoperative transsexual status. Both preoperative transsexuals reported having been passive partners in anal intercourse.

*Post*operatively, the male–to–female transsexuals were able to have "vaginal" sex with men, but they did not always do so. Only three of the eight *post*operative male–to–female transsexuals reported vaginal sex with men in the last twelve months, and only four out of the eight had *ever* done this. (One of the others had recently had genital surgery and said she planned to have vaginal intercourse.) Those who had *not,* either were worried about their attractiveness, had not yet met an acceptable partner, or defined themselves as "lesbian." The fear of AIDS also limited their acceptance of male partners.

Most of the transsexuals in 1988 still defined themselves as "bisexual" (eight of the eleven). They did not necessarily *behave* "bisexually" though, as only three of these eight reported having had sex with both male and female partners in the last year. And all of these three worked as prostitutes. Of the remaining eight transsexuals, five had *only* male partners in the last twelve months and three had *only* female partners. Overall, more of the transsexuals reported a decrease in the number of sexual partners they had compared to five years earlier. Seven of the eleven reported fewer male partners than five years ago, one more, and three the same. Six of the eleven reported fewer female partners than five years ago, three more, and two the same. About half of them had had no partners at all in the last year.

Regarding unconventional sex, in the last twelve months, five of the eleven had engaged in "threesomes" (though ten of the eleven had at some time experienced this), five of the eleven had gone to swing parties (eight of eleven at some time) and four of the eleven engaged in anonymous sex (six of eleven at some time), three of these with men only and one with both men and women. Finally, we were struck with the involvement of about half of the transsexuals in sadomasochistic practices.[2] Five reported such practices, one with men only (prostitute), one with women only, and three (one of whom was a prostitute) with both men and women as partners. Given that two of these five transsexuals were prostitutes, it is difficult to speculate on the meaning of SM to the transsexuals. The one transsexual with only women as SM partners had problems relating to men and strongly feared AIDS.

The second main factor that affected the transsexuals' sexual behavior was hormonal therapy. Estrogen can reduce the sex drive and atrophies the genital tissues. After about six to twelve months of continual treatment, it is difficult for the male–to–female transsexual to get an erection.[3] Six of the transsexuals had experienced a marked decrease in sex drive (four postoperative and two preoperative).

Tomi, a preoperative male–to–female transsexual who changed her self–label to "lesbian," reported that hormonal therapy reduced her high need for sexual release to a more comfortable level. In the last twelve months, she reported having three to four men as sex partners (less than "five years ago") and ten to fourteen women as partners (more than "five years ago") in accord with the direction of the change in her sexual preference (from "bisexual" to "lesbian"), yet she still had male partners. She also attended swing parties about once a month (which had not changed over the period). Her hormonal therapy, which reduced her felt need for orgasms, had helped her explore more her relationships with women.

Jolene was postoperative and also defined herself as "lesbian" when we interviewed her. Before estrogen treatments she said, "Sex ruled my life." She fantasized being female and having sex with men. Estrogen cut her sex drive tremendously, which she experienced as a relief, as she said it allowed her to see how sex dominated her life. Now she said she could relax and enjoy sex in ways that were not "compulsive." She said she enjoyed women more than men though she still had heterosexual fantasies. However, she had had only one female partner "in the last twelve months." Her sexual behavior was markedly reduced in frequency compared with five years earlier. As noted above, she attributed this change to hormonal therapy.

Leslie showed a similar pattern. Before estrogen treatments, she masturbated several times a day, but now did so about two to three times a week. She believed that her anxieties about being transsexual produced the strong sex drive and that the surgery had helped reduce these feelings. She said the estrogen had also played a role. She identified as "bisexual," but had had only male partners (ten to fourteen) in the last year. In line with her gender identity as a woman, she felt heterosexuality was important. However, she had been *post*operative for a year and during that time had not had vaginal intercourse. Her behaviors were limited to mutual masturbation, performing oral-genital sex, and receiving finger-anal stimulation.

Beth likewise reported a significant reduction in sex drive as a result of hormonal therapy. She now defined herself as "bisexual." She had had only one male partner in the last twelve months, but saw herself as more "heterosexual," seeking male partners so that "I can relate as a woman to men." She reported having vaginal intercourse less than once a month. Her only other sexual behavior was masturbating a male partner. Both of these activities were engaged in *less* frequently than five years earlier. Both Beth and Leslie, then, suggested that estrogen decreased their sex drive.

Of the remaining two transsexuals, Sammi did not mention the effects of hormonal therapy. This would be difficult to evaluate, anyway, as she was a prostitute and worked as such both pre- and postoperatively. Her reported 100 or more male partners during the past twelve months thus would appear to have nothing to do with her sex drive. Linda reported that her "libido [was] down" because she had recently received genital surgery. She was highly sexual preoperatively and intended to be so again when she recovered. Before surgery (while receiving hormonal treatment), she capitalized on having female breasts and male genitals. Such persons are extremely attractive to some males (and some females, too) so she was in demand in ways that other transsexuals were not (some transsexuals choose to remain celibate or hidden during their transition). Her experiences with men during this time, however, involved a lot of "ridicule" and "discrimination," she said. She had a lot of hostility toward men because of this and currently only had sex with her female partner. The fear of AIDS also added to this. The female partner was the same one she had before surgery. She reported ironically, "My relationship with the *same* partner was defined as okay when I was identified as a man but now it's defined as lesbian; and I'm the *same person!*"

A final point about the sexuality of the transsexuals we talked to is worth making. At the time of our research, there was a transsexual network in San Francisco; many transsexuals knew one another and offered information and support with regard to their specific problems. Because of social and economic discrimination too, transsexuals not only socialized together but sometimes lived together. All of this provided opportunities to have other transsexuals as sexual partners. We asked seven of our eleven transsexuals whether they had had sex partners who were transsexual. Six of the seven said they had been, or were in, such relationships, and these partners were just as likely to be preoperative as postoperative. For four of the six, the transsexual partners were only or mostly male-to-female. For the remaining two, they were about equally male-to-female and female-to-male.

Beth, who had not had a transsexual partner, said she knew a few cases of sexual relationships between transsexuals, but that these relationships had broken up after one or both partners completed their transition (surgery). Jolene also observed that some transsexuals had sexual relationships with other transsexuals during their sex change, but after their transition, "they blend away into society and want to become invisible." This seems fairly common; such relationships are usually seen as temporary by transsexuals who are seeking either a genetic female or heterosexual male as their ideal partner.[4] Thus, postsurgically, they wanted the final sign of "womanhood," a normal sex life. Again, in their words, they indicated wanting to "blend into the woodwork."

Relationships with other transsexuals thus seemed to have a built-in fragility, functioning primarily as a proving ground for sex and gender. As Sammi (who had relationships only with male-to-female partners) remarked, "Some of the transsexuals I associated with attempted to seduce other transsexuals into sex as a way of proving they were more feminine." Linda had "knowingly" only had sex once with a male-to-female preoperative transsexual when she was auctioned off at a Halloween party. The experience was negative because the partner was really only interested in men. "Never again," she said, would she seek transsexual partners.

Experimentation seemed to be the issue in the other cases too. Jolene had an eight-month relationship with a *female-to-male* transsexual (postoperative), who left her when he fell in love with a genetic female. She had sex with four male-to-female transsexual friends, relationships she referred to as "experiments." Leslie said that three years ago, she was having sex with both male-to-female and female-to-male transsexuals because of her "healthy libido." With preoperative male-to-female partners she received anal intercourse (she was preoperative herself at the time), and she also played a feminine role with female-to-male partners. In each case, she would imagine herself to be a female during sex. Serena had sex with only male-to-female transsexuals both when she and they were pre- and postoperative. She found this important in the preoperative phase because "men could get me freaked out" if they discovered that she had a penis. This was not a concern if their partners were other preoperative transsexuals. Postoperatively she did not have sexual relationships with other transsexuals because, she said, "I am more congruent physically with the gender I say I am." She now looked for sex just with genetic males, while females remained more important for satisfying her emotional needs.

We do not mean to say that all relationships with other transsexuals were necessarily short-lived or that they did not bring both sexual and emotional gratification. Tomi, for example, had relationships with both male-to-female and female-to-male transsexuals that she found satisfying. She said, "they are just folks like everybody else." However, the transsexuals in our study did seem to find it harder to "blend into the woodwork" than they wanted, and their reliance on other transsexuals for sexual and emotional sustenance appeared to have a more permanent character than many of them would admit.

In conclusion, then, a concern with gender identity was central to many of the changes experienced by the transsexual bisexual. Their sexual behavior changed in ways that validated or supported the coalescence of a particular gender identity. Further, some of the changes on the Kinsey dimensions played a part in changes in self-labeling. Fear of AIDS may have played a role, but it was mentioned less frequently than other factors as a cause of change. The other external factors that influence sexual preference also affected them, e.g., the influence of feminist ideology. So too did their own recognition of the independence of gender and sexual preference. Thus, even though the transsexuals changed their gender identity—notably from being a man to being a woman—this did not predict the resulting direction of their sexuality or how they eventually defined their sexual preference.

23

Changes in Sexuality

The emergence of AIDS in the United States had a profound effect on bisexuals. Not only were they lumped together with homosexuals as the original culprits in spreading the disease throughout the country, they were also seen as the main conduit of AIDS into an "innocent" hetero-sexual community. In addition, the bisexuals in our study had witnessed the AIDS-related deaths of lovers and many friends and were living in a city where AIDS warnings were constant.

More than anything else, the response to AIDS in San Francisco had been that people should practice "safe sex." We first look at how the meaning of sex had changed for bisexuals in the face of trying to be both "safe" and "sexual" at the same time. Then, we compare sexual patterns of our respondents in 1983 and 1988. If the meaning of sex had changed, we wanted to know if this was reflected in their actual sexual practices?

Safe Sex

We asked the general question: "How has your sex life changed in the face of the AIDS situation?" Almost *all* of the answers included some reference to safe sex. "Safe sex" has become a common term as society has struggled to control the AIDS crisis. Used by scientists and health officials, it refers to practices that supposedly reduce the chance of con-tracting the AIDS virus while a cure is being sought. But what of the population targeted by this message? How do they interpret it? Little is known about the *meaning* of "safe sex" to various groups. For example, what sexual acts do they perceive as risky? What constitutes a "reduction" in sexual partners? Are some categories of partners (e.g., heterosexuals, lesbians) safer than others? And, what lifestyles must be given up?

Certainly, our society has been bombarded with safe-sex messages. But as we know from research on the media, such messages are not necessarily received directly nor understood as intended. Rather, messages are received in a social context and filtered through social interaction to provide specific meanings. In our case, messages about AIDS are filtered through a society that is very homophobic and where those with AIDS are often seen as being responsible for their own predicament because of their lifestyle. The reaction of bisexuals to this is further affected by subcultural lore, individual experiences, and personal needs and desires. Thus, the concept of "safe sex," we argue, invariably involves a variety of meanings to those at risk as they search for a solution that allows them to be "safe" and "sexual" at the same time.

The Meaning of "Safe" in Safe Sex

Almost all of the bisexuals in our study knew someone who had died of AIDS. How does one avoid being a victim oneself? The only answer, short of sexual abstinence, is to practice "safe sex." But what does "safe" mean? Many people thought that there was an exhaustive, comprehensive set of rules that would protect them:

> AIDS has, at this time, caused me to study the process of the disease more and use safe-sex practices. [What do you mean by safe-sex practices?] I think there's eight of them. Well, condoms. Restricting, not taking in, bodily fluids. And I tend to have cuts on my hands, so I have to be very careful where I put my hands.   (M)

Others, however, seemed to rely less on formulas and were quite fatalistic:

> I never use safe sex with female partners. I hate condoms. Two or three years ago I decided to trust my intuition and maybe also ask a few questions. I said "damn it, what's going to happen will happen."   (M)

> It took me a while to realize I had to practice safe sex. I still don't 100 percent of the time. It's because I think no one gives a shit about us anyway—they'd wish we'd all die.   (M)

Common-sense probability calculations were sometimes made that led to feelings of despair about ever being able to evaluate the risks.

The risk is like walking out of your front door and getting hit by a car. You could, but you also have to live. I do interview partners, but I don't put a lot of credence in this. People don't tell all. They don't tell all to themselves, let alone prospective partners. They can't, they don't know all. They don't know who their partners' partners were. (F)

I found that I tended to get involved with men more carelessly. I don't want to do that any more. Part of it had to do with AIDS. Having a lot of sexual partners is like playing Russian roulette, though I practice safe sex. (F)

Some people based their calculations of infection on personal observations of what were assumed to be risky behaviors:

I don't know if I want to go back to sex parties. There's unsafe sex there with gay guys jumping from a male to a female partner. It's extremely painful to see the unsafe behavior at both the straight and mixed parties. Seeing women play unsafe at straight group parties makes me crazy. I can't believe they are at times more in denial than men. (F)

I go to the baths in Berkeley about once a month. It's still very active there. I see a lot of unsafe sex there—guys being fucked without rubbers. There's also a maze with glory holes and no cock sticking out has a rubber on it. (M)

For some people, seeing violations of the safe-sex contract highlighted the risk involved. Others who witnessed a lot of behaviors considered to be unsafe at the time expressed a sense of immunity:

I go over to the baths at Berkeley a lot. I have oral sex there, never anal sex. I've never seen anyone in the baths use condoms with oral sex. . . . I went ahead because I'm extremely oral. If oral sex caused AIDS, I would have been infected by now, that's my conclusion. (M)

It was not uncommon for certain "favorite" sexual behaviors to be considered exempt from risk despite their being considered a possible risk.

I do safe sex 80 percent of the time. There's a lot of areas they don't know are safe. I don't believe in the total abstinence definitions of safe sex. I'll take risks, like sucking cock without a condom. (M)

I do practice oral sex and I don't know if that's okay. But since neither of my partners sleeps with men, I don't have to practice safe sex. (F)

Most of those we surveyed, however, were doing all they could to reduce risk. A very common solution was to be "selective" in their choice of sexual partners. But feeling "safe" meant putting one's trust in the person chosen as well as being honest oneself, and honesty could never be assured.

Thus "selectivity" for the bisexuals covered a variety of methods to ascertain how trustworthy their partners were. Of course the greatest feeling of trust and safety was in a monogamous relationship:

I'm monogamous with my husband. How boring! I don't want to put myself at risk. That's it, that's enough. (F)

I've gone from occasional casual relationships to monogamous relationships. That pretty much sums it up. Sometimes the feelings are still there, but I have a definite concern about AIDS and other diseases. (M)

But trust and monogamy were not necessarily synonymous because some bisexuals had a definition of monogamy that did not involve sexual exclusivity. Known as "bi-monogamy," this involved being faithful to one opposite-sex *and* one same-sex partner. It could be for this reason, or a felt distrust, that a number of people who said they were monogamous also used other "safe-sex" practices:

I'm in a monogamous heterosexual relationship now. I've gone to safe-sex workshops and shared information with my partner. We practice safe sex even though we are monogamous, just as a caution. (M)

This strategy was probably wise given the following person's definition of monogamy:

AIDS has absolutely no effect on my sex life. I am basically monogamous and am usually involved with people who have had far fewer sexual partners than myself. I do not consider myself at risk even though I realize there certainly is a possibility I could have been infected along the way. (F)

Nonetheless, many people seemed to be seeking monogamy as a solution to the AIDS crisis:

I've stopped having casual sex. I've been looking for a [steady] partner, but not been finding one. I got discouraged because nothing occurred. (M)

Monogamy, however, was for some a reluctant choice, especially among those who accepted the earlier bisexual ideology that non-monogamy was the preferred lifestyle:

> I've chosen to stay in one long-term monogamous relationship. If AIDS didn't exist, I no doubt would have played around a lot more. He probably would have too.   (F)

> There has been deep emotional wounding among gay men. . . . It pushes them back into a self-hating stereotype of being wholly promiscuous. The straight media has said AIDS has made gay men grow up by making them monogamous—made them come to their senses. That's insulting. The gay media echoes it too . . . as they praise monogamy. "The party's over!"   (M)

If one was unable to achieve a monogamous relationship, "selectivity" took a different meaning. It involved putting trust in strangers or persons whose sexual lifestyle one did not fully know. Uncertainty thus increased as people tried to screen out those who might have had many partners, or did not use condoms, or engaged in high-risk sexual behaviors, or were intravenous drug users. Everyone seemed to know the importance of being selective in these terms.

> I'm less willing to have casual sex. AIDS brought to focus some ideas—that when you become sexually intimate with someone, you "marry" them for the rest of your life. So I've been more choosey about who I let into my space this way.   (M)

Trust, however, did not preclude mutual ignorance or a shared misconception about sex (*i.e.* my definition of "safe sex" is not the same as your definition).

Because of the uncertainty surrounding "selectivity," people had adopted a variety of strategies to give themselves a sense of security. The most common was not having sex with whole categories of persons, especially bisexual and homosexual men. This approach was consistent with the experiences of many women anyway. As they saw it, men did not care much about the consequences of sex before AIDS, and AIDS had not changed men's attitudes.

> Women always have had to be responsible. . . . I see a sliding attitude by men. They're leaving AIDS prevention up to their women partners too and not taking responsibility for their own lives.   (F)

Such selectivity, however, was based on the assumption that female partners were either having sex only with other women (who themselves had no male partners) or with heterosexual men (who were assumed to be wholly heterosexual or not having sex with bisexual women). Although the chains of inference were weak, they did provide a sense of security for many of the people we interviewed.

> I've put aside the interest in men I've had. If a feeling came up about this, I wouldn't act on it because it wouldn't be worth the risk.   (M)

> The only time you have to be concerned is with women who are involved in many other relationships. The woman I'm involved with now, in a secondary relationship, her primary relationship is also with a woman and I'd be surprised if she'd slept with a man. If they're lesbian then I don't worry about sleeping with them.   (F)

One person introduced his version of "responsible sex" as a method of protection, showing just how idiosyncratic the meaning of safe sex behavior could become:

> I'm extremely cautious about safe sex. Now I don't rule out everything that is technically unsafe. My roommate is straight and has had no gay experiences. With someone like him I would consider unprotected intercourse. That's how I distinguish between "safe sex" and "responsible sex." I see it as responsible sex even though it is technically unsafe. In Michigan I had unprotected sex with a straight woman which is technically unsafe, but I felt it extremely unlikely I'd get AIDS from her.   (M)

A global sense that "men are dangerous" was part of the social context too:

> AIDS has made me pull back from acting on my sexual feelings toward men. I get come-ons from gay men in the city, but not as many in the last three years. Propositions aren't as forthcoming from gays as the fear of AIDS has made them pull back.   (M)

This perception about men was reinforced when the bisexuals of either sex were themselves rejected or seen as risky by others because of their sexual desires for men.

> It's hard to date most lesbians because of my past involvement with men, especially my ex-lover who has AIDS. Some feel that I'm a risk to them.   (F)

> When the AIDS issue became known I chose not to participate in homosexual activities. I've also had less frequent heterosexual contacts. My potential partners became less available because of their fears—they knew I was bi and were afraid.   (M)

Other methods of "selectivity" involved searching for clues that a prospective partner was "safe," such as the following case, for a bisexual couple who were into swinging.

> I'm extremely selective and very attentive to what people say. There's two ways to meet people. The primary way is through swingers magazines, the second way is through private referrals. Whatever way, there is a lot of correspondence first. What people write is important prior to even meeting them. If they haven't been too sexually active or with too many partners, then we might meet with them. I look for consistency in what people say in the correspondence. (M)

Taking care over who one chooses as a sexual partner, again, did not ensure that one's partner was equally selective, or the partners of one's partner, and so on.

> It is important to me to know my partners and to know who their partners are. I feel very cautious about choosing a partner. And I feel a lot more anxiety about my husband's choice of partners. And I think my growing interest in monogamy is partly related to this. (F)

> Most of my secondary relationships are with people who have been in primary relationships for years. I feel safer. (F)

But despite all of these precautions, it was impossible to be sure how safe a given situation was:

> I'm hesitant meeting new people. I meet them at workshops, but you can never tell if they are using safe sex. Anyone with homosexual feelings is aware of safe sex. But the heterosexual crowd don't take it seriously. (F)

This is not to say, however, that the bisexuals in our study were always totally honest themselves. A few admitted they were not:

> I think heterosexual women are afraid of bisexual men. I needed to be honest that I was bisexual. I wouldn't blurt it out on a first date, but I'd get to know them and see how they'd react. (M)

Trust is always uncertain. Even restricting the universe of potential sexual partners was known not to be risk-free, although people felt more in control of events when doing this. A decrease in partners was common among the bisexuals in our study and it signaled the end of a particular sexual lifestyle—e.g., a lot of anonymous sexual partners or multiple partners in group sex. Some deliberately limited themselves to self–masturbation.

> I've been celibate for a year. That has a lot to do with AIDS. I'm extremely cautious about getting involved with someone. I'd get involved with someone I knew in the past because I'd know him and trust him if they told me they were safe. If I met someone new I would wait a while before I got involved with them sexually.   (F)

> My relationships outside my primary relationships have been significantly fewer.   (M)

> The only sexual contact I have now is with myself. The reason why is that casual sex is too risky because of AIDS.   (M)

> We don't go to swing parties any more. We don't go to bath houses, that kind of thing. We have a close circle of lovers, there's no more anonymous sex. We're just more cautious all around.   (F)

The Meaning of "Sex" in "Safe Sex"

What has happened to the meaning of "sex" during the AIDS crisis? First of all, the word "sex" was deconstructed. What was a life-affirming activity, a source of personal and social validation, was stripped of its wider meanings and became, first and foremost, a physical act constituting a prime route for a deadly virus. To a remarkable degree the "sex" in "safe sex" was focused primarily on the exchange of various "bodily fluids" regardless of the who, where, when, emotionality, passion, intimacy, and the like that gives meaning to sexuality. Not that these were absent, but they were secondary and were only considered important insofar as they were relevant to the issue of contagion. For many bisexuals, sex became equated with death.

> The concept that sperm is a deadly weapon has debilitated our society.   (M)

This climate led people to see themselves, and to see others viewing them, as a danger. In the words of two persons who tested positive for contact with the virus:

> I always practice safe sex. But I am uncomfortable with someone who's listed negative. I feel like I have a deadly disease—leprosy— and they may catch it. But there's a stronger connection with those who have tested positive.   (M)

You feel dangerous. But I have to keep it in perspective and hold on to my sexuality in the face of all the horror. I feel like giving it up at times though.   (M)

The atmosphere created by AIDS meant, for many, that the joy, the fun, and the spontaneity that accompanied sex was gone.

It's definitely put a damper on being sexually free and open. I'm inundated with the whole AIDS issue, since I know so many gay men. It places a mood on sexuality such that it's not easy, clear, or fun to the same degree it used to be.   (F)

I'm afraid. I'm less open to experimentation and flowing with the moment—less open to seeking adventure.   (M)

Even if death was not a preoccupation, the mechanics of safe sex were interpreted negatively. The use of condoms and other methods of "latex sex" were said to detract from sexual pleasure:

AIDS has definitely ruined my sex life. Condoms take all the fun out of fellatio and really make a penis look and smell like a rubber stick. Dental dams [latex between one partner's organ and the other's mouth] completely block sensation, smell, and taste. I have a lot less sex, and what I have isn't worth squat.   (F)

I hated condoms when I had to use them for birth control. I can't imagine anyone enjoying having to use them. Rubber dams are even worse. Cocks and clits are warm and moist and soft. Rubber gloves are cold and unyielding barriers to sensation.   (F)

Some types of favorite sex practices were also given up, usually with sadness:

For me, making the additional changes that safe sex requires wasn't difficult, with the exception of giving up fisting [vaginal]. I was much less willing also to change my feelings about bisexual men. I retrained myself to eroticize heterosexual men and, believe me, that took some doing.   (F)

I don't rim anymore [oral-anal sex]. I miss rimming. I'm committed to practicing safe sex with all my future sex partners.   (F)

And some said they had to forego the exploration of their sexuality, to put it on hold, because they felt it would be risky to do otherwise:

Anal sex is a fantasy that I've wanted to experience for a long time, but when I found a partner I felt comfortable exploring this with, the idea was too threatening due to the AIDS situation. Knowing that condoms are not foolproof, I didn't want to take the risk.   (F)

In general, among the bisexuals, we found a widespread lack of sexual satisfaction, a decreased sexual repertoire, and the fear that safe sex might not be all that safe no matter what the precautions taken:

I felt a sense of loss and mourning about just giving up sexual practices with men I enjoyed, even doing safe sex. Sucking with a rubber wasn't a turn-on. I was very much into oral sex. I had just started to enjoy receiving anal sex when AIDS came around. I felt real frustrated; even engaging in safe sex I felt anxious. What if the rubber broke? If someone came on me and I had a cut? Got to be not worth it.   (M)

Reluctance and disappointment were also based on less pragmatic reasons. Semen, for example, was interpreted as being an integral part of sexual relations:

In every article or book I've read that refers to why people don't use condoms, I have yet to read anyone who seems to know why. It's because ejaculation is part of the satisfaction for many people, men and women.   (M)

Sex becomes more complicated with condoms. I think there's some spiritual meaning in exchanging bodily fluids. That's gone when I wear a condom.   (M)

In any case sex for many came to involve a lot of work that it never did before, suggesting a change of meaning:

I will not do anything in the least unsafe with a secondary partner. I use gloves and condoms on a dildo with my primary partner. I will kiss and do oral sex with my primary partner. You have to think about sex a little more before you do it now. You have to buy things, you have to make sure you are supplied with rubber items. You can't just spontaneously slip your hand in, you have to go find a glove. You have to think about where your supplies are. You have to be prepared.   (F)

Lots of changes. A year and half ago when I broke up for the third time with the man I'd been dating, I moved back to San Francisco and had to confront the AIDS and safe-sex issue. So I had to have the test, do six months of absolutely safe sex, and have the test again. That was very difficult, a big adjustment because my sex life has not previously involved the use of condoms. I've had my tubes tied. I've relaxed a bit on the safe-sex issue, but I know a lot about it and I'm very selective.   (F)

The "work" that came to accompany sex was not only restricted to preliminaries like having the correct supplies, but required a new sexual etiquette. Safe sex to bisexuals meant a dialogue with partners in which past experiences, current partners, likes and dislikes, health status, and so on were discussed before sex occurred. Again this often detracted from the experience of sex since clear rules of etiquette did not exist, and asking too many questions could call into question a partner's integrity:

> People that we may have agreed to meet and swing with in the past, I'm now asking more difficult questions, and testing their patience more, and being extremely selective.   (M)

> I feel like every woman I go out with I have to explain my past and explain a lot about how AIDS is transmitted. I don't think it's changed who I've had a relationship with but it's slowed up the sexualization of a relationship.   (F)

At the same time, however, some of the bisexuals said that such preliminaries improved sex for them:

> AIDS is making people more thoughtful and that's good for relationships in general.   (M)

> I think it's [AIDS] taught me to communicate more with my sexual partners about what I like to do sexually and what I will not do.   (M)

> I use the AIDS situation. I'm shy sexually and cautious as a rule. I can use AIDS as an excuse to get to know somebody before having sex. In my mind I know I can use this excuse.   (F)

But all of these changes were considered oppressive for some of the bisexuals. Together they destroyed the meaning of sexual freedom.

> It [AIDS] makes me very leery. I'm very scared of AIDS. I'm afraid I'll become more repressed sexually again. It took me so long to come out of this once already.   (F)

Of course, there were others who simply gave up the attempt to practice safe sex altogether.

> I saw my current lover eight months before we had sex. We waited for sex till we had the antibody test. Then we used condoms. But

condom use was brief as he hated using them. We were both negative on the test so we stopped using them.   (F)

The previous case illustrates another element in the safe-sex situation —the test for the HIV antibodies. Testing negative was often seen as a signal that one could proceed safely. This was so even though the accuracy of the test and the interpretation of results were by no means definitive. For many people, however, the test did inject what appeared to be a scientific certainty into an uncertain situation. And overall, most respondents had continued their sexual lives.

We were concerned about safe sex at the early period. We had a hard time changing to safe sex. My sex was limited to oral sex before Billy, and then we got into anal sex. We didn't want to use rubbers for this anal thing. But we had to make a change. We still practice safe sex—we are both HIV positive.   (M)

What occurred for many was a reconstruction of sexuality whereby safe sex itself had become eroticized. One way of doing this, the most common, was to expand the meaning of sexuality to include or enhance behaviors that were once of less importance.

It has shifted the focus of most of my sexual activity from orgasms to the broader scope of pleasure.   (M)

The world has opened up—there are more options and discoveries. AIDS means everything is new again. Beyond barrier protection there are erotic potentials to safe sex. There is more permission to explore, more holding, touching, not just penetration. For example, rubbing cocks over the body, playing with latex, involvement in SM, play with fantasy especially. The sharing of this newness accounts for more intense orgasms.   (M)

Many respondents claimed to have rediscovered fantasies as a way of heightening sexual arousal.

I've been into more active fantasizing, and I've placed more reliance on pornography to satisfy my sexual urges.   (M)

There are some fantasies I no longer make reality. Like fantasies of anonymous sex—the fantasy of having sex with a lot of people blindfolded. That would be risky in today's world.   (F)

For many more, there was the discovery, or rediscovery, of the joys of self-masturbation:

I've invested in a world-class vibrator.   (F)

When I was first interviewed I had just begun to identify as bisexual and had not had sex with a man. Now my bisexuality is expressed in personal fantasies and masturbation.   (M)

I do a great deal of phone "jack off" as my sex outlet. I really like twenty-year olds; it's a real fetish.   (F)

Masturbation was especially favored by those who had AIDS or whose partners had AIDS:

He has AIDS, a full-blown case. I feel confident that we are performing the safest sex we can with the most pleasure and satisfaction. Most often I masturbate myself while he holds me. If he has the strength we have intercourse and that of course includes condoms, using a sponge and lubricant with nonoxynol 9.   (F)

New opportunities for mutual masturbation also appeared over five years. By 1988 there was a homosexual club that permitted only self- and mutual masturbation. It was well-lit and had TV monitors to ensure that no penetrative sex of any kind ever occurred. Another new institution in the sexual underground came to be known as jack- and jill-off parties or clubs, where both sexes masturbated themselves or each other (called a "jack-off party" if it was just for men).

I go to jack- and jill-off parties. Everything happens there except intercourse. . . . with anyone that's agreeable and most people are.   (M)

Others turned more to SM and bondage as avenues for sexual expression as these activities did not necessarily involve the exchange of bodily fluids.

I had to take most of my bi urges and find indirect ways to satisfy them—fantasy, kinky correspondence, nonintercourse SM activities with men.   (M)

I have stuck my head in the sand about AIDS. Meaning I have not done anything about it. I have never been tested. I don't screen potential partners. And I don't feel the need to yet. Fortunately, most of the nontraditional sex that I like, such as bondage, tends to be safe sex by its nature in that there is no exchange of fluids. There is no anal or vaginal intercourse. The actual sexual contact usually comes down to manual stimulation.   (M)

None of these changes in the meaning of sexual behavior is mutually exclusive. Often they blended together in different combinations:

> In general, I have less sexual intercourse with both males and females. More so with males than females though, more emphasis on nonintercourse. I have been expanding SM and finding new ways to be sexual: non-SM stuff—I did one "jack-off party" to see what it was like. Not too interesting. I use fantasy a lot more. And very extensive correspondence with gay males. That's how I meet the demands of my gay side.   (M)

SM could be an alternative to conventional sexual practices. However, people involved in the SM scene in San Francisco reported that the number of persons interested in such experimentation had not increased. Like most other sexual institutions in this particular sexual underground, though, the SM scene had changed in the face of AIDS. Practices like anal and vaginal fisting had declined because they were being defined as risk factors, and activities that involved cutting and piercing were being done very carefully if at all. One bisexual prostitute who specialized in these practices reported that the demand for her sexual services had increased because she was also a registered nurse and knew how to carry them out safely.

Indeed, any organ penetration or exchange of bodily fluids was now rare at SM parties. This was partially due to AIDS, but SM aficionados also used AIDS as an excuse to make such parties purely SM occasions. As one respondent explained it:

> We say this is an SM party. Go to a swingers' party if you want sex.   (M)

It would be tempting to suggest that the social control possible in group situations like SM and jack and jill-off parties and clubs would ensure compliance with safe-sex guidelines. In most instances this was probably the case (as in the homosexual club we observed). However, there were exceptions. As one woman said:

> I've organized jack and jill-off parties. The rules are no fucking and we provide latex gloves, rubbers—all under safe-sex guidelines. People sign a statement that they will follow these. They don't always, though. I've got depressed offering these parties.   (F)

The concern with the possible dangers of unfettered sexuality was reflected in the changes we discovered in specific sexual behaviors between 1983 and 1988.

Changes in Sexual Behavior

There were obvious questions that accounts about "safe-sex" raised: Did the need for "safe-sex" and the fear of AIDS lead to an overall decrease in the sexual desire of the bisexuals? To a greater selectivity about their partners? To a change in their sexual practices? We expected they would. In terms of sexual patterns, we assumed the most common changes would be a decrease in the *number* of partners, especially male partners, a decrease in group sex and anonymous encounters, and a decrease in "unsafe" behaviors such as fisting and anal intercourse. Conversely, AIDS and "safe-sex" we felt should lead to an increase in self-masturbation and manual sex. All of this would depend on how people interpreted the "safe-sex" message. But "safe-sex" was not necessarily defined in a clear or consistent way among the bisexuals we interviewed, and even if it was, following its dictates was often difficult. Thus, in this section we look at the *actual* changes that had occurred in sexual behavior by comparing what our respondents said they were doing in the last 12 months when interviewed in 1983 with what they said about their sexual behavior in the last 12 months when they were interviewed in 1988. (We will refer to these different periods simply by the year of the interview.) Some of the findings do not reach statistical significance because of the small sample size.

Number of Partners

Men showed a decrease in female partners (Table 23.1). Four percent had no female partners in 1983, compared to just over 20 percent in 1988. The number reporting five or more female partners dropped from a third in 1983 to only 14 percent in 1988.

For the women, the decline in the number of opposite-sex partners was even greater. The number with no male partners had risen from seven percent in 1983 to a third in 1988. And 44 percent had five or more male partners in 1983 as compared to about 15 percent in 1988.

The men also had somewhat fewer same-sex partners in 1988. Primarily, the proportion who had no male partners rose from around 14 percent to about a third. Similarly, the proportion reporting five or more such partners declined from about a half to a third.

There was little change in the number of same-sex partners for the women. This is consistent with the general assessment that men are riskier partners than women with regard to AIDS.

More relevant to "safe" sex messages is experiences with anonymous partners. Among the men, there was little change in having had an anonymous female partner in the past twelve months. There was a decrease, however, in having had an anonymous male partner: from two-thirds in 1983 to less than one-half in 1988.

For women, the percentage who had anonymous sex with a man fell by more than half. There was little change in their anonymous sex with other women.

Group Sex

There was no significant change in the percentage of men participating in threesomes or group sex (Table 23.1). The women changed more in both instances. Half had engaged in sexual threesomes in 1983, but less than a fifth had done so in 1988. For group sex, the number dropped from 37 percent to 11 percent. Apparently messages about "safe-sex" had affected women more than men.

Sexual Practices

Self-Masturbation

There was no significant change in the frequency of self-masturbation for the men (Table 23.2). Around 80 percent masturbated once a week or more in 1983 compared to 70 percent in 1988. The women showed a greater decrease, with the number masturbating once a week or more declining from 85 percent to 54 percent. This ran counter to our expectation that self-masturbation might increase to compensate for a decrease in other sexual activities. It may be that the disenchantment with sex and the AIDS crisis carried over to sex in general more than simply to sex with partners. In other words, the emotional effect was so profound for some as to lessen sexual desire overall.

Heterosexual Sex

The men showed the most apparent decrease in their frequency of performing masturbation, performing and receiving oral-genital sex, and coitus with a female. In 1983, twice as many of the men (around 60 percent) masturbated a woman at a frequency of once a month or more as in 1988 (one-third). Performing oral sex on a woman at the same frequency was down from over two-thirds to a third, and the frequency of receiving oral sex from a woman also declined from two-thirds to around 20 percent. Finally, about 80 percent of the men engaged in coitus once a month or more in 1983, compared to just over a half in 1988.

There was also a visible decrease among the men in performing and receiving finger-anal sex with a female (from around three-fifths and two-thirds on each variable in 1983 to about 40 percent in 1988 on both). The other forms of anal sex with a female partner were too infrequent to measure any change, as was the case with other forms of heterosexual activity (SM, enema, urine, or feces play). Indeed, in 1988, these behaviors had become even more rare.

The women decreased somewhat in performing and receiving masturbation of or by a male. Over 60 percent engaged in these behaviors once a month or more in 1983 as compared to around 40 percent in 1988. They showed a greater decrease in the frequency of oral-genital sex with men. About 80 percent performed fellatio once a month or more in 1983 as compared to over a half in 1988. Approximately three-quarters received oral sex from a man once a month or more in 1983 as compared to just over a third in 1988.

Coitus also decreased noticeably among the women: almost three-quarters engaged in coitus once a month or more in 1983 as compared to over half in 1988. Though less common than coitus at either time, there was a similar decrease in finger-anal sex, from around 60 percent to 37 percent (performing or receiving). Anal-receptive intercourse also decreased (about half answered "never" in the last twelve months for 1983 as compared with 70 percent for 1988). As noted above, too few of the women engaged in other forms of anal sex and unconventional heterosexual activity to measure any change.

Homosexual Sex

In terms of same-sex behaviors, the men changed little in the frequency of masturbating or being masturbated by a male partner. About a third reported this once a month or more in 1983 and 1988. There was a large decrease, however, in fellatio. About half of the men performed fellatio once a month or more in 1983 compared to only 14 percent in 1988. There was a decrease from 50 percent in 1983 to 21 percent in 1988 of those who had received oral-genital sex from a man.

Anal sex of all types decreased visibly as well for the men. Roughly 55 percent performed and 60 percent received same-sex finger-anal stimulation in 1983 as compared to roughly a third in 1988. Nearly 60 percent performed and two-thirds received same-sex anal intercourse in 1983 as compared to approximately a third and less than a fifth, respectively, in 1988. The most dramatic cutback in same-sex behaviors for the men was receiving anal sex.

For the women there was also limited change in masturbation with a same-sex partner. About a third masturbated or were masturbated by a woman once a month or more in both 1983 and 1988. There were more women in 1988 who said they stopped this behavior altogether compared to five years earlier (a 25 percent change with regard to performing masturbation and 12 percent in receiving it). There was less change in the frequency of oral-genital sex with a same-sex partner than among the men: almost half performed cunnilingus once a month or more in 1983 and about one-third in 1988. The corresponding figures for receiving cunnilingus from a woman were similar.

Decreases in performing or receiving finger-anal sex with same-sex partners were likewise evident for the women, from nearly two-thirds in

1983 to 80 percent in 1988 answering never. For both men and women, other homosexual activities were too infrequent to measure any change. Any sexual activities such as anal-fisting or enemas that involved potential unsafe sex did not increase.

It is worth re-emphasizing that masturbation with a same-sex partner for both men and women did not increase, especially given that manual sex has been widely touted as an important part of the safe-sex message—to compensate for decreases in other sexual outlets. Rather the trend was for it to decrease. Similarly, there was no change in self-masturbation for the men. In fact, the women decreased in their frequency of self-masturbation which perhaps indicated that the negative attitude toward sex outweighed anything else.

Overall the data suggest that the general effect of the AIDS situation had a far-reaching impact, reshaping sexual desire across the board more than particular sexual behaviors. What does stand out, however, is the decrease in anal-receptive intercourse among men, indicating that the message about what is ostensibly the most dangerous sexual behavior was clearly heard.

Safe Sex and AIDS

What the bisexuals considered to be "safe-sex" at the time (mainly the use of condoms and a decrease in anal and oral sex) was, overall, said to be practiced more with male partners than with female partners (Tables 23.3, 23.4, and 23.5). Approximately three-quarters of the men and over half of the women said they *always* practiced safe-sex with male partners; about one-third and under half, respectively, said this with regard to female partners. Safe sex was practiced less with primary partners (over a third said always), probably because many were now in monogamous relationships. More said they practiced safe sex with secondary partners (over half said always), and even more said they did so with casual and anonymous partners (approximately 85 percent said always).

Parallel to this was the care the bisexuals indicated they had taken in practicing safe sex. More said they were *very careful* in their safe-sex practices with male partners (70 percent of the women, 95 percent of the men) than they were with female partners (75 percent of the women, 80 percent of the men). Men who were sexual with men, then, were the most careful. Around two-thirds of the respondents said they were very careful in their practice of safe sex with their primary and secondary partners, and 90 percent or higher with casual and anonymous partners.

Approximately two-thirds of the people we interviewed had taken the HIV antibody test. Seventeen percent of the men tested positive and 6 percent of the women. We know that two of the respondents had developed AIDS and have since died. Also, since one-third of the respondents reported in 1988 that they had not been tested for HIV antibodies, we cannot conclude what proportion actually were exposed to the virus, nor how many have been since then.

To summarize, changes in the sexual patterns of bisexuals were congruent with changes we found in the meaning of sex. Nearly everyone was having less sex in 1988 than they were five years earlier. Certain "high-risk" activities (anal sex) had decreased, but "safer" practices (masturbation) had not increased to compensate overall. Rather than simply changing patterns of sexual activity, AIDS appears to have been an overall restraint to sexuality, changing the meaning of sex since the more hedonistic days of the early 1980s.

Since men were considered riskier partners than women, both sexes had cut back on the number of male partners. The bisexual men also showed a drop in the number of female partners, possibly because each was cutting back on overall sexual activity, or perhaps because bisexual men were considered a "high-risk" group and were therefore having trouble finding female partners. Hand in hand with these changes in the patterns and meaning of sex came significant shifts in the relationships that the people in our study considered most important in their lives.

24

Changes in Relationships

During our original round of interviews in 1983, we found that most of the bisexuals had a clear idea of the relationships they ideally wanted. The most common ideal was to have two primary relationships (one heterosexual and one homosexual). The second most commonly desired arrangement was one primary relationship (heterosexual) with a secondary (homosexual) partner. Some kind of group or communal arrangement, though not common, ranked third in preference. No matter what the arrangement, the underlying criterion was that these relationships had to be *nonmonogamous*. Did these ideals survive?

The Fate of Ideals over Five Years

Monogamous Ideals

Approximately 60 percent of the respondents said that they had changed their perception of what an ideal relationship would look like (Table 24.1). What stands out is the increased value given to monogamy. Many of those who said that their ideals had changed now considered a monogamous relationship the best possible arrangement.

AIDS played a major, but varied, role in the desire for monogamy. For some, AIDS was the direct and only factor:

> My ideal is to have just a female lover. I don't want to get involved sexually with men. It has to do with AIDS; it's like playing Russian roulette.   (F)

AIDS often interacted with other factors, however. The two most frequently mentioned were a desire for more "intimate relationships" and "getting older" or "more mature." In this respect, as we have already seen, AIDS might have been a catalyst for changes that would have occurred anyway. Regarding the former reason, one person who desired more intimacy said:

> I'd like to be in a monogamous, committed relationship with a female. Only a monogamous relationship is the most satisfying in terms of developing intimacy. And it's not safe to have a non-monogamous relationship because of AIDS. (M)

> AIDS made me look at the real emptiness I experienced having anonymous sex. (F)

Those who mentioned the passage of time cited an accompanying lack of energy, or simply becoming "more mature," as facilitating a change in their ideals in conjunction with AIDS:

> I attribute more of my ideal view to age rather than bisexuality. Now monogamy would be fine for me; it's less of a hassle. I want to feel more settled with one person. . . . It seems to me I have less time and energy to do things and nonmonogamy takes a lot of time and energy—juggling schedules, keeping your communications clear. . . . Part of it has to do with AIDS and the sexual climate—more people are inclined to be settled. I just pick up on that aura. (F)

> Commitment and monogamy—the desire for a committed relationship—have gained greater credibility for me because of maturity and aging. And I'm scared of AIDS. (M)

Some people in our study referred to the desire for intimacy and the effect of aging alone—without any reference to AIDS. They frequently mentioned the complexity of dealing with multiple relationships, especially as one got older.

> I've been concentrating on finding a supportive life-mate. This is more realistic for me. I could barely deal with the [one] relationship I have. I [have begun to] concentrate on more adult things, like my career. (F)

And we should not forget that bisexuality is considered deviant by the wider society, so that fears and social pressures often shaped ideals.

> With society as it is, I prefer to be married to a loving, gentle, intelligent, caring young man. If society was different I would not care at all. (F)

I prefer to live as a heterosexual because I don't want the stress of living in the closet. I'm an elementary school teacher and I was disowned by my family. (F)

Finally, we should acknowledge that sometimes ideals changed but without a specific resolution for the person involved.

My ideal changes from year to year. Five years ago, I was married and thinking of a dual relationship as my ideal. I've been wanting to focus on my husband and he switched around and wanted to be involved with two women. Now I've separated and I don't know what my ideal is. A monogamous relationship doesn't seem complete, but a relationship with two people is too confusing for me. (F)

Nonmonogamous Ideals

Monogamy had not become the only ideal, however. A considerable number of people who changed in their ideals moved toward the types of ideal relationships we found predominant in 1983. Some wanted two primary relationships—one heterosexual and one homosexual:

I'm clear now about wanting a committed relationship with a man and a woman even though I don't want to live with them. (F)

I've discovered that I would like to include a man as another primary relationship in my life, to emotionally expand. (M)

They provided the familiar reasons for holding such ideals—that bisexuality offers greater opportunities for psychological and sexual satisfaction. Other respondents wanted only *one* primary relationship with one or more outside partners.

I used to want to be in a primary relationship with a woman with a male relationship on the side. It's changed to the opposite now. I had this [the former] and it didn't work as women get jealous more often. (M)

Now I'd like to have in my ideal situation a male lover, but just female friends—perhaps sexual. Three years ago I stopped having intercourse with women and that freed me from a lot of performance pressures I've had all my life. (M)

Having positive relationship experiences with the same sex directly helped to change the criteria for ideal relationships:

This is my first relationship. I don't think I had a clear ideal before. I now have an open relationship [with a woman] with extremely good communication and a great deal of love. We're both free to have relationships with others.   (F)

I'm in a new relationship. I've accepted a lot of personal characteristics that were not on my list for an "ideal" partner. It took me a long time to admit that I love her, and I saw it could work with someone who previously wasn't my ideal.   (F)

Of course, for some of the people we interviewed, life changes could be sweeping—especially for those who had changed their sexual preference and the way they labeled themselves, resulting in a corresponding change in ideals.

I identified as bisexual in 1983 and as gay now. I used to go out with Jay [a female] one night and Gary [a male] the next. My ideal now is a primary nonmonogamous gay relationship. Transitioning from bi to gay would have to change my ideal!   (M)

There were others who decided they wanted families, but also wanted to remain nonmonogamous:

I'm from the Midwest and my family wonders why I haven't settled down. I couldn't be monogamous, I'd still want a male partner. But since 1983 I've had the desire for a family so I want a primary female partner.   (M)

I want a primary relationship with a woman and I think I even want to get married. I'd like her to be bi and open to bringing in other people.   (M)

Some kind of communal or multi-partner arrangement became the ideal for other respondents:

I didn't used to have an ideal; now I do. My ideal is a non-monogamous group marriage. My ideal would be four primary partners, two men and two women.   (F)

To be a member of a foursome, two males and two females, where the males relate sexually and the females relate sexually to each other and as male/female pairings do too.   (M)

As with other arrangements, however, real-life experiences modified communal ideals, leading to more modest goals.

I used to think group marriage, with both a male and a female, was ideal. Now it's become foggy because it's hard to achieve the ideal. Right now the idea is in flux. I moved into a house with a guy and some other people and it didn't work out socially. I found it wasn't that easy.   (F)

AIDS was not always a concern for those whose ideal was to have multiple partners. Some seemed to ignore the threat or deny its relevance. Thus, it did not always have an impact.

I've thought things out more and elaborated my philosophy on the subject [of polyintimacy] in spite of AIDS I guess.   (M)

More commonly people acknowledged AIDS and modified their ideals in the face of it. They did not always abandon nonmonogamy, however:

Due to the AIDS situation, I had to clarify what I was willing to do and not to do based on risk. I concluded I was not willing to be monogamous, but what I'd like would be a committed non-monogamous group marriage using safe sex. I was not willing to live my life based on fear.   (F)

My ideal was oriented toward multi-partner sex. Now my ideal is fewer partners and a less wide scope of sexual practices. AIDS influenced my sex practices and also my ideal.   (F)

We also asked a general question, "What is your current attitude toward nonmonogamy?" in order to see how this value had changed. After all, in 1983, 80 percent of the respondents said that they were nonmonogamous and their primary relationships were "open" ones.

Most often they still had a favorable attitude toward nonmonogamous relationships. The same kinds of reasons for this also still held: that nonmonogamy avoids depending on one person to fulfill all needs; that it removes the strain from relationships, emotionally and sexually; and that it allows for more exploration and fulfillment in one's sexual life.

By 1988 though, many had a negative attitude toward nonmonogamy. Some said AIDS was a reason, but again it was not the major one or the only one. Most, as noted previously, felt that nonmonogamy was too time consuming, took too much energy, or was too complicated. They also thought that it got in the way of developing love, trust, and more intimate relationships with a partner.

Some people gave heavily qualified or ambivalent answers that may have reflected the tension between the ideology of nonmonogamy and increasing social pressure for monogamy because of AIDS. One person said that he had practiced nonmonogamy because he felt pressured to be a role model for bisexuals. Others said that nonmonogamy was acceptable but only under stringent safe-sex conditions. Most, however, were willing to see nonmonogamy as possibly good for others but not for themselves.

Despite the AIDS situation, then, a considerable number of the bisexuals retained nonmonogamy *as a value* although perhaps not practicing it themselves. For those who had become monogamous, the shift seemed to be as much the result of having tried an open relationship as a fear of AIDS. A good number also remained ambivalent.

Ideals, then, whether linked to AIDS, particular experiences, the effects of aging, or time constraints, changed for many of our respondents. We turn next to the actual relationships that the people in our study were in and the changes that had occurred in them.

Changes in Primary Relationships

Over 60 percent of the bisexuals in the 1988 study were in a primary relationship—the same proportion as in 1983 (Table 24.2). Almost three-quarters of these relationships were heterosexual, again the same proportion as five years earlier.

Approximately three-quarters of the bisexuals, also however, reported a change in one or more primary relationships since 1983. Some people began new relationships, ended old ones, got divorced or married, were involved in a series of relationships, increased or decreased their number of simultaneous primary partners, and so forth.

We organized these changes into three categories: people who no longer had the primary relationship they had at the time of the original interview and were not in a new one; people who were in a primary relationship they had entered into since then; and people who reported a series of primary relationships or an increase or decrease in primary partners since 1983. In terms of these changes, about 40 percent of the people we re-interviewed were no longer in a primary relationship, close to a third were in a new one, and a third had a series or an increase or decrease in their number of primary partners.

What reasons did people give for the changes in their primary relationships? Basically, the same reasons were given for all of the changes, and most were reasons that heterosexuals and homosexuals would also offer for changing important relationships. Some important reasons, however, were linked to a bisexual lifestyle and to AIDS.

Conventional Reasons

Personality conflicts were the most frequently cited reasons for ending a relationship or moving through a series of them. Usually, these were said to be problems the partner had, not the respondent. The partner was usually described as unwilling or unready to change.

I'm no longer living with the man I was living with when I was interviewed before. I couldn't stand living with him. He was not willing to change and grow. I recently got into another primary relationship with a man three months ago.   (F)

I had two primary relationships. One with a woman broke up. She was too depressed and her therapist recommended she become single to resolve her personal problems. She was not ready to be in an adult relationship.   (F)

Five years ago I was involved with a male and a female, then I got into a relationship with a man, but that ended too. He was unhappy about his life, everything, but was unwilling to change anything.   (M)

Sometimes the alleged personality conflict was extreme, in one case involving violence.

I got rid of my battering husband. I found out that he was a closet drinker too. I'm much stronger now, but I'm not actively looking for a primary relationship. I don't want to take care of another man.   (F)

As with changing "ideals," other changes were related to the passage of time. Some people had taken on new commitments or interests that left them little time or energy for the type of primary relationship(s) they were in, or they had "matured" and wanted a different type of person or arrangement.

The only primary relationships I had up to that point were live-in relationships. Since then I've learned how to sustain primary relationships where we live apart. I attribute this to maturity. I used to have only one model for a primary relationship, which was living together in an exclusive relationship.   (M)

I ended an ongoing relationship with a woman. She was too demanding of my time. I now have a busy work and play schedule that doesn't allow for a lot of quality time for one person.   (M)

People in our study also changed primary relationships because they wanted more intimacy, giving up relationships that did not provide this and searching for those that would.

I had a primary relationship with a man and then with a woman, the one I'm in now. My bisexuality previously emphasized sex. Over the last couple of years, I've become more interested in women in a more emotional way. That's what I want.   (F)

Other reasons for changing primary relationships unrelated to sexual preference were: the partner died, the partner became ill, or the partner moved away.

Reasons Related to Sexual Preference

Among the reasons people offered for changing their primary relationships, a large number had to do in some way with sexual preference. Two major ones were first, problems associated with maintaining bisexual relationships; and second, problems that were connected to the AIDS situation. Both these reasons were offered with about equal frequency.

As we have seen, many of the bisexuals experienced their sexual preference as a changing phenomenon. Their feelings, affections, and behaviors were subject to continuous change, which caused problems for some relationships. Some of the people we spoke to went through changes in their bisexual desires which led them to reorganize their involvements.

> I now have a primary relationship with a female. I found I couldn't make a commitment to a man. I decided I couldn't be with men any more. (F)

> I left my primary male lover. We did not have a sexually satisfying relationship as I wanted to be with a woman primarily. I was always attracted to women more and basically felt I was more in a brother-sister relationship with my male lover. (F)

Sometimes the *partners* of the people we interviewed changed in their bisexuality:

> My first wife and I divorced because we realized her main interest in life was other women and even though she loved me a great deal she could not treat me as her primary partner. (M)

> Ultimately I found it satisfying to be with a self-defined straight male. In the past, the men I've been involved with who were bisexual were uneasy with that definition, experiencing pressure from their male partners and gay male friends. This made it difficult for them to be emotionally relaxed and open with me as their primary female partner. (F)

In other cases a mixture of changes related to the bisexuality of both partners hurt the relationship:

> I divorced my wife in 1986. The fact that I needed men on the side forced my ex-wife to have men on the side too. This just created too many problems for us. (M)

In addition to changes in sexual preference, there were problems with managing bisexual relationships themselves. Most frequently mentioned were the problems of dealing with nonmonogamy.

I had a relationship for about a year [with a man] that was supposed to be nonmonogamous. We negotiated around it being a monogamous relationship, but it was clear it wouldn't work.   (M)

I met my first woman lover in March 1984. The relationship was basically open. I could see other men, but not other women. It didn't last because of her jealousy. I had a date with a man and overslept and missed a date with her. She couldn't deal with her own jealousy. From there it went downhill.   (F)

Reasons Related to AIDS

Just as AIDS can poison the body, it can also poison relationships, either existing or potential ones. As in other instances, however, the effects of AIDS were complex. Sometimes people reacted with fear directly or quoted AIDS along with other relationship issues. As we have seen, the fear of AIDS led many people to reject men as partners:

I ended a primary relationship with a man and took up with a woman. I did try a relationship with a man I liked a lot, but AIDS was an unfortunate preoccupation.   (F)

When AIDS became epidemic, I shifted towards women because they were less risky. I've made a conscious effort to choose women and avoid the AIDS problem with men.   (M)

On the other side of the coin were those whose primary relationships were affected because *they* were considered a risk.

I had a primary relationship with a man and a woman for a year and a half. It broke up because the other guy had deep fears about AIDS.   (M)

The end of my relationship with a woman and the lack of relationships for a period may have been due in part to people's fear of AIDS.   (M)

And again, in other cases AIDS increased the desire for monogamy.

I've moved from an open relationship with a man to a closed relationship with another man. AIDS significantly affected my life in 1984. I stopped seeing my bisexual male lover because I was afraid of AIDS. I'm now seriously trying to get married.   (F)

Sadly, some of the people in our study had relationships in which they or their partners tested positive for HIV antibodies or who had AIDS itself. The strains on these relationships were evident:

> My longest relationship broke up because he was HIV positive. Much to my dismay, my love for him was compromised. The fear overwhelmed the love. . . . To take a gay lover now means the possibility of a horrible death—for both of us. This has thrown the worst pall over gay love that could ever be devised.   (M)

> Everything's changed. I have no primary partner right now. I don't see my ex–male lover very often. He has AIDS. We're still close, sometimes we have sex. If he wasn't sick, we could work out our relationship. He was gay-identified and an IV drug user the last part of our relationship—that's why I broke it off.   (F)

Changes in Secondary Relationships

Changes in secondary relationships were as complex as those in primary relationships, with people adding them, dropping them, experiencing changes in their composition or quality (Table 24.3). In 1988, about half of the people we interviewed said that they had a secondary relationship, the same proportion as in 1983. Approximately two-thirds of the people in the study also reported a change in their secondary relationship(s). Of these, over a half said that they now had no such relationship or fewer such relationships. Twenty percent had a secondary relationship that they did not have in 1983 or had increased the number of such relationships. The remainder mentioned some type of "qualitative" change—e.g., in having a same-sex or opposite-sex partner.

Of those who now had no, or fewer, secondary relationships, most spoke of AIDS as the reason. Some had direct experiences with AIDS that profoundly affected them.

> I have dispensed with them [secondary relationships]. Maybe if AIDS wasn't around I could play on that level. But dealing with people who are dying or whose partners have died takes too much energy, which earlier could have gone into playful secondary relationships. Everyone is in grief overload.   (M)

> I had two secondary relationships. One of them, a man, subsequently got AIDS and died. Since then I don't have any more secondary partners.   (M)

> There are more people who don't want to play with me because of my diagnosis [AIDS]. Only women who are SM identified and highly AIDS aware—not the majority of the female community—

will play with me. [What do you mean by "play"?] All sex, not just SM.   (F)

Others, as we saw in the case of primary relationships, cited AIDS in combination with other factors as reasons for change. The most frequent answer was that the AIDS risk had increased an already existing desire for a monogamous relationship. Many had found such a relationship, or were searching for one, making secondary relationships unimportant.

I am now monogamous. Five years ago I had a male lover. Now I want to be in a long-term secure relationship with my partner. Also I wouldn't want to take a risk with AIDS by becoming attracted to someone else.   (M)

I had multiple secondary relationships, but now I'm monogamous. It was due to AIDS. I was [also] seeking a long-term relationship with marriage and baby potential.   (F)

The move away from secondary relationships was not solely due to the fear of AIDS. As we have seen, many people also mentioned the desire for a deeper or a more fulfilling relationship with one person:

I no longer have secondary relationships. I asked myself why hadn't we become more deeply involved. Why was it just sexual? I decided I wanted more out of a relationship. [Were these changes related to the AIDS situation?] If I'm going to put my life in someone else's hands, I want it to be a committed person who I can trust.   (M)

Other people who now had no or fewer secondary relationships did not mention AIDS at all. Some of the people in primary relationships did not have the time to pursue or maintain secondary relationships:

Two years ago I had a secondary relationship with a woman and now I don't have any. I would like a man, but I'm just not meeting many men. I've been extremely busy and haven't had time.   (F)

These things could affect anyone's sexual relationships regardless of sexual preference. Bisexuality itself, though, because it usually involved more than one relationship, produced its own problems with respect to secondary relationships.

I went from having a relationship with a lady to nothing. She couldn't stand my husband, so that was it. She wouldn't share my time and couldn't handle me having another relationship. And I think she felt threatened by a man. She was originally my primary partner and then I met my husband so she became secondary.   (F)

I don't have secondary relationships. If I had a relationship with someone else it would have to be two primary relationships. I'm interested in a team relationship where everyone could be equal and working together. (F)

Some bisexuals did not mention an increase or decrease in the number of secondary partners but said that the tenor of the secondary relationships they sought had changed. About half of these people referred to AIDS as a reason. Many had become more selective:

These relationships have tended to be more ongoing and slower to develop. I'm more interested in knowing the person behind the sex organs. And I'm more selective because of AIDS. (M)

Unlike a large proportion of swingers we're not interested in short-term relationships any more, so our relationships tend to last. We haven't met any new people for over two years because of AIDS. I'm extremely selective, I screen out people who seem too risky. (M)

Some of this selectivity involved a difference in the sex of the secondary partner—a preference for women over men because of the risk of AIDS.

I tend to be involved with females more now because of AIDS. Before I had mainly male secondary relationships. (M)

But not all such selectivity had to do with the fear of AIDS. Feelings toward men or women changed for other reasons too:

I'm more aware of my ability to love a same-sex partner. My experience has made me more comfortable caring about a female partner. (F)

I'm seeing more women. In the past five years I've gotten more angry with men because of their lack of emotional involvement. (M)

Finally, as with primary relationships, some changes in secondary relationships had nothing to do with AIDS or bisexuality *per se* but rather occurred as a result of the passage of time, unique experiences, or a reduction in opportunities.

I didn't use to have a secondary relationship, now I do. I don't know if I'm going through a midlife crisis since we've [she and her primary partner] been together for nine years and the sex is getting boring. I think that happens. (F)

In general, then, AIDS appeared to have as much of an effect on secondary relationships as on primary relationships.

Changes in Ground Rules

Most of those who had primary relationships in 1983 were non-monogamous and most of them had adopted some type of ground rules to deal with jealousy. Fewer had ground rules in their secondary relationships as jealousy did not seem to be a prime concern here. These rules were not always fixed; pragmatic concerns or the changing nature of relationships did lead to their alteration. In 1983, the threat of death from AIDS was not an issue. AIDS added a new dimension to ground rules. Previously, sexually transmitted diseases were the main factor in the ground rules respondents had for secondary relationships. Now, the fear of death made such a concern the major one in any type of sexual relationship.

Primary Relationships

About two-thirds of the people we re-interviewed reported changing ground rules in their primary relationships since 1983 (Table 24.4). There were three major changes: about 40 percent of the respondents now had a rule of monogamy, just over a quarter had rules for safe sex, and about a third had rules for making their relationship an open one.

AIDS was a major reason for adopting rules about monogamy, either alone or together with a general concern about the sex of the outside partner.

> We went from a specific written agreement that we could see outside partners to an explicit verbal agreement to be monogamous. This is health-related—because of AIDS.   (F)

> When I was married we had ground rules—a specific night out for both of us, a time to be back home, call if late, etc. I'm pledged to monogamy now. I couldn't get involved with anyone else now emotionally because of AIDS. I've lost too many friends.   (M)

> My last boyfriend and I lived in different cities and saw each other only on weekends. We had an understanding we could have different partners in between. When AIDS became epidemic, I shifted towards women because they were less risky and accepted monogamy.   (M)

As noted previously, some entered into monogamy reluctantly:

> I have agreed to be monogamous. I've never agreed to this before. I believe he's honest and it seems important to him. Anyway, I guard my [HIV] negative status fiercely.   (F)

Again, there were also non-AIDS-related reasons for monogamous ground rules. To some it just made a relationship easier. As one woman said, any other rule than monogamy is "too complicated for me emotionally." For others, monogamy was looked on in the traditional way, as a sine qua non for a relationship, especially a marital one:

> When we decided to get married we made up a list of ground rules. We will be monogamous unless we both discuss it prior. We wanted to avoid disagreements about what being married means.   (F)

> Marriage is based on monogamy. I'm willing to make a commitment at this point in my life. Refusing to do so is copping out on the relationship.   (M)

Almost as many people in our study changed their ground rules to precautions concerning safe-sex as adopted rules about monogamy. All of them directly mentioned AIDS as a reason. These people took AIDS into account, yet did not feel that they had to sacrifice their open relationships:

> Before, I didn't make demands about my partners' sexual behavior or safe sex. Now I expect them to know if they're antibody positive or negative and to do only safe sex.   (F)

> I took recommendations for "safe sex" and made them the rules for our sexual activities. I got an AIDS test and it came back positive, but it was a *false* positive. I had to wait two weeks for the recheck. That's a real experience! Everyone I was involved with was scared. It really brought home to me the importance of safe sex.   (M)

Despite the AIDS situation, some changed their ground rules to allow a more open relationship, often because they always had wanted an open relationship and now had met a partner who would support them in this. Or they had negotiated this with an existing partner:

> We have an open relationship now. A couple of years ago I wanted an open relationship and my primary said no, she wanted monogamy. Then she fell in love with an older woman and I said go ahead—so we had an open relationship.   (F)

> When I was with a woman, she and I were completely monogamous. She never would have tolerated me having other partners. My new partner is a man and he doesn't care. He is not threatened by my bisexuality and wouldn't be jealous if I had sex with another man or woman.   (F)

Secondary Relationships

In 1983, about two-thirds of those with secondary relationships said they had no ground rules with those partners. Those who did mentioned precautions against sexually transmitted diseases as being the most common rule. Of those who reported a change in ground rules for a secondary relationship, the most frequent change was the adoption of a rule to reduce the risk of AIDS:

> I'd prefer they had a primary relationship also so that I know who they were sleeping with. Before we do anything now we talk about it. It's just good to be clear and honest.   (F)

> I definitely will not do unsafe sex with a secondary partner. I tell them about my primary relationship. Before, I didn't share that information. I also got a lot more practice in what it took to have a secondary relationship. I had to weigh the risks of what was it worth to me to have fun on the side. It was not worth unsafe sex.   (F)

These findings echo those in Chapter 21. Even though the fear of AIDS was an important reason for changes between 1983 and 1988, it was not always mentioned as a reason, the most important reason, or the only reason. Many relationships changed for conventional reasons.

Mundane reasons, however, coupled with the fear of AIDS, did result in a very different picture of bisexual relationships in 1988. Instead of searching for a pair of primary relationships (one with each sex) or a primary and secondary relationship, monogamy had become the ideal for many of the people in our study. The proportion who said they were in primary and secondary relationships had not changed, but the ground rules for those relationships were different. Earlier most of the ground rules appeared to be designed, directly or indirectly, to prevent jealousy in the primary relationship. Now AIDS was a major factor, and the new ground rules stressed monogamy and safe sex. Despite the threat of AIDS, some of the people we re-interviewed had opened up their relationships since 1983, but the major trend seemed to be toward closing them in favor of monogamy.

25

Adapting to a New World

Because of AIDS, bisexuals confronted a different world in 1988 than they had in 1983. Earlier, their major concern had been dealing with how others reacted to their sexual behavior. Now they also had to address the growing belief that bisexuals were carriers of HIV. Not only did this reinforce the conception that many lesbians had about bisexuals; it also increased the fear of those heterosexuals who saw bisexuality as a conduit for AIDS between the homosexual and heterosexual worlds. This chapter examines the reaction of the bisexuals in our study to these problems. Specifically, we examine whether they became more secretive about their bisexuality, whether they felt increasing hostility from heterosexuals and homosexuals, and whether they became less self-accepting in this new environment.

Changes in Disclosure

We asked the bisexuals whether they had become more or less wary about disclosing their bisexuality to others since 1983 (Table 25.1). Forty percent said that they had become more wary, just under 20 percent that they had become less wary, and about 40 percent that they had not changed. About a quarter of those who changed attributed this change to AIDS.

Becoming More Secretive

When asked a follow-up question, whom they had become more wary of, our respondents most often said "everyone" or "most people." The

273

next most common answer was "employers" or "work associates." The relative lack of family or friends mentioned was due, as we have seen, to the fact that respondents had already disclosed themselves to these groups. On the other hand, people can change their jobs frequently and the general social climate toward bisexuality can change.

Three times as many men as women cited AIDS as a reason for being more wary. The major issue here was that those we interviewed saw other people as believing that bisexuals were disease carriers:

> I don't disclose as much because I'm afraid of triggering people's paranoia about AIDS.   (M)

> There's a myth in the lesbian community that bi women are carrying AIDS into their community. This makes it harder to disclose.   (F)

> Because of AIDS people have taken their homophobia completely out of the closet now. And I don't want to take on the monster of homophobia right now in my life.   (F)

> AIDS has hurt the bi community really bad. . . . The bi male is scapegoated as the door for AIDS into the heterosexual community.   (M)

As important as the AIDS reason for becoming more secretive was the fear of discrimination, especially in employment. Respondents were at pains to point out that even without the AIDS situation, stereotypes of bisexuals persisted and often provided a basis for discrimination. Sometimes personal experiences of this kind had made them more wary:

> I've learned the hard way. It's injured my career and limited my opportunities and options. I used to be "out." It has nothing to do with AIDS.   (F)

> I've gotten more wary because I've gotten some bad reactions from people when I tell them I'm bi. They didn't believe me; they don't think that bisexuality exists.   (F)

> I am a school teacher. When I taught in San Francisco I came out to almost every teacher. In Arizona [where she now works] I can't do this. A lot of teachers make jokes, which intimidates me.   (F)

Becoming Less Secretive

Most of those becoming less secretive said that they now disclosed their bisexuality to "everyone." The next most common answer was family and friends. Few mentioned AIDS as a reason for becoming less secretive. The most important reason was increased self-acceptance:

I've become more comfortable with my bisexuality. Now I speak in front of large groups about being bi. [How did this happen?] Through a lot of therapy, a lot of life, and getting to the point where I didn't need other people's approval about who I was.　(F)

I'm more secure in my own sexuality so I'm less influenced by other people's negative responses. I'm not willing to lie now.　(F)

Others became less secretive as a "political" gesture or to be a role model for other bisexuals:

It's important for people who feel comfortable with their bisexuality to be out. It's important that bisexual people be recognized as having our own sexual orientation. If my coming out and telling my story helps just one person, then it's worth it.　(F)

The above woman was one of the few people who mentioned AIDS as a reason for becoming *less* secretive. She saw it as her mission to tell people that bisexuals were not disease spreaders:

Time magazine called bisexuals the pariahs of society—that we were spreading a gay disease to the straight community. That's just not true. People fear bisexuality because of negative stereotypes surrounding AIDS and bisexuals. I hope to dissolve those stereotypes.　(F)

Finally, changing patterns of social contact also led to disclosure. People tended to disclose when they shared strong bonds:

I've been relating more with sexual minorities—as straights aren't so great. It's easier to disclose to the sexually unconventional.　(F)

I'm no longer with the man I was living with. He was a private person, and wary about others finding out I was bi, because it might affect him and his work.　(F)

As I've gotten to know people longer our friendships don't hinge on whether or not they know. It just hasn't become a problem.　(F)

Relationships with Heterosexuals and Homosexuals

As we have seen, in 1983 bisexuals reported serious problems with both heterosexuals and homosexuals. In particular, they reported the most discomfort in the company of heterosexual males and lesbians. We wondered if things were different in 1988. To what extent did AIDS account for any changes? Were people who shared the homosexual stigma more supportive of each other? Were heterosexuals more condemnatory? (Table 25.2)

Changes in Relationships with Homosexual Men

A third of the people in our study—both men and women—said that their relationships with homosexual men had become better since 1983. About 10 percent said that they had worsened. The rest, over half, reported no change.

Those who saw an improvement in their relationships with homosexual men most often attributed this to the AIDS situation. For some this was born of compassion or an increased closeness and intimacy as they saw gay men deal with the crisis:

> I met many gay men sick with AIDS as a nurse and my relations with them are best of all. Because of their caring for each other, I am a better, more loving person.   (F)

> I was in a positive antibody group for bi and gay men. I really have grown to love and admire gay men. Their love for other men is just like what my parents' love for each other was.   (F)

> I began talking more to gay men because of AIDS. It opened opportunities to talk about other things. It helped in my . . . relationships with them in terms of increasing our communication skills.   (F)

Many people, though, talked about how AIDS presented a shared crisis for both bisexuals and homosexuals and created a more cohesive sense of community between them:

> I could pass as straight if I wanted to and the fact that I don't is appreciated by gays as a sort of solidarity. All gays and bi's feel there's a siege mentality now because of AIDS in that everyone is pulling together.   (M)

> I think the bisexual community has been acknowledged more by the gay and lesbian communities. This is a consequence of gay and bisexual males being lumped together as one high-risk group for HIV infection. It has provided an issue around which many other differences dwindle in importance.   (M)

Aside from AIDS, some of the bisexuals' relationships with homosexual men improved as they came to terms with their own bisexuality and achieved some measure of self-acceptance. This sometimes involved a global self-acceptance:

> I communicate more and better now. I am very loved in my life. I am out there, not hiding myself in the least—what you see is what you get.   (F)

Or specifically, other times it emerged from an acceptance of their bisexuality and standing up for themselves:

> I have been more open as a bisexual and have asked that I be respected and treated as okay.   (M)

> I don't apologize for it any more. I take the position that I know as much about being with the same sex as they do.   (F)

Of course, less conflict came with a change in the way they defined their sexual preference:

> Well now I identify as gay and this has pretty much eliminated the conflict.   (M)

> It's changed because I now define as lesbian. . . . There's no sexual tension involved now and there was before when I was bi.   (F)

Finally, the passage of time alone sometimes led to improved relationships, especially when it led to increased interaction between bisexuals and homosexuals.

> I'm with a lot more gay men today because I work in a gay business. They all know I'm bi. My relationships have improved because I didn't know many gay men before and now I do.   (F)

> The only problem I ever had was with gay men, usually out of ignorance or lack of experience [with bisexuals]. Lately it seems that gay men have gotten educated about bisexuality or even tried it themselves.   (M)

Ironically, those who felt that their relationships with homosexual men had worsened mentioned AIDS about as often as did those who said relations had improved. Some blamed gay men personally, saying that they were currying favor with heterosexuals by buying into the belief that bisexuals were responsible for further spreading the HIV virus:

> Gays have ditched us in their efforts to minimize their own blame [for AIDS] from heterosexuals. They've joined the straights in blaming the bi's.   (M)

On the other hand, ironically some bisexuals were avoiding homosexual men because they saw them as disease carriers:

> I stay away from them because I'm afraid of getting sexually involved because of AIDS.   (M)

> I'm very wary of sexual relationships with gay men because of
> AIDS—even when doing safe sex.   (M)

The reactions toward homosexual men in light of the AIDS situation
could be very diverse:

> Gay men claim that AIDS is a gay disease, but it isn't. They operate
> as if they own it; it's as if nobody else is devastated by it.   (F)

> So many of my gay men friends have died because of AIDS. It's
> depressing. The others talk about nothing but AIDS and death. I
> avoid them now. They make me feel guilty somehow for being alive
> and healthy.   (F)

As both the above and following case show, the sheer horror of the AIDS
situation could hurt relationships with homosexual men:

> There's impacted grieving from losing so many men. I've lost twen-
> ty to thirty close male friends. So there are big blocks in getting close
> to men. There's the association between gay love and seeing gay men
> die, which is more than I can deal with. My relations with gay men,
> therefore, have become dysfunctional.   (M)

Other reasons for worsening relationships, familiar from the 1983
interviews, revolved around what bisexuals still saw as stereotypes held
about them:

> I'm sick of them putting down bi's. They still have the same old
> stereotypes.   (F)

> I've had a couple of instances recently. One gay man said he didn't
> believe there were such things as bisexuals. Another man thought
> that a bi meeting held in a gay space [building] was an intrusion. I'm
> more aware now of some attitudes I hadn't previously been aware of,
> especially that "bi's don't exist"!   (F)

Changes in Relationships with Homosexual Women

Over a third of the people we interviewed said that their relationships
with lesbians had improved since 1983, and only a few said that they had
worsened. Almost 60 percent indicated no change. These figures are
similar to changes in relationships with homosexual men. Bisexual men
and women were about equally likely to say that relationships had improved, which was unexpected given that twice as many women as men
reported having problems with homosexual women in 1983.

The bisexuals most often cited AIDS as a reason for improved relationships with homosexual women. As with homosexual men, AIDS had created an increased sense of community:

> I have seen the lesbian community come to the aid of the gay male community. I think the women are more open and understanding than the male community. When I hear the term "bitchiness," I think in terms of the gay male community, not the female lesbian community. (M)

> In general, lesbians are working more with gay men, providing for them, because AIDS has so crippled the gay male population. And the men are welcoming the women more into their lives. Lesbians are more accepting of my personal relationships with men. (F)

> They now consider me part of their family. Men and women are pulling together because of AIDS. (M)

Increased interaction with lesbians, independent of the AIDS situation, had also improved relationships:

> Since the closing of the Bi Center, I have found myself spending more time in the gay and lesbian communities and have developed stronger ties there. (M)

> I've had more chances to meet gay women and make friends with them. I seem to be running into less strident and angry gay women now. (F)

One theme in these responses, absent among the men, was an increasing solidarity based on gender: women began to identify as women first and as bisexual or lesbian second.

> There seems to be more tolerance towards bisexuals. I seem to have more in common with them [lesbians]. I guess being a woman puts me more in touch with them. (F)

> I don't see myself as different from them any more. I used to think they would ostracize me for being sexual with men and now this doesn't matter to me because I identify with what's the same about us, which is we both love women. And a lot of lesbians love men also, even with the lesbian label. (F)

As is the case for male homosexuals, the most common non-AIDS reason cited for better relationships with lesbians was increased acceptance by the people we interviewed of their own bisexuality, or a general increase in self-acceptance:

My increased level of comfort with my own sexuality. One of my primary relationships was with a former lesbian, and many of her friends were still militant lesbians. I learned to get along with them.   (M)

I learned to be more assertive and to explain bisexuality better, so they question their assumptions about it.   (F)

Some of the bisexual men too had reduced their fear of women in general:

I tend to fall in love with women now. I met some wonderful ones. I'm less threatened by women now having gotten to known them better.   (M)

And, of course, there were those whose relationships with female homosexuals had improved because they later defined themselves as gay or lesbian:

I defined myself as bisexual for about ten years and during that time it was very difficult to be among lesbians. Now I define as lesbian and feel a lot more acceptance in the lesbian community.   (F)

Because now I've defined as lesbian. When I defined as bi, they didn't like me as much because they thought I'd leave a woman for a man and they don't like to sleep with a woman who's having sex with a man—it turns them off.   (F)

Those who said that their relationships with homosexual women had worsened most often attributed this to AIDS, and most of those evaluations came from women. The reasons were similar to those of homosexual men; that bisexuals were a medical threat:

Because lesbians, as a result of the AIDS crisis, are more frightened of anyone who touches sperm. I've had some lesbian women, when I tell them I'm bisexual, just get up and walk away. They see me as a potential disease spreader.   (F)

They use the excuse that bi's are more likely to spread AIDS. They were previously hostile to bi's, but now it's worse.   (F)

I've been bashed by lesbians. Just hear them putting down bi's—they're conduits for AIDS, bi women will leave them for men, bi women claim heterosexual privilege, and so on.   (F)

As this last comment illustrates, negative relationships were attributed to the persistence of stereotypes about bisexuals as well as the perceived risk of AIDS.

Changes in Relationships with Heterosexual Men

In 1983, the bisexuals reported a great deal of difficulty with heterosexual men. Five years later, about a sixth of our respondents said their relationships with heterosexual men had improved. A similar proportion, however, said that relationships had worsened. Two-thirds saw things as about the same.

No one mentioned AIDS as a factor in improved relationships with heterosexual men. Rather, most of those who claimed an improvement attributed this to the increased acceptance of their own bisexuality or an improved self-acceptance in general:

> I'm assertive about saying I'm bisexual and not backing down in front of straight men.   (F)

> Two years ago I couldn't relate comfortably with them. Now I feel more comfortable about myself and I look to have more heterosexual friends.   (M)

> I'm more confident of my own masculinity so I'm less threatened by them.   (M)

The second most common reason for better relationships with heterosexual men was increased interaction and more positive social experiences with them:

> My career has passed the point of bitter competition with the men in my field. I'm more relaxed with them and socialize with them more.   (F)

> In general, I've found the straight men I've come out to, to be very secure with themselves and are not threatened by my sexuality.   (M)

Those who said their relationships with heterosexual men had become worse most often gave a reason related to AIDS. The main one was that heterosexual men saw bisexuals as disease carriers from the homosexual to the heterosexual world:

> They [heterosexual men] say we're not at risk and the bi's will contaminate us.   (F)

They're uncomfortable about bi's spreading AIDS because I've had sex with multiple partners. I'd be defined as an AIDS risk.   (F)

Straight men feel more threatened now as they see bisexuals as a potential source of AIDS. There has always been homophobia among straights and this fuels their alienation toward bi's.   (M)

Again, stereotypes about bisexuals were still seen to exist and were viewed along with AIDS as a reason for worsened relationships.

Changes in Relationships with Heterosexual Women

About 20 percent of our respondents said that their relationships with heterosexual women had improved since 1983; another 16 percent, that they had worsened. The vast majority, over 60 percent, reported no change. Bisexual men, especially, noted that the situation had become worse (a quarter versus less than ten percent of the bisexual women).

As in the case of relationships with heterosexual men, no one said that AIDS had led to improved relationships with heterosexual women. Again, most attributed the improvement to the acceptance of their own bisexuality or their increased self-acceptance.

I'm more confident of myself and I'm less threatened by them.   (M)

I feel more comfortable with myself and have sought more straight friends. I have less fear and more attraction to straight women, a feeling of being on an equal basis with them.   (M)

As previously, the second most common reason for improved relationships was increased social interaction and more positive social experiences with heterosexual women:

I've had more interest and involvement with heterosexual women. I haven't met many bisexual women since I dropped out of the Bi Center.   (M)

I have more close relationships now with [heterosexual] women. I have more positive contacts.   (F)

Most of those who said that their relationships with heterosexual women had worsened mentioned AIDS as a reason, and those reasons were similar to those we found in the case of their evaluation of heterosexual men—that bisexuals were seen as disease carriers.

They are afraid to become sexual and intimate with us because they think that's how you get AIDS.   (M)

They are afraid because of the AIDS situation. There's distrust for bisexual men who they think bring AIDS into their heterosexual lives.   (M)

It seems fair to say that had the AIDS situation *not* occurred, increased disclosure, interactions, and self-acceptance might have made for an even greater improvement in relationships with heterosexuals. Ironically, however, it took the AIDS situation to bring bisexuals and homosexuals more together.

Accepting a Bisexual Identity

Even given the problems the bisexuals faced, in 1983 we found that 80 percent of the bisexual respondents had no regrets whatsoever about their sexual preference, and among those who did, the regrets were minor (Table 25.3). Most accepted their sexual preference and felt positive about it. Their major problems had more to do with society's negative attitudes toward bisexuality. We especially wanted to know whether more had regrets over their bisexuality in 1988 than in 1983 because of AIDS.

We found no significant change. Twenty percent of the people we interviewed in 1988 still said they regretted their bisexuality, but almost all of them said that their regrets were not strong. The major reason for these regrets had not changed: being bisexual made life complicated because of society's reaction.

It would be easier to swim in the same direction as the rest of the fish. There are moments that it would be neat to have marriage and a house with a white picket fence.   (M)

I regret that I have experienced some job discrimination in the past. I also regret that my bisexuality has made some women wary of me and scared some, though certainly not all, away.   (M)

A new reason did appear in 1988, however. Added to social prejudice, the next most frequently mentioned reason for regrets had to do with fear of AIDS:

Accepting my bisexual behavior, I feel, has indirectly put me in the fast lane for AIDS, from my past. In the old days I got it on with anonymous bi men at sex parties. If I had not explored my bisexuality, I wouldn't be living through this fear, wondering if the antibodies are going to show up.   (F)

There's an increased risk of dying!   (M)

Given the relatively small number of people who regretted their bisexuality and the number of these who offered AIDS as a reason, it does not seem that AIDS played a major role in shaking their acceptance of their sexual preference. This would seem to pay tribute to the potential that the bisexual identity holds for adapting to radical changes in the environment.

26

Conclusions:
Understanding Bisexuality

We began the book by posing a riddle: how can current theories of exclusive homosexuality, with their emphasis on fixed biological characteristics, explain bisexuality? It did not seem feasible to us that these theories could do justice to the complexity of a sexual preference in which people engage in sex with both same- and opposite-sex partners and often change this mix over their lives. If current theories cannot explain these patterns, how exactly do our general conceptions of sexual preference need to be rethought?

Our answer both emphasizes and augments the early contribution of Alfred Kinsey. On the one hand, we believe bisexuality is a universal human potential. And we recognize, as Kinsey did, the danger of converting types of behaviors into types of people—of calling, for example, all persons engaging in sexual activity with both men and women, "bisexuals," as if they were a homogeneous group. But in other ways we go beyond Kinsey. As sociologists we pay more attention to the social factors that foster or inhibit a bisexual potential. We do not ignore, as he did, the social identities that people adopt to make sense of their sexuality. Indeed, we look at the interaction of sexual behaviors and identities in great detail, as well as the emotional dimensions of sexual preference that embed sex in a web of intimate relationships. Finally, we are able to show how people's sexual preferences can change. Unfortunately, it was the AIDS crisis that helped us here.

Our research concentrated on one type of sexual preference and on a particular group in one particular place over a short period of time, so any generalizations must be made with care. At the same time, the group studied is a natural group rather than a "sample," the place is San Francisco, which is a veritable laboratory for sexual experimentation, and the time period involves a serious reexamination of sexuality occasioned by the emergence of a deadly disease. For these reasons, it is at least possible to suggest some of the ingredients of a more general understanding of sexual preference.

Pathways to Sexual Preference

Although we do not have detailed accounts about the early lives of people who became "bisexual," we were able to gain some insights into the ways in which early sexual experiences are related to subsequent sexual behavior.[1]

Beginning with the belief that persons are born not only with a bisexual potential but the potential to eroticize—learn to give sexual meaning to—many things, what stands out in our research is the relationship between one's earliest sexual feelings and behaviors and one's subsequent sexual preference. In the case of heterosexuals and homosexuals, sexual preference generally followed the direction of the earliest sexual experiences they reported (opposite-sex and same-sex respectively), even though many of them had sexual experiences later that were discordant with their sexual preference, and even though many of the heterosexuals still had homosexual feelings and homosexuals heterosexual feelings. This seems to suggest a continuing bisexual *potential* in many persons throughout their lives, regardless of the sexual identity they eventually adopt.

Bisexuals generally completed their heterosexual development first (their opposite-sex attractions and behaviors preceded their same-sex ones), but in a way that did not close the door to further development in a homosexual direction. Both future heterosexual and homosexual adults seemed to develop a clear pathway relatively early (in adolescence). Bisexuals, on the other hand, often engaged in homosexual behavior for the first time later and self-identified as bisexual in their twenties and beyond.

At this point, then, we can say that early sexual attractions and experiences do seem predictive of later sexual preference. At the same time, regardless of their future sexual preference, many persons did show evidence of a bisexual potential in that their early experiences involved a mix of sexual feelings and behaviors toward both sexes (and this persisted into adulthood for many of them).

No theory of sexual preference should ignore the mundane feature of sexual pleasure. Unfortunately, many of them do.[2] We believe that sexual pleasure in its various forms is ordinarily the main reason people have sex. The role of pure physical pleasure seems much clearer for men. Men, in all three preference groups in our research, had their first *homo-*

sexual experience much earlier than women. Men thus learn early that sexual pleasure is possible with both sexes, and that given the greater difficulty of getting female partners, other men may be acceptable substitutes. This accounts for why there seemed to be a more genital focus to the same-sex behavior of the bisexual men.

Learning the Gender System

If a mix of sexual attractions toward *both* men and women is not uncommon, why does it remain part of the sexuality of some persons but become buried in others? As people develop sexually they not only learn about the pleasure that their genitals bring but also what this pleasure is supposed to mean. Much of the meaning of sexuality is conveyed to people as they learn about gender. This is done through what we call the "traditional gender schema"—a cognitive framework that organizes a person's perception of what characteristics are or should be associated with each sex. Obviously, the gender schema teaches us socially appropriate sexual expression. As one sociologist states:

> When we do begin having sex in our society, our beliefs about women/men strongly influence who we have sex with, what sexual things we do, where and when we will have sex, the reasons we agree, and the feelings we have.[3]

Thus, "many societies, including our own, treat an exclusive heterosexual orientation as the *sine qua non* of adequate masculinity and feminity."[4] Not only should a man want to have sex with women, he *must* do so in order for society to consider him a man. Sometimes this bias of our society makes its way into scholarship, and the resulting theories of sexuality have suggested that persons who are not exclusively heterosexual are deficient in their gender learning (e.g., a woman is homosexual because she has not fully learned the feminine qualities and roles of our society).

We cannot agree with this. We found that the people we studied held a very traditional view of gender. They *had* learned the qualities and roles our society assigns to men versus women. Learning what gender is all about may be no different for most persons who end up with different sexual preferences. What seems more important is what they *do* with that learning as they put their sexualities together.

The gender schema—the map we carry in our head to process ideas about males and females—gives meaning to what differences between men and women can or should be eroticized. These socialized desires that bring us *sexual* pleasure thus can be considered *gendered* pleasures. It is gendered bodies we initially discover, first our own and then those of the opposite sex. We may not need too much help from social learning to discover that genitals produce sexual pleasure, but it is social learning (via the gender schema) that gives meaning to this pleasure. Notably, we learn that only pleasures derived from opposite-sex bodies should be entertained. This then can become generalized as a whole panoply of socially constructed gender differences—personality, demeanor, clothing, speech styles, comportment, carriage, and the like, all of which are eroticized to produce conventional heterosexual desire. All these we learn as society directs us toward those qualities which are appropriately and consistently "masculine" or "feminine."

Variations in sexual preference involve variations in the extent to which individuals *actualize* what they have learned about gender in putting together their sexuality, rather than some deficiency in learning. We argue that persons initially learn the desirable aspects of *both* sexes and that they can produce gendered pleasure in both directions. To become conventionally heterosexual, however, requires unlearning, disattending to, or repressing desirable features of one's own gender that could be or have been eroticized. Homosexuality would result in a similar way by disattending to the erotic qualities of the opposite gender. Thus heterosexuality and homosexuality develop in similar ways. The major difference is that, given the institutionalized pressures toward learning heterosexuality, homosexuals probably have a greater understanding of heterosexuality than heterosexuals have of homosexuality—one is more easily avoided than the other. As evidence, in our data, more homosexuals have had heterosexual experiences, especially the women, than vice versa.

Following this line of argument, bisexuality then emerges we suggest as a *result of failing to unlearn, or rediscovering both same and opposite-sex gendered pleasures*. Its complexity results from the ability to combine gender attributes together in a variety of ways. Putting together sexuality in this fashion can be far from easy. It is not surprising that we came across a great deal of confusion in the sexual histories of many bisexuals and homosexuals, and a considerable number of heterosexuals.

But still the question remains as to exactly how bisexuals combine gender attributes in constructing their sexuality. We maintain that this involves the development of what we call an "open gender schema" that reconstructs the traditional gender schema in a particular way.

Open Gender Schemas

Bisexuals do learn the traditional gender schema—what is erotic about both sexes. They are thus successfully socialized. But they fail to successfully suppress the erotic aspects of their own gender. We would argue that instead of organizing their sexuality in terms of the traditional gender schema, bisexuals do so in terms of an "open gender schema," a perspective that *disconnects gender and sexual preference,* making the direction of sexual desire (toward the same or opposite sex) *independent* of a person's own gender (whether a man or a woman).

Because of this disconnection a number of things follow. The acquisition of an open gender schema can explain why some persons activate their bisexual potential and some do not. Many persons "add on" homosexuality to an already developed heterosexuality because they do not close off their ability to eroticize same-sex characteristics. This openness also accounts for why change seems so much more prevalent among bisexuals. In putting together their sexuality, they seem to be able to build upon and respond to a wider range of gendered signals than heterosexuals and homosexuals do. It is this last fact that draws attention to the myriad environmental factors that can affect sexual preference. An open gender schema allows persons to be extremely *adaptable* in meeting their sexual needs as their situation changes.

Homosexuals also learn traditional gender views. They too develop an open gender schema, probably before bisexuals do. But rather than remain at this stage they go on and develop a schema where gender and sex preference are *reconnected*. For them, however, this is *only* with persons of the *same* gender. Thus in many ways homosexuals and heterosexuals are closer to each other than they are to bisexuals.

Producing Open Gender Schemas

Theoretically, the question that follows is how open gender schemas are produced. There may be a variety of ways this occurs. It should be pointed out that the openness of such schemas can also be *relative*—the more open the schema, the more bisexual in sexual preference people become. Unfortunately, we do not have the data to describe how such perspectives come about, so what follows must of necessity be speculative.

The central process involves the ability to cognitively disrupt the connection between gender and sexual preference. For example, many hustlers (male prostitutes who may consider themselves "heterosexual" but have men as clients) deny the implications of their homosexual behavior for their sexual identity by behaving sexually in ways that protect their identity as "men," e.g., engaging only in the "active" or "masculine" sexual role.[5] This form of bisexuality is very common among male prisoners[6] and among males in certain ethnic groups.[7] The important feature of this is that it is supported by a network of male peers. The adoption of an open gender schema in these cases is transitory, because most men who engage in homosexual sex in prison seem to revert to exclusive heterosexuality upon their release.[8]

These manifestations of bisexuality in which the sexual component is paramount are characteristic of men. For women, the disconnection between gender and sexual preference seems to follow a different route. It is not the pursuit of sex that is the central issue but rather the pursuit of intimacy. Following gender scripts for heterosexuality brings women to heterosexual sex, but men do not always satisfy their emotional needs for intimacy and closeness. Their bisexuality may center around a close relationship with another woman in which there is little or no sex. Thus the implication for "sexual preference" is less direct—some of our female respondents emphasized that their behavior was bi-intimate rather than bisexual. This is not to say that women are not interested in same-sex genital behavior: the married woman who engages in homosexual sex at swing parties can come to like it, but her marital status helps her to adopt an open gender schema that is situational and does not necessarily carry any implications for what she feels is her true sexual preference.[9]

Open gender schemas can be produced in more dramatic ways. The transsexual bisexuals among our respondents showed this to be the case. These were people who had a lifelong struggle with the gender schema as they tried to achieve a gender identity that was inconsistent with their biological sex. They came to know the gender schema firsthand as it affected their sexual preference (e.g., "heterosexuality" validating their change in gender). Of necessity, then, they had to undergo a cognitive reorganization concerning gender in the process of discovering their sexuality. Bisexuality was not a necessary outcome but one that made sense. Most persons do not develop an open gender view so consciously, but many do. A great many women, for example, come to bisexuality as a result of their involvement in the women's movement, which leads them to closely analyze gender. They learn that to be a woman does not necessarily mean also being heterosexual.

The *conscious* elaboration of an open gender schema characterizes most of the bisexual respondents. All had adopted a bisexual identity. This often was related to their membership in the Bisexual Center. Here they were provided with an ideology that taught them that they need not necessarily confine their sexuality to the opposite sex and that exploring bisexuality was natural and a way to personal growth. In effect, it taught them that they were *bisexuals*—persons with a *permanent* open gender schema.

Finding a Sexual Identity

Sexual identities—naming oneself or being named in terms of the sex of the partner one chooses—are crucial to sexual preference. They give meaning to a person's sexual feelings and behaviors by defining these as signs that the individual is a special type of person—in our culture a "heterosexual," "homosexual," or "bisexual." For persons dealing with the confusions that dual attractions can bring, a sexual identity can *stabilize* sexual preference. If you know what you *are*, it organizes what you *do*. And it allows for social support from others who identify similarly. Sexual identities provide the social "cement" which sets sexual preference in place.

If bisexuality is a universal potential, then adopting the sexual identity of "heterosexual" ("straight") or "homosexual" ("gay" or "lesbian") can restrict a person from becoming "bisexual." That is, people who adopt an exclusive sexual identity may not even think about entering into sexual relations with both sexes, because it would violate their sense of who they are. This leads them to interact socially primarily with like-minded others, further reinforcing their sexual identity.

Nonetheless, we were struck by the similarities in the histories given by those who eventually labeled themselves "bisexual" and those who eventually labeled themselves otherwise. There are no watertight compartments between many "heterosexuals," "bisexuals," and "homosexuals," but rather overlaps. Thus, in their sexual profiles, the "somewhat mixed" heterosexual and "somewhat mixed" homosexual types were quite similar to two of our bisexual types. Although the overlap is not large, enough exists to raise the possibility that for some people in the three groups, traffic at the border could raise questions of identity. For such people, the *identity* they adopt explains their subsequent lives more clearly than does their behavior.

Bisexuals found it impossible to make sense of their sexuality by adopting either a heterosexual or homosexual identity. On the other hand, because the bisexual identity as a social category is not well defined or readily available to them, many experienced confusion in coming to grips with their sexuality and defined themselves as "bisexual" at a relatively late age. We believe that the study of sexual identities—where they come from, how they are put together, how they are disseminated, how they are different among different cultures and groups, and how they change over time—is indispensable to any theory of sexual preference.[10] Equally important is understanding how individuals relate to these social categories. For example, many of the "bisexuals" in our study believed that they had to have regular sexual relations with *both* men and women to be bisexual. Ending up in an exclusive relationship, as many of them did, often called into question their identity as bisexual. Not being sexually active with both sexes contemporaneously seemed to some of them to breach the prevalent social definition of "bisexuality."

Even after having adopted an identity, then, the bisexuals were often faced with continued uncertainty, an absence of closure about who they were that admits that further change may be possible. Exclusive relationships were not the sole reason for this uncertainty. Some failed to receive social support, or yielded to the demands of both the heterosexual and homosexual world to "get off the fence," and so forth. To accept the bisexual identity with some comfort, then, is often to accept the continued uncertainty that this choice implies.

In all of this we should not lose sight of the fact that persons who share the same identity label are very different. There are "bisexualities" just as their are "homosexualities"[11] and, indeed, "heterosexualities." We saw five general types of bisexuality. The most common was the heterosexual-leaning type, those for whom, we have argued, homosexuality is an "add on." Few of the people we studied were stereotypically bisexual, the "pure type," scoring exactly in the middle of the

Kinsey scales. But the broader mid type was relatively common. The homosexual-leaning type was uncommon, suggesting that perhaps most people come to bisexuality from an initial exclusive heterosexuality.

Of course there are different sexual profiles for those who define themselves as "heterosexual" and "homosexual" too; neither is a pure category. But the variance in these two categories is slight compared with that for bisexuality. "Bisexual" appears to be a more expansive sexual identity, sheltering a larger variety of persons who may have less in common in terms of their sexual preference (e.g., the profile of their Kinsey scale scores) than in the label they adopt. Another important finding is that people who are *not* at the ends of the Kinsey scale may experience dramatic changes in their feelings and behaviors without a corresponding change in sexual identity. Those at the ends—the more exclusive sexualities—are perhaps more loath to change or experiment with their sexuality because of the implications this may have for their sexual identities. The distances between the categories of the Kinsey scale do not, therefore, have a uniform meaning.

At this point, then, we add the notion of *sexual identity* to fill out the process of becoming bisexual. We can say that this involves the choice of a particular perspective (identity) from which to make sense of one's sexual feelings and behaviors. Which sexual identity one chooses in turn has important consequences for one's continuing sexual preference.

Sexual Opportunities

A person may have developed an open gender schema and may seek gendered pleasure. And perhaps the person has organized these desires through the adoption of a sexual identity. But something else must be added to fully explain sexual preference—an appreciation of the socially structured ways in which these desires can be realized.[12] This not only explains how bisexuality itself may be "done" but also the range of bisexualities that can exist. These factors we refer to as sexual opportunities.

By far the most pervasive of these structures is the gender system itself, which produces "gendered situations"—differential constraints and opportunities for sex depending on whether a person is male or female. This explains why bisexuality can take a different pattern for men and women.

San Francisco, as a whole, provides extensive sexual opportunities in the institutions of the "sexual underground." Most widespread and organized of all institutions in the "sexual underground" is the homosexual subculture, more specifically the gay *male* subculture—not only bars, clubs, organizations, etc., but also covert phenomena like certain restrooms, cruising areas, etc., that facilitate sexual interactions between men. Male economic power and male command of public space makes their sexual opportunities far more extensive than those for women. And it is this range of opportunities for male homosexual sex that bisexual men had available to them. Lesbian subcultures are less extensive and public than gay subcultures, so bisexual women had fewer opportunities for meeting same-sex partners. Thus the "add-on" of sexual behaviors with the same sex may be more common among men than women simply because of different ranges of opportunity.

Heterosexual opportunities, *in general,* may be no greater or fewer in San Francisco than other American cities, so they may not be as important for bisexuals as homosexual opportunities. The heterosexual opportunity structure did, however, intersect with the sexual underground to produce additional heterosexual and bisexual opportunities. In San Francisco this took two major forms: swing clubs and SM parties. Swing clubs provided opposite-sex opportunities for men and access to both male and female partners for women. And the SM community would organize a "bisexual night" every now and then, which allowed for same- and opposite-sex partners for both men and women.

For many people in our study, sexual opportunities were organized around an "open" marriage or marital-like relationship. The partner was considered "primary," the other relationship(s), "secondary." The primary relationship was likely to be heterosexual and the secondary relationship homosexual. Most bisexual arrangements were concurrent, and therefore based on nonmonogamy, but nonmonogamy was restricted by ground rules that protected the relationship from jealousy. Jealousy seemed more rife when someone's "extramarital" partner was of the same sex as the partner in the primary relationship. This meant that additional partners were more likely to be homosexual than heterosexual. Opportunities to meet other bisexuals were mostly limited to the Bisexual Center. The major role of the center, however, was not to facilitate opportunities for relationships but to provide a supporting ideology for those who chose a bisexual identity.

The most obvious constraints surrounding the expression of bisexuality come from society at large, which sees such a preference as a form of deviance. Thus bisexuals reported "coming out" to others much as homosexuals did, testing the waters first with those they trusted and avoiding those whose reaction they suspected would be negative. Because the homosexual subculture was and remains extensive in San Francisco, the homosexual can live a life circumscribed almost entirely by other homosexuals. Bisexuals in the 1980's lacked such a subculture, which, even if it were more extensive, would not circumscribe their lives—because bisexuals by definition retain their commitment to the heterosexual world. As a result bisexuals were *less* likely to be "out" than were homosexuals. Secrecy thus played a greater role in bisexuals' lives.

Another constraint on bisexual opportunities came from the homosexual world. Lesbians especially were negative toward bisexual women, who in turn found a source of expected partners and support to be less accessible than they had thought and hoped. In response to such stigmatizing, bisexuals emphasized the need for political organization and efforts to educate others about being bisexual. The Bisexual Center exemplified one way in which new sexual identities can emerge, be disseminated and reinforced.

Ultimately sexual opportunities do affect the character of one's sexual preference. We can see this first in the mix of partners found for the people in our study. Men had more same-sex partners and women more opposite-sex partners. That is, partners for *both* were more likely to be *men*—because men are more easily available for sex. The institutions of the gay subculture provided a steady supply of easily available male sex partners for men, and other institutions of the sexual underground provided willing male partners for women. Bisexual women had a greater number of *total* partners than did heterosexual and homosexual women because of their access to sexual partners from heterosexual, bisexual, and homosexual institutions. Bisexual men, however, did *not* have more total partners than *homosexual* men; the homosexual world operates well in providing a large number of male sexual partners.

Any general view of sexual preference, then, must take sexual opportunities into account. Obviously opportunities can change—a new gay bar may open, a bath house may close, a swing house may change its frequency of parties or who it lets in, a marriage may end, an ideology may change. All these changes alter the ability of people to meet prospective partners or gain other types of support for their sexual preference.

The range of sexual opportunities available to each group demonstrates most clearly the role of the social environment in shaping the form that sexual preference takes. But larger external forces can have an even greater impact.

Social Change and Sexual Preference

Any theory of sexual preference must be able to explain changes in the direction of a person's sexuality. Sexual preference is not totally immutable, and the direction of a person's sexual and romantic feelings and sexual behaviors are subject to both short-term flux and long-term modifications.

How do we account for changes in sexual preference? This is a crucial question because it goes far toward explaining the nature of sexual preference itself. Our research has shown that a considerable number of factors mediate changes across the Kinsey scale. All of these are part of the social environment.

At the lowest level, face-to-face interaction, we find the most immediate determinants of change: entering or leaving an exclusive relationship, deepening an existing one, increasing or decreasing social interaction with heterosexuals or homosexuals, receiving social support from bisexuals.

At the next level, social institutions, change is caused by those factors that reflect what we have called a sexual opportunity structure—those institutions that facilitate or limit sexuality. Among these were organizations like the Bisexual Center that provided an ideology, and especially a sexual identity, that supported change.

The final level, the sociocultural, has been relatively ignored in theories of sexual preference. Here we find large-scale historical changes in the cultural meanings of sex and gender, the development of new moralities, the emergence and disappearance of social categories, and so on. For us, the most pressing factor in the large-scale social environment that affected the features of sexual preference was the emergence of the AIDS epidemic.

AIDS had a powerful impact on sexual preference, as many respondents changed their sexual behavior in an attempt to protect themselves from the virus. It also meant a change in sexual relationships toward a monogamous arrangement. AIDS also changed the sexual environment. Sex that had been a positive, hedonistic, phenomenon was transformed into something to be done "safely," if at all. Casual and anonymous sex decreased as did participating in sex with multiple partners. The sexual opportunities available to the people we studied changed radically, as indicated by the reductions in sexual partners and frequencies in 1988 compared with 1983. All these changes produced changes in the direction of sexual feelings and behaviors for bisexuals. Some called their bisexual identity into question. But few rejected the identity.

Despite the evidence of change among the people in our study, even among those who were exclusively heterosexual or homosexual, we are not suggesting that changing one's sexual preference is an easy matter. The unsuccessful attempts by many psychiatrists to change "homosexuals" into "heterosexuals" is a case in point. But to the degree that bisexuality is a universal potential, a person with a homosexual preference might be able to "add on" a heterosexual interest without the primary homosexual interest being extinguished. The same holds for a person with a heterosexual preference who might add on a homosexual interest but still retain his or her primary heterosexual attractions. This pays homage to the power of early sexual conditioning. Still this would not be the only thing that prevents a complete change. We must also note the power of sexual identities. Traditional labels of "heterosexual" and "homosexual" organize a person's sexuality in an exclusive direction. For example, considering oneself as *being* a "homosexual" cuts off entertaining any opposite-sex attractions. And social interaction with other homosexuals and possibly immersion in the homosexual world makes the availability of any heterosexual opportunity structure irrelevant.

The most change in the direction of sexual preference in the short run was reported by the bisexuals. And they reported the most extensive change. We think two things account for this. First, the adoption of an open gender schema made them more open to the effects of environmental changes. There were simply more ways available to meet their sexual and emotional needs. Second, the bisexual identity signifies a nonexclusive sexual preference. Unlike heterosexuals and homosexuals, such an identity means that change is an accepted part of life when it comes to partner preference. They may "add on" or "take off" partners of either sex according to their particular needs. The bisexual identity is also less politicized than the homosexual identity. For example, many homosexuals see a political advantage in defining homosexuality as a condition into which one is born. That is, if homosexuality is beyond one's conscious choice then this should add support to gay and lesbian demands for civil rights. Whether homosexuality is innate or not, the *belief* that it is certainly retards any attempts by homosexuals to alter their sexual preference.

Toward a Theory of Sexual Preference

We cannot provide a complete explanation of how sexual preference is formed at this point. For a start, we can only theorize about sexual preferences based on gender. Although those are the most common sexual preferences, there are other bases for partner choice that are more important to some persons (e.g., a partner's age or race). More important, the myriad factors at work seem to defy easy generalization. But some of these factors do appear to hang together enough for us to sketch the framework for a theory. We propose the following components.

Bisexual Potential   All individuals seem to have the potential to be sexually attracted to both sexes. This potential may remain throughout a person's life and may be activated at any time.

Initial Sexual Experiences   The kinds of sexual experiences one has, especially during the adolescent years, are important in directing or reinforcing the course of one's future sexual preference. Although we would hesitate to claim that there is a "critical period" for initial sexual experiences to occur and take hold, the *order* in which these experiences with the same and opposite sex occurs and the feelings derived from them does seem related to one's sexual preference.

Traditional Gender Schema   Despite the existence of a bisexual potential, bisexuality is suppressed by society's conventional way of thinking about gender. Meaning is given to sexual feelings and behaviors by directing them exclusively toward the opposite sex, and we learn to assume a necessary link between gender and sexual preference. This traditional view, enforced through a variety of powerful social institutions, ensures that the most common sexual preference will be heterosexuality. In the course of sexual socialization, however, we learn what is erotic about *both* genders. Variations away from heterosexuality occur when we fail to unlearn or relearn these aspects of our own gender.

Open Gender Schemas   Bisexuality occurs when people adopt an open gender schema, a mode of thinking and perception opposed to traditional views of gender. Central to this is the understanding that gender and sexual preference are independent. People can adopt this nontraditional view in many ways—indirectly through the continuation of pleasurable same-sex experiences, or directly through ideological analyses of sex and gender, especially through any radical questioning of gender that shows it as a social construction rather than a fixed biological fact. When people learn that gender and sexual preference do not necessarily go together, they are more likely to entertain having sex with both men and women.

Sexual Identity   Sexual identities are social constructions that serve as convenient simplifications of the diversity of sexual experience. They give social meaning to our sexual feelings and behaviors and thus stabilize particular gender schemas. The existence of a set of sexual identities in any culture at any historical time will determine the nature of sexual preference, as people shape their lives to fit them and in turn are shaped by what these identities direct. Thus the existence and predominance of exclusive sexual identities like "heterosexual" and "homosexual" prevent consideration of dual sexual attractions and act as blocks to sexual experimentation. The bisexual identity is one that is only just emerging in our society. Its relative lack of clarity and lack of dissemination throughout our society mean that only a minority of persons who behave bisexually have adopted it. Still emerging and taking shape, it seems to explain less about bisexuality than the homosexual identity does for homosexuality. But it is increasingly part of the politics of sexual identity in the United States, whereby various groups engage in social conflict over the meaning of different sexualities.[13]

Sexual Opportunities   However much a person recognizes a bisexual potential, develops an open gender schema, or adopts a particular sexual identity, little sexual behavior will follow without the existence of sexual opportunities. These opportunities are a set of social organizational forms that facilitate sex by providing ways to meet partners, places to have sex, justifying ideologies, and the like. For heterosexuals these would include things like singles bars and dances, dating services, marital institutions, and so on. For homosexuals there are gay bars and many clubs and institutions in the homosexual subculture. Bisexuality is usually fostered by participation in both the heterosexual and homosexual opportunity structures. Institutions like the Bisexual Center that cater to bisexuals per se are rare, which can affect the adoption of a bisexual identity and sometimes bisexual behavior.

The Social Environment   Sexual opportunities are clearly part of the social environment, but sexual preference unfolds within a broader set of social influences as well. Anything in the social environment that affects the development of open gender schemas, sexual identities and the relationships that support them, and the structure of sexual opportunities can affect sexual preference. This is why understanding a sexual preference like bisexuality seems so complex. Because bisexuals have the most "open" of open gender views, the least restrictive sexual identity, and (in San Francisco at least) an extensive sexual opportunity structure, the direction of their sexual preference is subject to many influences. Taking these broader social influences into account is important because it allows us to look beyond purely interpersonal social factors to wider sociohistorical events such as the AIDS crisis.

Epilogue

The closing of the Bisexual Center ended an important chapter in the attempt to provide social support for a bisexual identity. It would be wrong, however, to link the fate of bisexuals in San Francisco with that of the Bisexual Center. Despite its demise, the people it brought together and the ideas it espoused continue to influence other institutions that form the basis of a viable bisexual community in the Bay Area. Today there are educational groups such as the Bay Area Bisexual Network (BABN) with its Bisexual Speaker's Bureau and magazine *Anything that Moves*. There are also purely social groups like Bi-Friendly, and BiPOL, a bisexual, lesbian, and gay political action group. And there are groups associated with Queer Nation, an organization of younger homosexuals who include bisexuals in their membership.

What does the future hold for the bisexual identity in the United States? It appears that bisexuals are where homosexuals were in the early 1960s. They are known to exist as a type of individual, although, as we have pointed out, this knowledge was unfortunately disseminated in connection with the AIDS epidemic so that bisexuals are often regarded as "disease carriers" by both homosexuals and heterosexuals. As their sexuality is seen as more preferential than essential, they also receive less sympathy than do homosexuals. On the other hand, we are starting to see the development of bisexual organizations, as we have described, and the emergence of leaders. This will help dissipate the "invisibility" of bisexuals and allow more positive messages about the bisexual identity to appear, especially if national coordination is successful.

Homosexual groups in the early 1960s also existed in well-organized pockets, and national organization was beginning (e.g., the North American Conference of Homophile Organizations, NACHO). However, it took specific historical events—the civil rights movement, the Vietnam War, the rejuvenation of the women's movement—to galvanize homosexuals into forming the gay liberation movement with its far-reaching consequences for society and the homosexual identity.

Larger historical events that can affect the bisexual identity of course cannot be foreseen. No one saw the AIDS crisis coming. We can only predict changes on the basis of the more mundane and gradual developments we are currently seeing. One positive sign for bisexuals is the acceptance of bisexuality by younger homosexuals. Just as younger homosexual groups (e.g., Gay Activists Alliance) confronted older ones (e.g., Mattachine Society) about what it meant to be homosexual in the late 1960s, so do younger homosexual groups (e.g., Queer Nation) confront older ones today. This time an important issue is the exclusivity of sexual preference. Older homosexuals are most likely to emphasize exclusivity of same-sex partners as the basis of a homosexual identity, since this is the identity they had to fight for. Younger homosexuals, however, can approach an already-won identity and experiment with it. Such experimentation includes bisexuality. Thus, the "add on" that characterizes bisexuality could increasingly be a heterosexual "add on" from the homosexual end of the Kinsey scale.

Whether such behavioral changes are in fact occurring and whether they will lead to identity changes (i.e., such persons calling themselves "bisexual") is not the central issue. More important is the fact that phenomena like the "Queer Movement" exist. This suggests an inclusiveness among young persons today, gathering together those who combine sex and gender in unconventional ways. The bisexual movement is stronger today because its younger members are part of the Queer Movement, which has networks all over the country. How distinct particular identities will remain in this inclusive movement is hard to say. It could be subject to the factional infighting and breakoffs that often characterize radical movements. And the bisexual identity could be subject to a narrower, "politically correct" definition as has characterized the homosexual identity.

We do not think that this will happen. There is simply no one right way to be "bisexual" that can easily be captured by a social movement. It is an identity that includes swingers, transsexuals, and many others—all united by a desire to escape traditional boundaries of sex and gender. Challenges to such boundaries seem increasingly possible in our society, which can only benefit the bisexual movement. It is more difficult to speculate on the future of heterosexuality, since "heterosexual" is rarely a politicized identity subject to debate. The questions feminists have raised about the boundaries of gender, however, have made heterosexual roles more flexible. A more serious challenge to accepted notions of sexuality seems difficult in the restrictive sexual climate of the early 1990's. This climate is itself in part a consequence of AIDS, which long ago put dampers on those who would travel along the Kinsey scale.

Notes

Chapter 1

1. *New York Times* (August 20, 1991).
2. *New York Times* (December 17, 1991).
3. *New York Times* (December 18, 1991).
4. Robert Pool, "Evidence for Homosexuality Gene." *Science* 261 (1993), pp. 291–92. Dean H. Hamer, Stella Hu, Victoria L. Magnuson, Nan Hu, Angela M. L. Pattatucci, "A Linkage Between DNA Markers on the X Chromosome and Male Sexual Orientation." *Science* 261 (1993), pp. 321–27.
5. Alfred C. Kinsey, Wardell B. Pomeroy, and Clyde E. Martin, *Sexual Behavior in the Human Male* (Philadelphia: W. B. Saunders Company, 1948); Alfred C. Kinsey, Wardell B. Pomeroy, Clyde E. Martin, and Paul H. Gebhard, *Sexual Behavior in the Human Female* (Philadelphia: W. B. Saunders Company, 1953).
6. Kinsey, *Sexual Behavior* (1948), p. 661.
7. John H. Gagnon, "The Management of Erotic Relations with Both Genders." Paper presented at the CDC Workshop on "Bisexuality and AIDS," Atlanta, GA, 1989. See also John H. Gagnon, "Disease and Desire," in Stephen R. Graubard (ed.), *Living with AIDS* (Cambridge, MA: MIT Press, 1990).
8. For a comparison between those who define themselves as "bisexual" and those who do not, see Janet Lever, David E. Kanouse, William H. Rogers, Sally Carson, and Rosanna Hertz, "Behavior Patterns and Sexual Identity of Bisexual Males," *Journal of Sex Research* 29 (1992), pp. 141–62. They conclude that men with a bisexual identity have more bisexual sexual activity than those who do not self-define as "bisexual" but that there are few other differences associated with the identity.

Chapter 2

1. Martin S. Weinberg, Colin J. Williams, and Douglas W. Pryor, "Telling the Facts of Life: A Study of a Sex Information Switchboard," *Journal of Contemporary Ethnography* 17(1988):131–63.

2. All were in their mid-thirties or older and had previous experience conducting interviews. About ten hours of group-based training was provided to them. First, the interview schedule was reviewed item by item to familiarize the volunteers with the instrument itself as well as the basic purpose of the study. Second, instruction was given about how to use neutral probes to request more detail from respondents or to clarify vague words and phrases. Third, a formal mock interview was performed as a case example. Finally, each interviewer conducted a practice interview with one of the researchers present.

The training did not end with the instruction sessions. Each interview was carefully reviewed by the research team after it was completed and tips were provided to volunteers about how to improve their work. Reviews were routinely completed within twenty-four hours and before the volunteer was assigned to another interview. If a glaring mistake was found in an interview—for instance, if a question had been mistakenly skipped or an answer was extremely unclear—then the respondent was called back over the phone for the needed information. These reviews also allowed the volunteers to ask about any problems they might have had with their own interviewing.

3. As indicated in the previous chapter, the room where one member of the research team lived was converted into a field headquarters. Many interviews were completed at a card table in this room. Sometimes scheduling required that two or more interviews be done simultaneously. In this case, a couch in one of the meeting rooms or a table in a back veranda were used. Some of the interviews could not be conducted at the Bisexual Center. For example, a few respondents lived east of Berkeley and did not want to travel into San Francisco for the interview. Thus two volunteers who lived in this area conducted the interviews in their own homes or at the homes of respondents. Also, one couple who lived close to the Bisexual Center provided us with the use of their home for interviews on a few occasions when space limitations required it. All interviews, regardless of the setting, took place in privacy.

4. A closed (or closed-ended) question is one in which the person's responses are limited to specified alternatives (e.g., "Do you consider yourself to be bisexual, heterosexual, or homosexual?"). An open (or open-ended) question is one that asks for a response from the person with no structure provided for the reply (e.g., "What has it been like being bisexual during the AIDS crisis?").

Chapter 3

1. Vivien C. Cass, "Homosexual Identity Formation: Testing a Theoretical Model." *Journal of Sex Research* 20 (1984), pp. 143–167; Eli Coleman, "Developmental Stages of the Coming Out Process." *Journal of Homosexuality* 7 (1981/2), pp. 31–43; Barbara Ponse, *Identities in the Lesbian World: The Social Construction of Self* (Westport, CT: Greenwood Press, 1978).

2. Martin S. Weinberg, Colin J. Williams, and Douglas Pryor, "Telling the Facts of Life: A Study of a Sex Information Switchboard." *Journal of Contemporary Ethnography* 17 (1988), pp. 131–163.

3. Donald Webster Cory and John P. Leroy, *The Homosexual and His Society* (New York: The Citadel Press, 1963), p. 61.

Chapter 4

1. Alfred C. Kinsey, Wardell B. Pomeroy, and Clyde E. Martin, *Sexual Behavior in the Human Male* (Philadelphia: W. B. Saunders Company, 1948).

2. See David P. McWhirter, June Reinisch, and Stephanie Sanders, *Homosexuality/Heterosexuality: The Kinsey Scale and Current Research* (New York: Oxford University Press, 1990).

3. See Philip Blumstein and Pepper Schwartz, "Intimate Relationships and the Creation of Sexuality," in Barbara Reisman and Pepper Schwartz (eds.), *Gender in Intimate Relationships* (Belmont, CA: Wadsworth, 1989), pp. 120–29.

4. A. MacDonald, "Bisexuality: Some Comments on Research and Theory," *Journal of Homosexuality* 6 (1981), pp. 21–35, leveled this criticism at Weinberg and Williams for classifying as "homosexuals" individuals who were less than exclusively so (a 6) on the Kinsey scale. See Martin S. Weinberg and Colin J. Williams, *Male Homosexuals: Their Problems and Adaptations* (New York: Oxford University Press, 1974), Chapter 15.

5. In response to MacDonald's "Bisexuality: Some Comments" criticism above, despite the fact that respondents in our 1974 survey ranked themselves as less than exclusively homosexual on the scale, they all identified themselves as homosexual even to the extent of being members of a homosexual political and social organization. This identification enabled Weinberg and Williams to understand homosexuality through a combination of a person's "identity" along with his own numerical Kinsey ratings, yielding some insight into what lies behind categories of sexual identity.

6. This is referred to as the "identity construct" approach. See Kenneth Plummer, *The Making of the Modern Homosexual* (Totowa, NJ: Barnes and Noble, 1981).

7. Alan P. Bell and Martin S. Weinberg, *Homosexualities: A Study of Diversity Among Men and Women* (New York: Simon and Schuster, 1978).

8. Michael Shively and John DeCecco, "Components of Sexual Identity," *Journal of Homosexuality* 3 (1977), pp. 41–48. Klein and his coauthors added further dimensions. For example, they include sexual attraction, social preference (e.g., the sexual preference of friends), self-identificaton, and having a heterosexual or homosexual lifestyle. They provide a total of seven different dimensions in the Klein Sexual Orientation Grid. A summated score from these dimensions thus includes numerous factors. Unfortunately, the problem with such scores is that the dimensions usually must be disentangled in a data analysis to show how factors relate to one another—for example, between the sexual attraction score and the adoption of a heterosexual or homosexual lifestyle score. Klein and his colleagues found self-identification to be the best predictor for placement on the summated grid, which supports our decision to use self-identification as our

starting point. Fritz Klein, Barry Sepekoff, and Timothy J. Wolf, "Sexual Orientation: A Multi-Variable, Dynamic Process," *Journal of Homosexuality* 11 (1985), pp. 35–49.

9. This is congruent with DeCecco's definition of sexual orientation: "The individual's physical sexual activity with, interpersonal affection for, and erotic fantasies about members of the same or opposite biological sex." John DeCecco, "Definition and Meaning of Sexual Orientation," *Journal of Homosexuality* 6 (1981), pp. 51–67.

10. John DeCecco and Michael Shively, "From Sexual Identity to Sexual Relationships: A Contractual Shift," *Journal of Homosexuality* 9 (1983/4), pp. 1–26.

11. Klein et al., "Sexual Orientation." The predictive value of self-identification was especially true for people who labeled themselves "bisexual."

12. Diane Richardson, "The Dilemma of Essentiality in Homosexual Theory," *Journal of Homosexuality* 9 (1983/4), p. 85. See also Barbara Ponse, *Identities in the Lesbian World: The Social Construction of Self* (Westport, CT: Greenwood Press, 1978); Thomas Weinberg, "On 'Doing' and 'Being' Gay: Sexual Behavior and Homosexual Male Self Identity," *Journal of Homosexuality* 4 (1978), pp. 143–56.

13. See Klein et al., "Sexual Orientation"; DeCecco and Shively, "From Sexual Identity"; Philip Blumstein and Pepper Schwartz, "Bisexuality in Men," *Urban Life* 5 (1976), pp. 339–58, and "Bisexuality: Some Social Psychological Issues," *Journal of Social Issues* 33 (1977), pp. 30–45.

Chapter 5

1. Sandra Bem, "Gender Schema Theory: A Cognitive Account of Sex Typing," *Psychological Review* 88 (1981), pp. 354–364.

2. This can have unintended consequences. According to William DuBay: "In our society, a male can never be male enough, and this in itself can create a sense of longing and aspiration that triggers an erotic response, even among the most masculine, towards those perceived as meeting a higher standard. Our society emphasizes individualism and hero worship and at the same time bans the emotions often caused by those beliefs." William Dubay, "New Directions for Gay Liberation: Getting Beyond the Labels." Unpublished MS. 1992, p. 3.

3. R. A. Laud Humphreys, *Tea Room Trade* (Chicago: Aldine Publishing Company, 1975); Albert J. Reiss, "The Social Integration of Queers and Peers," *Social Problems* 9 (Fall 1961), pp. 102–20; Joan K. Dixon, "Sexuality and Relationship Changes in Married Females Following the Commencement of Bisexual Activity," in Fritz Klein and Timothy J. Wolf (eds.), *Bisexualities: Theory and Research,* (New York: The Haworth Press, 1985), pp. 115–33.

Chapter 6

1. In that all but one of these respondents is a male-to-female transsexual, the chapter will be written from this perspective with little to say about female-to-male transsexuals.

2. In addition to the transsexuals, we refer to three other respondents (female bisexuals) whose primary *partners* were transsexual. These partners were all male-to-female transsexuals (two *pre*operative and one *post*operative).

3. S. J. Kessler and W. McKenna, *Gender: An Ethnomethodological Approach* (New York: John Wiley, 1978).

4. Harry Benjamin, *The Transsexual Phenomenon* (New York: The Julian Press, 1966).

5. As Kessler and McKenna, "Gender" (p. 15) put it: "attaching one of these gender-based labels [homosexual or heterosexual] to someone depends on the gender attributions made about *both* partners (e.g., that one is male and the other is female). The gender attribution determines the label 'homosexual' or 'heterosexual' *but the label itself does not lead to a gender attribution* [our emphasis]."

6. E.g., Wardell Pomeroy, "The Diagnosis and Treatment of Transvestites and Transsexuals," *Journal of Sex and Marital Therapy* 1 (Spring 1975), pp. 215–224.

Chapter 7

1. Martin S. Weinberg and Colin J. Williams, "Gay Baths and the Social Organization of Impersonal Sex," *Social Problems* 23 (December 1975), pp. 124–136.

2. Philip Blumstein and Pepper Schwartz, *American Couples* (New York: William Morrow and Company, Inc., 1983), esp. pp. 295–98.

3. See Margaret Nichols, "Doing Sex Therapy with Lesbians: Bending a Heterosexual Paradigm to Fit a Gay Lifestyle," in *Lesbian Psychologies: Exploration and Challenges,* ed. The Boston Lesbian Psychologies Collective (Urbana and Chicago: University of Illinois Press, 1987), pp. 97–125.

4. See Alan Bell and Martin S. Weinberg, *Homosexualities: A Study of Diversity Among Men and Women* (New York: Simon and Schuster, 1978), pp. 85–86, 93–94.

5. Blumstein and Schwartz, *American Couples,* pp. 240–43.

6. Martin S. Weinberg, Colin J. Williams, and Charles Moser, "The Social Constituents of Sadomasochism," *Social Problems* 31 (April 1984), pp. 379–89.

7. We report two statistics about their current situation: the incidence of the problem—whether it was or was not considered a problem—and the seriousness of the problem—combining the categories of "much" and "very much" of a problem.

Chapter 9

1. Michael Ross, *The Married Homosexual Man* (London: Routledge and Kegan Paul, 1983).

2. Timothy J. Wolf, "Marriages of Bisexual Men," in *Bisexualities: Theory and Research,* eds. Fritz Klein and Timothy J. Wolf (New York: The Haworth Press, 1985), pp. 135–48.

3. Eli Coleman, "Bisexual Women in Marriages," *Journal of Homosexuality* 11 (1985), pp. 87–99, and "Interaction of Male Bisexuality and Marriage," *Journal of Homosexuality* 11 (1985), pp. 189–209; Jean S. Gochros, "Wives' Reactions to Learning that their Husbands are Bisexual," *Journal of Homosexuality* 11 (1985), pp. 101–113.

4. For the processes involved, see Eli Coleman, "Developmental Stages of the Coming Out Process," *Journal of Homosexuality* 7 (1981/2), pp. 31–43; Frederick W. Bozett, ed., *Gay and Lesbian Parents* (New York: Praeger, 1987).

5. Gochros, "Wives' Reactions."

6. Coleman, "Bisexual Women."; John J. Brownfain, "A Study of the Married Bisexual Male: Paradox and Resolution," *Journal of Homosexuality* 11 (1985), pp. 173–88; David R. Matteson, "Bisexual Men in Marriage: Is a Positive Homosexual Identity and Stable Marriage Possible?" *Journal of Homosexuality* 11 (1985), pp. 149–171; J. D. Latham and G. D. White, "Coping with Homosexual Expression Within Heterosexual Marriages: Five Case Studies," *Journal of Sex and Marital Therapy* 4 (1978), pp. 198–212.

7. Cf. Matteson, "Bisexual Men"; Michael W. Ross, "The Married Homosexual Man"; B. Maddox, *Married and Gay: An Intimate Look at a Different Relationship* (New York: Harcourt Brace Jovanovich, 1982).

8. For similar problems among lesbian families cf. Sally Crawford, "Lesbian Families: Psychosocial Stress and the Family-Building Process," in The Boston Lesbian Psychologies Collective (eds.), *Lesbian Psychologies: Explorations and Challenges,* (Urbana: University of Illinois Press, 1987), pp. 195–214.

Chapter 11

1. Ritch C. Savin-Williams, *Gay and Lesbian Youth: Expressions of Identity* (New York: Hemisphere Publishing Company, 1990).

2. See Dank (1971) on this for homosexual identities. Barry M. Dank, "Coming Out in the Gay World," *Psychiatry* 34 (May, 1971), pp. 180–97. Certainly we have seen this earlier during the period of confusion experienced by our respondents. How this confusion differs for bisexuals and homosexuals and how each deals with it will be further explored in the next section of the book.

3. See Savin-Williams, *Gay and Lesbian Youth,* p. 112.

Chapter 12

1. To simplify the process of completing the questionnaire, we used a self-mailing design, see Earl Babbie, *The Practice of Social Research* (Belmont, CA: Wadsworth Publishing Company, 1992), p. 264. The questionnaire was folded and stapled with the name and address of the respondent and a return address facing out. When the respondent was finished, it could then be folded back the other way with only our address visible. One major drawback to the mailing procedure concerned the use of return postage. Because of limited funds, the respondents in the first mailing to two organizations were asked to pay for return postage themselves. Certainly this discouraged participation. It is hard to know who might have dropped out as a consequence: people with the most radical or conservative sexual lifestyles who were reluctant to provide the information requested, those with lower incomes who did not want to pay for postage, people who were extremely busy and who felt inconvenienced? Return postage, through the use of a business reply mail permit, was provided for the second mailing to the above two organizations, the one-time mailings to one of the other organizations, and the hand-circulated questionnaires. Thus return postage was provided with at least one mailing to three of the four participating organizations.

2. The questionnaire data were coded and entered into the computer by three research assistants who worked together as a team. One of the research assistants, the same graduate student who assisted with the quantitative coding on the interviews, monitored the other two assistants on a case-by-case basis. The three coders were also closely supervised by the coauthors. Once the data were entered, basic cross-tabulations and tests of statistical significance were run on the data broken down two ways, first by sex and then by sexual preference.

3. John H. Gagnon, "Science and Politics of Pathology," *Journal of Sex Research* 23 (February, 1987), pp. 120–21.

Chapter 13

1. This is supported by other studies. See Eli Coleman, "Developmental Stages of the Coming Out Process." *Journal of Homosexuality* 7 (1981/2), pp. 31–43; Richard R. Troiden, *Gay and Lesbian Identity: A Sociological Analysis* (New York: General Hall, Inc., 1988); and Fritz Klein, Barry Sepekoff, and Timothy J. Wolf, "Sexual Orientation: A Multi-Variable Dynamic Process." *Journal of Homosexuality* 11 (1985), pp. 35–49.

2. See John H. Gagnon and William Simon, *Sexual Conduct: The Social Sources of Human Sexuality* (Chicago: Aldine, 1973), Chapter 2.

3. Adrienne Rich, "Compulsory Heterosexuality and Lesbian Existence," *Signs* 5 (Summer, 1980), pp. 631–60.

Chapter 14

1. We also included those who were a 3 on two dimensions and a 1 or a 5 on the other. The third combination was a mixture of a 2 and a 4 with at least one or both. In each case, all three rankings were within 2 scale points and the combined mean was between 2 and 4.

2. Included as well were a few people who scored a 3 on one dimension, 1 on a second dimension, and 2 or less on the third (e.g., 321, 311) and who thus appeared to be predominantly heterosexual. The mean value of all three scale measures together was less than or equal to 2.

3. The most common combinations for each type in the typology, and the combinations that overlap in the three sexual preferences, are presented in Table 13.3.

4. Also included were a few cases where the respondents were a 3 on one dimension, a 5 on another dimension, and a 4 on a third (e.g., 453, 354, 435) and who thus appeared overall to be more homosexual in their sexual preference. The average profile score was greater than or equal to 4 in every instance.

Chapter 15

1. Rebecca Shuster, "Sexuality As a Continuum: The Bisexual Identity," in The Boston Lesbian Psychologies Collective (eds.), *Lesbian Psychologies: Explorations and Challenges* (Urbana and Chicago: University of Illinois Press, 1987), pp. 56–71.

2. Joan K. Dixon, "Sexuality and Relationship Changes in Married Females Following the Commencement of Bisexual Activity," in Fritz Klein and

Timothy J. Wolf (eds.), *Bisexualities: Theory and Research,* (New York: Haworth Press Inc., 1985), pp. 115–33.

3. Such changes require specific time referents in order to be understood. For example, Klein, Sepekoff and Wolf rate changes in sexual preference through three referents: past (undefined) and present (as defined by the preceding year) and the individual's ideal choice (future). Their results, for example, show that self-identified bisexuals add on significant increases in homosexuality from past to present and as an ideal. See Fritz Klein, Barry Sepekoff, and Timothy J. Wolf, "Sexual Orientation: A Multi-Variable Dynamic Process," *Journal of Homosexuality* 11 (1985), pp. 35–49.

4. Respondents answered by selecting two values from the Kinsey scale that best approximated the beginning and end of the largest change they had experienced in sexual feelings. We again derived net change scores for each group, totaling the percentage who experienced a change in their sexual feelings that was of two (or more) scale points, and a smaller subgroup who experienced a change of at least three scale points.

Chapter 16

1. See Martin S. Weinberg and Colin J. Williams, *Male Homosexuals: Their Problems and Adaptations* (New York: Oxford University Press, 1974), Chapter 12 on same-sex kissing among males.

2. Lower homosexual frequencies were also found by Kinsey. See Alfred C. Kinsey, Wardell B. Pomeroy, and Clyde E. Martin, *Sexual Behavior in the Human Male* (Philadelphia: W. B. Saunders Company, 1948), pp. 632–34 for possible reasons for this difference.

3. Most who reported a problem rated its severity as "a little." Thus, significant differences usually were only between "a little" and "not at all."

Chapter 17

1. This is true even when controlling for the number of relationships they actually had.

2. This can be shown another way by taking the variable that measured whether or not respondents were nonmonogamous outside of their involved relationships and controlling for the total number of involved relationships with either the same or the opposite-sex. This approach yields five types of relationships: closed-single, open-single, closed-multiple, open-multiple, and finally a residual category of uncoupled respondents. The open-multiple relationship type emerges as the most common one for the bisexual men and women. The closed-single relationship type was the most common one among the heterosexual men and women and the homosexual women. The homosexual men tended to have either a closed-single structure on an open-multiple one. Thus some basic differences in the structure of relationships by sexual preference are apparent.

3. None of the above findings were affected when we controlled for age.

Chapter 18

1. Erving Goffman, *Stigma: Notes on the Management of Spoiled Identity* (Englewood Cliffs, N.J.: Prentice-Hall, 1965), p. 3.

Chapter 19

1. Douglas Shenson, "When Fear Conquers: A Doctor Learns About AIDS From Leprosy." *New York Times Magazine* (February 28, 1988), p. 48.

2. *AIDS and KS Foundation Newsletter* 1 (May, 1983).

3. Public Health Service, Centers for Disease Control and Prevention, *HIV/AIDS Surveillance Report* (February, 1992), pp. 8, 17.

4. *Morbidity and Mortality Weekly Report* (June 5, 1981), pp. 250–52.

5. J. Seale, "How to Turn a Disease into VD," *New Scientist* 1461 (1985), pp. 38–41.

6. Sander Gilman, "AIDS and Syphilis: The Iconography of Disease," in Donald Crimp (ed.), *AIDS: Cultural Analysis, Cultural Activism* (Cambridge, MA: MIT Press, 1988), p. 90.

7. *Time* (September 6, 1982).

8. Simon Watney, "The Spectacle of AIDS," in Donald Crimp (ed.), *AIDS: Cultural Analysis, Cultural Activism* (Cambridge, MA: MIT Press, 1988), p. 77.

9. *Newsweek* (August 12, 1985).

10. *New York Times* (February 14, 1988).

11. *U.S. News and World Report* (January 12, 1987).

12. *New York Times* (February 14, 1988).

13. Watney, "The Spectacle of AIDS," p. 37.

14. Watney, "The Spectacle of AIDS," p. 23.

15. Frances Fitzgerald, "A Reporter at Large: The Castro," *New Yorker* (July 21, July 28, 1986).

16. *New York Times* (February 14, 1988).

17. Noted in *Newsweek* (August 12, 1985).

18. Noted in Katie Leishman, "Heterosexuals and AIDS," *Atlantic Monthly* (February, 1987) pp. 39–58.

19. Quoted in the *New York Times* (May 28, 1985).

20. *San Francisco Examiner* (February, 1989).

21. Jan Zita Grover, "AIDS: Keywords," in Donald Crimp (ed.), *AIDS: Cultural Analysis, Cultural Criticism* (Cambridge, MA: MIT Press, 1988), p. 21.

22. Grover, "AIDS: Keywords," p. 21.

23. Leishman, "Heterosexuals and AIDS," p. 48.

24. Quoted in Leishman, "Heterosexuals and AIDS," p. 49.

25. *Newsweek* (July 13, 1987).

26. There were no significant differences between these sixty-one persons and the complete group of one hundred in their answers to the questions asked in 1983.

27. All told, both instruments together contained 128 close-ended and 35 open-ended questions. The open-ended replies were all recorded by hand as the respondents talked. Pacing techniques were once again employed.

28. Under careful daily supervision, two research assistants, the graduate student who went to San Francisco and an undergraduate student with an extensive background in quantitative methods, coded and entered the closed-ended variables into the computer. The data for the repeat respondents were then merged with the closed-ended data from the 1983 interview study. Again, cross-tabulations and statistics were run on the repeat follow-up cases with breakdowns by sex. There were cross-tabulations between matching time-one and time-two variables and tests of statistical significance run to compare the distributions with

a model of chance. The open-ended coding was done as it was in the 1983 interview study. Frequencies were again recorded with breakdowns by sex.

Chapter 20

1. Such effects have been noted, e.g., Hirschfeld's study of the effects of World War I on sexual behavior. Magnus Hirschfeld, *The Sexual History of the World War* (New York: The Panurge Press, 1934).
2. *New York Times* (September 22, 1985).

Chapter 22

1. Anne Bolin, *In Search of Eve: Transsexual Rites of Passage* (South Hadley, MA: Bergin and Garvey, Inc., 1988).
2. Also noted by Bolin, *In Search of Eve*, pp. 167–72.
3. See Harry Benjamin, *The Transsexual Phenomenon* (New York: The Julian Press, 1966), p. 49; Deborah H. Feinbloom, *Transvestites and Transsexuals: Mixed Views* (New York: Delcorte Press, 1976), p. 27.
4. See Bolin, *In Search of Eve*, p. 298.

Chapter 26

1. This was emphasized, of course, by Kinsey. See Alfred C. Kinsey, Wardell B. Pomeroy, and Clyde E. Martin, *Sexual Behavior in the Human Male* (Philadelphia: W. B. Saunders Company, 1948).
2. For example, DeCecco (1987) sees many current theories of homosexuality as centering around the belief that homosexuals are not "true" men or women because they are in some way deficient in their gender roles. The theories being referred to seem far removed from the *sexual* aspect of sexual preference. John DeCecco, "The Two Views of Meyer-Bahlburg: A Rejoinder," *Journal of Sex Research* 23 (1987), pp. 123–27.
3. John H. Gagnon, *Human Sexualities* (Glenview, IL: Scott Foresman, 1974), p. 59.
4. Sandra Bem, "Gender Schema Theory: A Cognitive Account of Sex Typing," *Psychological Review* 88 (1981), p. 361.
5. Albert J. Reiss, "The Social Integration of Queers and Peers," *Social Problems* 9 (Fall 1961), pp. 102–20.
6. Wayne S. Wooden and Jay Parker, *Men Behind Bars: Sexual Exploitation in Prison* (New York: Plenum Press, 1982).
7. Joseph M. Carrier, "Mexican Male Bisexuality," in Fritz Klein and Timothy J. Wolf (eds.), *Bisexualities: Theories and Research* (New York: Haworth Press, 1985).
8. Edward Sagarin, "Prison Homosexuality and Its Effect on Post-Prison Sexual Behavior," *Psychiatry* 39 (1976), pp. 245–57.
9. Joan K. Dixon, "Sexuality and Relationship Changes in Married Females Following the Commencement of Bisexual Activities," in Fritz Klein and Timothy J. Wolf (eds.), *Bisexualities: Theories and Research* (New York: Haworth Press, 1985), pp. 115–33.

10. This is the project of what is called "social constructionism" in current sex research. See Edward Stein (ed.), *Forms of Desire: Sexual Orientation and the Social Constructionist Controversy* (New York: Garland Publishing Inc., 1990).

11. Alan P. Bell and Martin S. Weinberg, *Homosexualities: A Study of Diversity Among Men and Women* (New York: Simon and Schuster, 1978).

12. Maureen Mileski and Donald J. Black, "The Social Organization of Homosexuality," *Urban Life and Culture* 1 (July, 1972), pp. 187–202.

13. For a recent view on this see Paula C. Rust, "The Politics of Sexual Identity: Sexual Attraction and Behavior Among Lesbian and Bisexual Women," *Social Problems* 39 (November 1992), pp. 366–86.

Appendix A: Tables for the 1983 Interview Study

The following abbreviations are used in the tables for the interview study:

BIM	bisexual men
BIW	bisexual women
OS	opposite-sex partners
SS	same-sex partners

In order to do the chi-square (χ^2) test of significance, we sometimes had to combine adjacent categories. When such collapsing was done we note this under the respective table. We accept a difference as being statistically significant when the probability of its being due to chance is equal to or less than five in a hundred ($p \leq .05$).

CHAPTER 2: BISEXUALS IN SAN FRANCISCO

Table 2.1   Demographic and Background Data, 1983 Interview Sample

Age

	1983		
	Men	Women	Combined
19-24	2.0%	15.9%	8.6%
25-29	16.3	25.0	20.4
30-34	28.6	15.9	22.6
35-39	24.5	20.5	22.6
40-44	16.3	9.1	12.9
45-49	8.2	2.3	5.4
50+	4.1	11.4	7.5
	(n = 49)	(n = 44)	

Collapsed top two rows and fifth and sixth rows. $x^2 = 9.702$ df = 4 p = .047.

Education

	1983		
	Men	Women	Combined
High school graduate	0	9.1%	4.3%
Some college	22.9%	27.3	25.0
College graduate	35.4	34.1	34.8
Master's	31.3	18.2	25.0
Professional degree (MD, PhD, LLD, EdD)	10.4	11.4	10.9
	(n = 48)	(n = 44)	

Collapsed top two rows. $x^2 = 3.016$ df = 3 p = .400.

Income level

	1983		
	Men	Women	Combined
< $5,000	6.1%	9.1%	7.5%
5,000-9,999	18.4	25.0	21.5
10,000-14,999	10.2	22.7	16.1
15,000-19,999	10.2	18.2	14.0
20,000-24,999	10.2	9.1	9.7
25,000-29,999	18.4	4.5	11.8
30,000-39,999	12.2	4.5	8.6
40,000-49,999	6.1	2.3	4.3
50,000+	8.2	4.5	6.5
	(n = 49)	(n = 44)	

Collapsed top two rows and fifth through seventh rows. $x^2 = 9.194$ df = 4 p = .059.

Race/Ethnicity

	1983		
	Men	Women	Combined
White	95.9%	93.2%	94.6%
Black	4.1	4.5	4.3
Hispanic	0	2.3	1.1
	(n = 49)	(n = 44)	

Collapsed bottom two rows. $x^2 = .015$ df = 1 p = .903.

Currently employed

	1983		
	Men	Women	Combined
Yes	89.8%	86.4%	88.2%
No	10.2	13.6	11.8
	(n = 49)	(n = 44)	

$x^2 = .036$ df = 1 p = .858.

Religious background

	1983		
	Men	Women	Combined
Catholic	18.4%	20.9%	19.6%
Protestant	44.9	39.5	42.4
Jewish	22.4	18.6	20.7
Other	14.3	20.9	17.4
	(n = 49)	(n = 43)	

$x^2 = .978$ df = 3 p = .807.

Religiosity

	1983		
	Men	Women	Combined
Not at all	72.9%	68.2%	70.7%
A little	14.6	22.7	18.5
Moderate	4.2	4.5	4.3
Strong	6.3	2.3	4.3
Very strong	2.1	2.3	2.2
	(n = 48)	(n = 44)	

Collapsed bottom three rows. $x^2 = 1.141$ df = 2 p = .573.

Political preference

	1983		
	Men	Women	Combined
Very conservative	2.0%	2.3%	2.2%
Conservative	4.1	0	2.2
Moderate	8.2	11.4	9.7
Liberal	44.9	27.3	36.6
Very liberal	30.6	34.1	32.3
Radical	10.2	25.0	17.2
	(n = 49)	(n = 44)	

Collapsed top three rows. $x^2 = 5.018$ df = 3 p = .177.

Degree of acceptance of feminist ideology

	1983		
	Men	Women	Combined
Not at all	6.1%	2.3%	4.3%
A little	4.1	2.3	3.2
Moderate	46.9	25.0	36.6
Strong	40.8	59.1	49.5
Radical	2.0	11.4	6.5
	(n = 49)	(n = 44)	

Collapsed top two rows and bottom two rows. $x^2 = 7.200$ df = 2 p = .030.

Distance from residence to San Francisco

	1983		
	Men	Women	Combined
More than a day	0	2.3%	1.1%
4 hours drive	4.1%	0	2.2
50 miles	8.2	4.7	6.5
In the Bay area	87.8	93.0	90.2
	(n = 49)	(n = 43)	

Collapsed top three rows. $x^2 = .247$ df = 1 p = .637.

CHAPTER 3: BECOMING BISEXUAL

Table 3.1   Sexual Identity

Currently in transition

	Men	Women	Combined
Yes	10.2%	9.1%	9.7%
No	89.8	90.9	90.3
	(n = 49)	(n = 44)	

$x^2 = .033$ df = 1   p = .856.

Someday define yourself as lesbian/gay or heterosexual

	Men	Women	Combined
Yes	42.2%	39.0%	40.7%
No	57.8	61.0	59.3
	(n = 45)	(n = 41)	

$x^2 = .091$ df = 1   p = .763.

Direction of change in sexual definition

	Men	Women	Combined
Lesbian/gay	21.1%	31.3%	25.7%
Heterosexual	15.8	6.3	11.4
Either direction	63.2	62.5	62.9
	(n = 19)	(n = 16)	

Collapsed two top rows. $x^2 = .097$ df = 1   p = .763.

Probability of change in sexual definition

	Men	Women	Combined
Not probable	78.9%	56.3%	68.6%
Somewhat probable	15.8	25.0	20.0
Very probable	5.3	18.8	11.4
	(n = 19)	(n = 16)	

Collapsed bottom two rows. Fisher's exact p = .141.

Table 3.2   Sexual Behavior

Someday behave exclusively homosexual or heterosexual

	Men	Women	Combined
Yes	85.1%	78.0%	81.8%
No	14.9	22.0	18.2
	(n = 47)	(n = 41)	

$x^2 = .732$ df = 1   p = .392.

Direction of change in sexual behavior

	Men	Women	Combined
Lesbian/Gay	10.0%	21.9%	15.3%
Heterosexual	42.5	9.4	27.8
Either direction	47.5	68.8	56.9
	(n = 40)	(n = 32)	

Collapsed top two rows. $x^2 = 2.465$ df = 1   p = .123.

Probability of change in sexual behavior

	Men	Women	Combined
Not probable	41.0%	40.6%	40.8%
Somewhat probable	38.5	25.0	32.4
Very probable	20.5	34.4	26.8
	(n = 39)	(n = 32)	

$x^2 = 2.260$ df = 2   p = .323.

Table 3.3   Confusion

Presently confused about bisexuality

	Men	Women	Combined
Yes	24.5%	22.7%	23.7%
No	75.5	77.3	76.3
	(n = 49)	(n = 44)	

x^2 = .002   df = 1   p = .967.

Ever confused about bisexuality

	Men	Women	Combined
Yes	83.8%	55.9%	70.4%
No	16.2	44.1	29.6
	(n = 37)	(n = 34)	

x^2 = 5.350   df = 1   p = .021.

CHAPTER 4: BISEXUAL TYPES

Table 4.1   Mean Ages (Standard Deviations) for the Development of Sexual Preference

	Men		Women		t	p
	Mean	(S.D.)	Mean	(S.D.)		
First heterosexual attraction	11.7	(4.79)	11.6	(4.58)	.10	.923
First heterosexual experience	17.3	(5.96)	14.7	(4.73)	2.22	.029
First homosexual attraction	13.5	(6.77)	16.9	(8.61)	2.09	.039
First homosexual experience	16.3	(8.52)	21.4	(9.14)	2.56	.012
Labeled themselves bisexual	27.2	(9.32)	26.8	(8.56)	.18	.854

Table 4.2   Kinsey Scale Rankings for the Men

	Sexual feelings	Sexual behaviors	Romantic feelings
0: Exclusively heterosexual	0	6.3%	14.6%
1: Mainly heterosexual with a small degree of homosexuality	4.1%	27.1	25.0
2: Mainly heterosexual with a significant degree of homosexuality	44.9	18.8	12.5
3: Equally heterosexual and homosexual	22.4	14.6	25.0
4: Mainly homosexual with a significant degree of heterosexuality	22.4	18.8	12.5
5: Mainly homosexual with a small degree of heterosexuality	6.1	14.6	10.4
6: Exclusively homosexual	0	0	0
	(n = 49)	(n = 48)	(n = 48)

Table 4.3 Kinsey Scale Rankings for the Women

		Sexual feelings	Sexual behaviors	Romantic feelings
0:	Exclusively heterosexual	0	14.3%	0
1:	Mainly heterosexual with a small degree of homosexuality	6.8%	11.9	7.0%
2:	Mainly heterosexual with a significant degree of homosexuality	18.2	31.0	14.0
3:	Equally heterosexual and homosexual	40.9	23.8	44.2
4:	Mainly homosexual with a significant degree of heterosexuality	29.5	9.5	23.3
5:	Mainly homosexual with a small degree of heterosexuality	4.5	7.1	7.0
6:	Exclusively homosexual	0.0	2.4	4.7
		(n = 44)	(n = 42)	(n = 43)

Table 4.4 Kinsey Scale Scores: Men Versus Women

Kinsey Dimension	Men		Women		t	p
	Mean	(S.D.)	Mean	(S.D.)		
Sexual feelings	2.816	(1.03)	3.068	(0.97)	1.21	.231
Sexual behaviors	2.563	(1.57)	2.333	(1.51)	.70	.484
Romantic Affections	2.271	(1.58)	3.233	(1.15)	3.28	.001
	(n = 49)		(n = 44)			

Table 4.5   Bisexual Types

	Scale Values			(Number of men)	Percent of men	(Number of women)	Percent of women
	Sexual feelings	Sexual behaviors	Romantic feelings				
Pure Type	3	3	3	(1)	2%	(7)	17%
	3 on each dimension						
Mid-Type	2-4	2-4	2-4	(14)	30	(16)	38
	With a 3 on at least one dimension[a]						
Hetero-sexual-leaning type	0-2	0-2	0-2	(21)	45	(8)	19
	0-2 on each dimension[b]						
Homo-sexual-leaning type	4-6	4-6	4-6	(7)	15	(7)	17
	4-6 on each dimension[c]						
Varied Type	0,3,3	0,3,5	1,4,3, etc.	(4)	9	(4)	10
	Not classified in above and at least 3 points between any two dimensions						
				(n = 47)		(n = 42)	

Note: Because of low expected frequencies, we only compared the "pure type" and the "heterosexual leaning type" where the major differences between the men and women lie. $x^2 = 7.017$ df = 1 p = .009.

[a]Also included is one person who ranked a 3 on two dimensions and a 1 on the third.
[b]Two persons included had a 3 on one dimension.
[c]Four persons included had a 3 on one dimension.

CHAPTER 7: SEXUAL ACTIVITIES

Table 7.1   Sexual Partners

Total in Lifetime

	Men[a]		Women[b]	
	Opposite-sex Partners	Same-sex Partners	Opposite-sex Partners	Same-sex Partners
0	2.0%	2.0%	0	4.5%
1-2	10.2	4.1	2.3%	6.8
3-4	4.1	4.1	0	15.9
5-9	20.4	12.2	15.9	31.8
10-14	12.2	10.2	9.1	18.2
15-24	18.4	12.2	9.1	2.3
25-49	12.2	14.3	18.2	4.5
50-99	4.1	12.2	11.4	13.6
100-249	10.2	8.2	18.2	0
250-499	4.1	6.1	2.3	0
500-999	2.0	8.2	6.8	2.3
1000 +	0	6.1	6.8	0
	(n = 49)	(n = 49)	(n = 44)	(n = 44)

Collapsed table: "0-9" vs. "10-14" vs. "15-24" vs. "25-49" vs. "50-100+."
[a]$x^2 = 5.791$ df = 4 p = .218; [b]$x^2 = 19.440$ df = 3 p = .000.
BIM vs. BIW (OS): $x^2 = 9.535$ df = 4 p = .049; BIM vs. BIW (SS): $x^2 = 19.169$ df = 4 p = .001.

Total sex partners (past twelve months)

	Men[a]		Women[b]	
	Opposite-sex Partners	Same-sex Partners	Opposite-sex Partners	Same-sex Partners
0	8.2%	8.2%	9.1%	22.7%
1-2	42.9	24.5	25.0	29.6
3-4	18.4	12.2	18.2	27.3
5-9	16.3	24.5	27.3	11.4
10-14	4.1	6.1	6.8	4.6
15-24	4.1	4.1	4.5	2.3
25-49	6.1	4.1	6.8	2.3
50-99	0	10.2	0	0
100-249	0	4.1	0	0
250-499	0	2.0	0	0
500+	0	0	2.3	0
	(n = 49)	(n = 49)	(n = 44)	(n = 44)

Collapsed Table: "0-9" vs. "10-500+." [a]$x^2 = 2.872$ df = 1 p = .093; [b]$x^2 = 1.444$ df = 1 p = .234. BIM vs. BIW (SS): $x^2 = 5.348$ df = 1 p = .021; BIM vs. BIW (OS): $x^2 = .262$ df = 1 p = .628.

Frequency of casual sex (past 12 months)

	Men	Women	Combined
Never	22.4%	23.3%	22.8%
< 1 month	30.6	44.2	37.0
1-3 month	30.6	16.3	23.9
1-3 week	16.3	14.0	15.2
4 + week	0	2.3	1.1
	(n=49)	(n=43)	

Collapsed bottom two rows. $x^2 = 3.113$ df = 3 p = .385.

Gender of casual sex partners (past 12 months)

	Men	Women	Combined
All women	5.3%	12.1%	8.5%
More women	21.1	21.2	21.1
Equal men and women	15.8	15.2	15.5
More men	28.9	33.3	31.0
All men	28.9	18.2	23.9
	(n = 38)	(n = 33)	

Collapsed top two rows. $x^2 = 1.240$ df = 3 p = .743.

Frequency of anonymous sex (past 12 months)

	Men	Women	Combined
Never	44.9%	67.4%	55.4%
< 1 month	32.7	20.9	27.2
1-3 month	18.4	7.0	13.0
1-3 week	4.1	2.3	3.3
4 + week	0	2.3	1.1
	(n = 49)	(n = 43)	

Collapsed bottom three rows. $x^2 = 4.795$ df = 2 p = .093.

Gender of anonymous partners (past 12 months)

	Men	Women	Combined
All women	3.7%	7.1%	4.9%
More women	3.7	7.1	4.9
Equal men and women	7.4	28.6	14.6
More men	33.3	21.4	29.3
All men	51.9	35.7	46.3
	(n = 27)	(n = 14)	

Collapsed top three rows. $x^2 = 3.953$ df = 2 p = .147.

Number of threesomes (past 12 months)

	Men	Women	Combined
0	46.9%	50.0%	48.4%
1	20.4	18.2	19.4
2-3	16.3	20.5	18.3
4-7	10.2	2.3	6.5
8 +	6.1	9.1	7.5
	(n = 49)	(n = 44)	

Collapsed bottom two rows. $x^2 = 0.731$ df = 3 p = .864.

Number of swinging or group sex experiences (past 12 months)

	Men	Women	Combined
0	58.3%	65.9%	62.0%
1	14.6	18.2	16.3
2-3	6.3	2.3	4.3
4-7	12.5	4.5	8.7
8 +	8.3	9.1	8.7
	(n = 48)	(n = 44)	

Collapsed bottom three rows. $x^2 = 1.720$ df = 2 p = .435.

Table 7.2   Sexual Frequencies (past 12 months)

Self-masturbation

	Men	Women	Combined
None year	2.0%	4.5%	3.2%
< 1 month	2.0	0	1.1
1-3 month	14.3	15.9	15.1
1-3 week	40.8	54.5	47.3
4 + week	40.8	25.0	33.3
	(n = 49)	(n = 44)	

Collapsed top three rows. $x^2 = 2.719$ df = 2 p = .262.

Masturbate partner with goal of orgasm

	Men[a]		Women[b]	
	Opposite-sex partners	Same-sex partners	Opposite-sex partners	Same-sex partners
None year	22.5%	20.4%	18.2%	29.6%
< 1 month	20.4	44.9	27.3	36.4
1-3 month	40.8	30.6	25.0	27.3
1-3 week	16.3	0	25.0	4.6
4 + week	0	4.1	4.6	2.3
	(n = 49)	(n = 49)	(n = 44)	(n = 44)

Collapsed bottom two rows. $^a x^2 = 8.862$ df = 3 p = .035; $^b x^2 = 8.055$ df = 3 p = .047.
BIM vs. BIW (OS): $x^2 = 4.005$ df = 3 p = .265; BIM vs. BIW (SS): $x^2 = 2.100$ df = 3 p = .557.

Being masturbated by partner with goal of orgasm

	Men[a]		Women[b]	
	Opposite-sex partners	Same-sex partners	Opposite-sex partners	Same-sex partners
None year	30.6%	20.4%	18.2%	25.0%
< 1 month	40.8	46.9	27.3	43.2
1-3 month	24.5	26.5	25.0	25.0
1-3 week	4.1	0	25.0	4.6
4 + week	0	6.1	4.6	2.3
	(n = 49)	(n = 49)	(n = 44)	(n = 44)

Collapsed bottom two rows. $^a x^2 = 1.340$ df = 2 p = .515; $^b x^2 = 8.320$ df = 3 p = .043.
BIM vs. BIW (OS): $x^2 = 10.380$ df = 3 p = .016; BIM vs. BIW (SS): $x^2 = 2.100$ df = 3 p = .557.

Mouth on partner's genitals

	Men[a]		Women[b]	
	Opposite-sex partners	Same-sex partners	Opposite-sex partners	Same-sex partners
None year	20.4%	8.2%	9.3%	25.0%
< 1 month	16.3	38.8	11.6	36.4
1-3 month	26.5	28.6	25.6	22.7
1-3 week	36.7	22.5	46.5	15.9
4 + week	0	2.0	7.0	0
	(n = 49)	(n = 49)	(n = 43)	(n = 44)

Collapsed bottom two rows. $^a x^2 = 17.589$ df = 3 p = .000; $^b x^2 = 17.593$ df = 3 p = .000.
BIM vs. BIW (OS): $x^2 = 3.667$ df = 3 p = .299; BIM vs. BIW (SS): $x^2 = 5.248$ df = 3 p = .162.

Partner's mouth on your genitals

	Men[a]		Women[b]	
	Opposite-sex partners	Same-sex partners	Opposite-sex partners	Same-sex partners
None year	18.4%	10.2%	13.6%	25.0%
< 1 month	20.4	34.7	9.1	38.6
1-3 month	28.6	26.5	27.3	25.0
1-3 week	32.7	22.5	43.2	11.4
4 + week	0	6.1	6.8	0
	(n = 49)	(n = 49)	(n = 44)	(n = 44)

Collapsed bottom two rows. [a]$x^2 = 3.128$ df = 3 p = .382; [b]$x^2 = 20.265$ df = 3 p = .000.
BIM vs. BIW (SS): $x^2 = 9.495$ df = 3 p = .025; BIM vs. BIW (OS): $x^2 = 4.011$ df = 3 p = .264.

Vaginal intercourse

	Men	Women
None year	14.3%	13.6%
< 1 month	16.3	15.9
1-3 month	30.6	20.5
1-3 week	32.7	38.6
4 + week	6.1	11.4
	(n = 49)	(n = 44)

Collapsed bottom two rows. $x^2 = 1.598$ df = 3 p = .663.

Perform anal intercourse (men only)

	Opposite-sex partners	Same-sex partners
None year	71.4%	44.9%
< 1 month	16.3	28.6
1-3 month	8.2	16.3
1+ week	4.1	10.2
	(n = 49)	(n = 49)

Collapsed bottom two rows. $x^2 = 7.180$ df = 2 p = .030.

Receive anal intercourse

	Men	Women
	Same-sex partners	Opposite-sex partners
None year	32.7%	63.6%
< 1 month	40.8	25.0
1-3 month	18.4	6.8
1 + week	8.2	4.5
	(n = 49)	(n = 44)

Collapsed bottom two rows. $x^2 = 9.201$ df = 2 p = .010.

Perform anal stimulation with fingers

	Men[a]		Women[b]	
	Opposite-sex partners	Same-sex partners	Opposite-sex partners	Same-sex partners
None year	42.9%	38.8%	50.0%	75.0%
< 1 month	20.4	28.6	25.0	20.5
1-3 month	24.5	26.5	13.6	4.5
1 + week	12.2	6.1	11.4	0
	(n = 49)	(n = 49)	(n = 44)	(n = 44)

Collapsed bottom three rows. $^a x^2 = 0.042$ df = 1 p = .846; $^b x^2 = 4.848$ df = 1 p = .031.
BIM vs. BIW (SS): $x^2 = 10.920$ df= 1 p = .001; BIM vs. BIW (OS): $x^2 = .232$ df = 1 p = .647.

Receive anal stimulation with fingers

	Men[a]		Women[b]	
	Opposite-sex partners	Same-sex partners	Opposite-sex partners	Same-sex partners
None year	36.7%	34.7%	47.7%	77.3%
< 1 month	30.6	32.7	25.0	15.9
1-3 month	20.4	30.6	15.9	6.8
1 + week	12.2	2.0	11.4	0
	(n = 49)	(n = 49)	(n = 44)	(n = 44)

Collapsed bottom two rows. $^a x^2 = .061$ df = 2 p = .970; $^b x^2 = 9.362$ df = 2 p = .010.
BIM vs. BIW (SS): $x^2 = 17.152$ df = 2 p = .000; BIM vs. BIW (OS): $x^2 = .668$ df = 2 p = .716.

Perform oral-anal stimulation

	Men[a]		Women[b]	
	Opposite-sex partners	Same-sex partners	Opposite-sex partners	Same-sex partners
None year	69.4%	67.3%	68.2%	81.8%
< 1 month	12.2	20.4	15.9	11.4
1-3 month	10.2	10.2	9.1	6.8
1 + week	8.2	2.0	6.8	0
	(n = 49)	(n = 49)	(n = 44)	(n = 44)

Collapsed bottom two rows. $^a x^2 = 1.614$ df = 2 p = .456; $^b x^2 = 2.48$ df = 2 p = .291.
BIM vs. BIW (OS): $x^2 = .309$ df = 2 p = .859; BIM vs. BIW (SS): $x^2 = 2.535$ df = 2 p = .285.

Receive oral-anal stimulation

	Men[a]		Women[b]	
	Opposite-sex Partners	Same-sex Partners	Opposite-sex Partners	Same-sex Partners
None year	75.5%	61.2%	63.6%	77.4%
< 1 month	12.2	22.4	15.9	14.0
1-3 month	12.2	14.3	11.4	8.6
1 + week	0	2.0	9.1	0
	(n = 49)	(n = 49)	(n = 44)	(n = 44)

Collapsed bottom two rows. $^a x^2 = 2.488$ df = 2 p = .290; $^b x^2 = 2.58$ df = 2 p = .279.
BIM vs. BIW (OS): $x^2 = 1.654$ df = 2 p = .448; BIM vs. BIW (SS): $x^2 = 2.793$ df = 2 p = .253.

Table 7.3   SM, Sex of Partners, and Dominant-Submissive Roles

	Men	Women	p
SM activity in past 12 months	28.6% (n = 49)	27.3% (n = 44)	.931[a]
Partners* Men or Women	28.6	41.6	
Men and Women	71.4	58.3	.387[b]
SM Dominant OS*	92.9	91.7	1.000[b]
Frequency < 1 month	83.7	90.9	.392[b]
SM Submissive OS*	92.9	100.0	1.000[b]
Frequency < 1 month	92.8	84.1	.580[b]
SM Dominant SS*	57.1	50.0	.977[a]
Frequency < 1 month	93.9	100.0	1.000[b]
SM Submissive SS*	64.3	50.0	.742[a]
Frequency < 1 month	98.0	100.0	1.000[b]
	(n = 14)	(n = 12)	

*In past 12 months. [a]x^2 probability (df = 1). [b]Fisher's exact probability.

Table 7.4   Incidence of Various Unconventional Behaviors (Past 12 Months)

	Men		Women	
	Opposite-sex partners	Same-sex partners	Opposite-sex partners	Same-sex partners
Urinating on partner	10.2%	6.1%[a]	11.4%	6.8%[h]
Urinated on by partner	14.3	8.2[b]	15.9	2.3[i]
Drinking partner's urine	10.2	2.0[c]	6.8	0.0[j]
Anally fisting partner	8.2	16.3[d]	9.1	4.5[k]
Anally fisted by partner	12.2	8.2[e]	4.5	0.0[l]
Gave enema to partner	4.1	8.2[f]	6.8	4.5[m]
Received enema from partner	4.1	6.1[g]	9.1	2.3[n]
Feces play	0	0	0	0
	(n = 49)	(n = 49)	(n = 44)	(n = 44)

Men
[a]OS vs. SS: x^2=.136   df=1 p=.716.
[b]OS vs. SS: x^2=.410   df=1 p=.533.
[c]OS vs. SS: x^2=1.598   df=1 p=.207.
[d]OS vs. SS: x^2=1.520   df=1 p=.221.
[e]OS vs. SS: x^2=.111   df=1 p=.745.
[f]OS vs. SS: x^2=.176   df=1 p=.682.
[g]OS vs. SS: x^2=.000   df=1 p=1.000.

Women
[h]OS vs. SS: x^2=.137   df=1 p=.715.
[i]OS vs. SS: x^2=3.438   df=1 p=.068.
[j]OS vs. SS: x^2=1.380   df=1 p=.246.
[k]OS vs. SS: x^2=.714   df=1 p=.417.
[l]OS vs. SS: x^2=.512   df=1 p=.483.
[m]OS vs. SS: x^2=.000   df=1 p=1.000.
[n]OS vs. SS: x^2=.848   df=1 p=.373.

Table 7.5   Sexual Problems (Past 12 Months)

Difficulty finding suitable partner

	Men[a]		Women[b]	
	Opposite-sex partners	Same-sex partners	Opposite-sex partners	Same-sex partners
Not at all	44.9%	22.9%	45.5%	12.5%
A little	8.2	14.6	13.6	27.5
Somewhat	20.4	25.0	15.9	15.0
Much	18.4	16.7	18.2	22.5
Very much	8.2	20.8	6.8	22.5
	(n = 49)	(n = 48)	(n = 44)	(n = 40)

Collapsed bottom four rows. [a]x^2 = 4.286 df = 1 p = .041; [b]x^2 = 9.366 df = 1 p = .004.
BIM vs. BIW (OS): x^2 = .024 df = 1 p = .883;   BIM vs. BIW (SS): x^2 = .968 df = 1 p = .333.

Dissatisfaction with sexual frequency

	Men[a]		Women[b]	
	Opposite-sex partners	Same-sex partners	Opposite-sex partners	Same-sex partners
Not at all	30.6%	28.6%	36.4%	7.3%
A little	18.4	22.4	27.3	22.0
Somewhat	26.5	31.7	15.9	24.4
Much	16.3	8.2	13.6	17.1
Very much	8.2	6.1	6.8	29.3
	(n = 49)	(n = 49)	(n = 44)	(n = 44)

Collapsed bottom four rows. [a]x^2 = .000 df = 1 p = 1.00; [b]x^2 = 9.666 df = 1 p = .004.
BIM vs. BIW (SS): x^2 = 5.960 df = 1 p = .016;   BIM vs. BIW (OS): x^2 = .135 df = 1 p = .718.

Difficulty sharing own sex needs with partner

	Men[a]		Women[b]	
	Opposite-sex partners	Same-sex partners	Opposite-sex partners	Same-sex partners
Not at all	34.7%	40.8%	25.0%	30.0%
A little	30.6	28.6	29.5	42.5
Somewhat	22.4	14.3	29.5	12.5
Much	8.2	12.2	9.1	2.5
Very much	4.1	4.1	6.8	12.5
	(n = 49)	(n = 49)	(n = 44)	(n = 40)

Collapsed bottom four rows. [a]x^2 = .174 df = 1 p = .685; [b]x^2 = .072 df = 1 p = .791.
BIM vs. BIW (OS): x^2 = .626 df = 1 p = .446;   BIM vs. BIW (SS): x^2 = .698 df = 1 p = .422.

Partner difficulty telling you sex needs

	Men[a]		Women[b]	
	Opposite-sex partners	Same-sex partners	Opposite-sex partners	Same-sex partners
Not at all	26.5%	35.4%	25.6%	25.0%
A little	32.7	31.3	48.8	30.0
Somewhat	34.7	29.2	14.0	25.0
Much	6.1	2.1	11.6	12.5
Very much	0	2.1	0	7.5
	(n = 49)	(n = 48)	(n = 43)	(n = 40)

Collapsed bottom four rows. $^a x^2 = 0.529$ df = 1 p = .477; $^b x^2 = 0.037$ df = 1 p = .856.
BIM vs. BIW (OS): $x^2 = .018$ df = 1 p = .896;   BIM vs. BIW (SS): $x^2 = .677$ df = 1 p = .429.

Failure to meet partner's needs

	Men[a]		Women[b]	
	Opposite-sex partners	Same-sex partners	Opposite-sex partners	Same-sex partners
Not at all	42.9%	40.8%	31.8%	25.6%
A little	34.7	36.7	47.7	41.0
Somewhat	20.4	14.3	13.6	15.4
Much	2.0	4.1	4.5	15.4
Very much	0	4.1	2.3	2.6
	(n = 49)	(n = 49)	(n = 44)	(n = 39)

Collapsed bottom four rows. $^a x^2 = .000$ df = 1 p = 1.000; $^b x^2 = .142$ df = 1 p = .709.
BIM vs. BIW (OS): $x^2 = .779$ df = 1 p = .395;   BIM vs. BIW (SS): $x^2 = 1.602$ df = 1 p = .207.

Partner's failure to meet your needs

	Men[a]		Women[b]	
	Opposite-sex partners	Same-sex partners	Opposite-sex partners	Same-sex partners
Not at all	38.8%	14.3%	23.3%	23.1%
A little	30.6	53.1	27.9	41.0
Somewhat	24.5	24.5	27.9	17.9
Much	6.1	6.1	18.6	5.1
Very much	0	2.0	2.3	12.8
	(n = 49)	(n = 49)	(n = 43)	(n = 39)

Collapsed bottom four rows. $^a x^2 = 6.334$ df = 1 p = .012; $^b x^2 = 0.059$ df = 1 p = .810.
BIM vs. BIW (OS): $x^2 = 1.887$ df = 1 p = .177;   BIM vs. BIW (SS): $x^2 = .615$ df = 1 p = .449.

Difficulty becoming and staying aroused

	Men[a]		Women[b]	
	Opposite-sex partners	Same-sex partners	Opposite-sex partners	Same-sex partners
Not at all	47.9%	60.4%	38.6%	55.0%
A little	41.7	22.9	25.0	27.5
Somewhat	6.3	12.5	20.5	7.5
Much	2.1	2.1	13.6	10.0
Very much	2.1	2.1	2.3	0
	(n = 48)	(n = 48)	(n = 44)	(n = 40)

Collapsed bottom four rows. [a]x^2 = 1.049 df = 1 p = .307; [b]x^2 = 1.646 df = 1 p = .199.
BIM vs. BIW (OS): x^2 = .487 df = 1 p = .491; BIM vs. BIW (SS): x^2 = .087 df = 1 p = .773.

Failure to orgasm

	Men[a]		Women[b]	
	Opposite-sex partners	Same-sex partners	Opposite-sex partners	Same-sex partners
Not at all	57.1%	61.2%	43.2%	39.0%
A little	24.5	18.4	18.2	34.1
Somewhat	4.1	12.2	20.5	4.9
Much	12.2	4.1	11.4	4.9
Very much	2.0	4.1	6.8	17.1
	(n = 49)	(n = 49)	(n = 44)	(n = 41)

Collapsed bottom four rows. [a]x^2 = 0.042 df = 1 p = .846; [b]x^2 = 0.018 df = 1 p = .896.
BIM vs. BIW (OS): x^2 = 1.292 df = 1 p = .261; BIM vs. BIW (SS): x^2 = 3.560 df = 1 p = .062.

Reach orgasm too slowly

	Men[a]		Women[b]	
	Opposite-sex partners	Same-sex partners	Opposite-sex partners	Same-sex partners
Not at all	67.3%	65.3%	30.2%	60.0%
A little	18.4	20.4	25.6	10.0
Somewhat	4.1	8.2	23.3	15.0
Much	10.2	2.0	14.0	7.5
Very much	0	4.1	7.0	7.5
	(n = 49)	(n = 49)	(n = 43)	(n = 40)

Collapsed bottom four rows. [a]x^2 = .000 df = 1 p = 1.000; [b]x^2 = 6.276 df = 1 p = .013.
BIM vs. BIW (OS): x^2 = 11.178 df = 1 p = .000; BIM vs. BIW (SS): x^2 = .087 df = 1 p = .773.

Reach orgasm too fast

	Men[a]		Women[b]	
	Opposite-sex partners	Same-sex partners	Opposite-sex partners	Same-sex partners
Not at all	65.3%	71.4%	84.1%	87.5%
A little	14.3	24.5	6.8	5.0
Somewhat	16.3	4.1	4.5	0
Much	2.0	0	2.3	0
Very much	2.0	0	2.3	7.5
	(n = 49)	(n = 49)	(n = 44)	(n = 40)

Collapsed bottom four rows. $^a x^2 = 0.189$ df = 1 p = .675; $^b x^2 = 0.018$ df = 1 p = .896.
BIM vs. BIW (OS): $x^2 = 3.348$ df = 1 p = .072;   BIM vs. BIW (SS): $x^2 = 2.498$ df = 1 p = .120.

Lack of orgasm by partner

	Men[a]		Women[b]	
	Opposite-sex partners	Same-sex partners	Opposite-sex partners	Same-sex partners
Not at all	57.1%	55.1%	63.6%	46.3%
A little	24.5	36.7	29.5	29.3
Somewhat	12.2	6.1	2.3	14.6
Much	6.1	0	0	2.4
Very much	0	2.0	4.5	7.3
	(n = 49)	(n = 49)	(n = 44)	(n = 41)

Collapsed bottom three rows. $^a x^2 = 7.941$ df = 2 p = .019; $^b x^2 = 5.453$ df = 2 p = .069.
BIM vs. BIW (OS): $x^2 = 2.784$ df = 2 p = .254;   BIM vs. BIW (SS): $x^2 = 4.504$ df = 2 p = .107.

Partner reaches orgasm too slowly

	Men[a]		Women[b]	
	Opposite-sex partners	Same-sex partners	Opposite-sex partners	Same-sex partners
Not at all	57.1%	53.1%	46.5%	66.7%
A little	22.4	38.8	34.9	25.6
Somewhat	12.2	8.2	16.3	5.1
Much	6.1	0	0	0
Very much	2.0	0	2.3	2.6
	(n = 49)	(n = 49)	(n = 43)	(n = 39)

Collapsed bottom three rows. $^a x^2 = 4.779$ df = 2 p = .094; $^b x^2 = 3.865$ df = 2 p = .153.
BIM vs. BIW (OS): $x^2 = 1.787$ df = 2 p = .422;   BIM vs. BIW (SS): $x^2 = 1.148$ df = 1 p = .286.

Partner reaches orgasms too fast

	Men[a]		Women[b]	
	Opposite-sex partners	Same-sex partners	Opposite-sex partners	Same-sex partners
Not at all	85.7%	58.3%	41.9%	97.4%
A little	10.2	25.0	27.9	2.6
Somewhat	2.0	14.6	16.3	0
Much	0	0	11.6	0
Very much	2.0	2.1	2.3	0
	(n = 49)	(n = 48)	(n = 43)	(n = 39)

Collapsed bottom four rows. [a]$x^2 = 7.738$ df = 1 p = .008; [b]$x^2 = 26.661$ df = 1 p = .000.
BIM vs. BIW (OS): $x^2 = 17.531$ df = 1 p = .000; BIM vs. BIW (SS): $x^2 = 15.895$ df = 1 p = .000.

Feel sexually inadequate

	Men[a]		Women[b]	
	Opposite-sex partners	Same-sex partners	Opposite-sex partners	Same-sex partners
Not at all	49.0%	57.1%	45.5%	40.0%
A little	26.5	24.5	31.8	35.0
Somewhat	18.4	12.2	13.6	15.0
Much	6.1	6.1	2.3	7.5
Very much	0	0	6.8	2.5
	(n = 49)	(n = 49)	(n = 44)	(n = 40)

Collapsed bottom four rows. [a]$x^2 = 0.369$ df = 1 p = .559; [b]$x^2 = 0.081$ df = 1 p = .780.
BIM vs. BIW (OS): $x^2 = .017$ df = 1 p = .898; BIM vs. BIW (SS): $x^2 = 1.949$ df = 1 p = .171.

Difficulty maintaining love for partner

	Men[a]		Women[b]	
	Opposite-sex partners	Same-sex partners	Opposite-sex partners	Same-sex partners
Not at all	57.1%	58.3%	45.5%	76.0%
A little	20.4	16.7	27.3	12.5
Somewhat	14.3	8.3	6.8	7.5
Much	8.2	8.3	13.6	0
Very much	0	8.3	6.8	5.0
	(n = 49)	(n = 48)	(n = 44)	(n = 40)

Collapsed bottom four rows. [a]$x^2 = 0.008$ df = 1 p = .933; [b]$x^2 = 6.731$ df = 1 p = .010.
BIM vs. BIW (OS): $x^2 = .843$ df = 1 p = .374; BIM vs. BIW (SS): $x^2 = 2.007$ df = 1 p = .166.

Table 7.6   Sexually Transmitted Diseases

Ever contracted STD

	Men	Women	Combined
Yes	61.2%	59.1%	60.2%
No	38.8	40.9	39.8
	(n = 49)	(n = 44)	

$x^2 = .044$ df = 1 p = .834.

Disease status at time of interview

	Men	Women	Combined	x^2	df	p
Herpes, Active	2.0%	6.8%	4.3%	1.329	1	.249
Herpes, Inactive	6.1	13.6	9.7	1.514	1	.219
Gonorrhea	0	2.3	1.1	1.514	1	.219
Syphilis	0	2.3	1.1	1.509	1	.219
Urethral Infection	6.1	2.3	4.3	.879	1	.349
Venereal Warts	4.1	4.5	4.3	.012	1	.912
Other STD	4.1	11.4	7.5	1.805	1	.179
AIDS	0	0	0			
	(n = 49)	(n = 44)				

Times in life contracted STD

	Men	Women	Combined
0	39.6%	41.9%	40.7%
1	14.6	20.9	17.0
2-3	18.8	16.3	17.6
4-7	16.7	4.7	11.0
8+	10.4	16.3	13.2
	(n = 48)	(n = 43)	

$x^2 = 4.443$ df = 4   p = .349.

Ever contracted particular diseases

	Men	Women	Combined	x^2	df	p
Herpes	10.2%	18.2%	14.0%	1.231	1	.267
Gonorrhea	24.5	22.7	23.7	.040	1	.842
Syphilis	10.2	4.5	7.5	1.106	1	.293
Urethral Infection	24.5	15.9	20.4	1.067	1	.303
Venereal Warts	24.5	11.4	18.3	2.754	1	.097
Other STD	40.8	43.2	41.9	.053	1	.817
	(n = 49)	(n = 44)				

Increase in STDs affected sexual patterns

	Men	Women	Combined
Not at all	12.2%	45.5%	28.0%
A little	24.5	29.5	26.9
Somewhat	22.4	11.4	17.2
Much	12.2	2.3	7.5
Very much	28.6	11.4	20.4
	(n = 49)	(n = 44)	

$x^2 = 18.431$ df = 4   p = .001.

CHAPTER 8: SIGNIFICANT OTHERS

Table 8.1   The Ideal Arrangement

Ever had their ideal sociosexual arrangement

	Men	Women	Combined
Yes	30.6%	34.1%	32.3%
No	69.4	65.9	67.7
	(n = 49)	(n = 44)	

x^2 = .019 df = 1 p = .892.

Length of time ideal arrangement lasted

	Men	Women	Combined
< 6 months	46.7%	57.1%	51.7%
6 months - 4 years	33.3	28.6	31.0
5+ years	20.0	14.3	17.2
	(n = 15)	(n = 14)	

Collapsed bottom two rows. x^2 = .037 df = 1 p = .847.

Table 8.2.   Current Relationships

Current significant involvement

	Men	Women	Combined
Yes	67.3%	70.5%	68.8%
No	32.7	29.5	31.2
	(n = 49)	(n = 44)	

x^2 = .010 df = 1 p = .926.

Number of significant partners

	Men	Women	Combined
1	57.6%	45.2%	51.6%
2	30.3	29.0	29.7
3	0	3.2	2.0
4 +	12.1	22.6	16.7
	(n = 33)	(n = 31)	

Collapsed bottom three rows. x^2 = .552 df = 1 p = .470.

Table 8.3   Most Significant Relationship

Sex of most significant partner

	Men	Women	Combined
Male	24.2%	80.6%	51.6%
Female	75.8	19.4	48.4
	(n = 33)	(n = 31)	

x^2 = 18.164 df = 1 p = .000.

Partner's sexual preference

	Men	Women	Combined
Gay	16.2%	6.9%	11.7%
Lesbian	3.2	3.4	3.3
Bisexual	35.5	51.7	43.3
Straight	45.2	37.9	41.7
	(n = 31)	(n = 29)	

N.B.: Collapsed top two rows. Still doesn't meet minimal expected frequencies for a x^2 test. No clear basis for collapsing into a 2x2 table.

Uses "primary" and "secondary" labels

	Men	Women	Combined
Yes	78.8%	73.3%	76.2%
No	21.2	26.7	23.8
	(n = 33)	(n = 30)	

$x^2 = .045$ df = 1 p = .832.

Label for most significant relationship

	Men	Women	Combined
Primary	84.6%	81.0%	83.0%
Secondary	0	9.5	4.3
Can't tell	15.4	9.5	12.8
	(n = 26)	(n = 21)	

Collapsed bottom two rows. $x^2 = .000$ df = 1 p = 1.000.

Location where respondent met most significant partner

	Men	Women	Combined
Friends, conventional parties, dinners, or get-togethers	12.9%	10.0%	11.5%
School, work, recreation group	48.4	50.0	49.2
Bars, sex related clubs, parties, or get-togethers	12.9	10.0	11.5
Rap, support, personal growth, or workshop groups	25.8	30.0	27.9
	(n = 31)	(n = 30)	

Collapsed top two rows as more conventional and bottom two rows as less conventional.
$x^2 = .000$ df = 1 p = 1.000.

Residential arrangement

	Men	Women	Combined
Lives with partner	42.4%	51.6%	46.9%
Does not live with partner	57.6	48.4	53.1
	(n = 33)	(n = 31)	

$x^2 = .236$ df = 1 p = .627.

Sleeping arrangement

	Men	Women	Combined
Shares bed	85.7%	100.0%	93.3%
Does not share bed	14.3	0	6.7
	(n = 14)	(n = 16)	

Fisher's exact p = .209.

Frequency of visits (if not living together)

	Men	Women	Combined
< 1 month	5.3%	0	2.9%
1-3 month	10.5	26.7%	17.6
1-3 week	68.4	46.7	58.8
4 + week	15.8	26.7	20.6
	(n = 19)	(n = 15)	

Collapsed top two rows. x^2 = .164 df = 2 p = .440.

Frequency of sexual activity with most significant partner

	Men	Women	Combined
Never	3.0%	6.7%	4.8%
< 1 month	3.0	0.7	1.8
1-3 month	15.2	10.0	12.7
1-3 week	57.6	56.7	57.1
4 + week	21.2	20.0	20.6
	(n = 33)	(n = 30)	

Collapsed top three rows and bottom two rows. x^2 = .000   df = 1   p = 1.000.

Respondent's evaluation of sex life with most significant partner

	Men	Women	Combined
Positive	72.7%	63.3%	68.3%
Okay	27.3	20.0	23.8
Negative	0	16.7	7.9
	(n = 33)	(n = 30)	

Collapsed bottom two rows. x^2 = .280   df = 1   p = .597.

Length of relationship

	Men	Women	Combined
< 6 months	24.2%	22.6%	23.4%
1 year	6.1	6.5	6.3
2-4 years	27.3	35.5	31.3
5-9 years	27.3	22.6	25.0
10-19 years	12.1	9.7	10.9
20 + years	3.0	3.2	3.1
	(n = 33)	(n = 31)	

Collapsed last three rows. x^2 = .520 df = 2 p = .773.

Occurrence of separations or breakups

	Men	Women	Combined
Yes	30.3%	19.4%	25.0%
No	69.7	80.6	75.0
	(n = 33)	(n = 31)	

$x^2 = .521$ df = 1 p = .470.

Length of time respondent expects relationship to last

	Men	Women	Combined
< 6 months	0	6.5%	3.2%
6 months-< year	6.3%	9.7	7.9
1-2 years	9.4	12.9	11.1
3-5 years	12.5	0	6.3
6-10 years	0	6.5	3.2
> 10 years	71.9	64.5	68.3
	(n = 32)	(n = 31)	

Collapsed top four rows. $x^2 = .127$ df = 1 p = .721.

How upset if relationship broke up

	Men	Women	Combined
Not at all	3.0%	6.7%	4.8%
A little	6.1	13.3	9.5
Moderately	18.2	20.0	19.0
Very much	72.7	60.0	66.6
	(n = 33)	(n = 30)	

Collapsed top three rows. $x^2 = .644$ df = 1 p = .422.

Thinks bisexuality affects ability to sustain long-term relationships

	Men	Women	Combined
Affects it	18.2%	38.7%	28.1%
Does not affect it	81.8	61.3	71.9
	(n = 33)	(n = 31)	

$x^2 = 2.394$ df = 1 p = .122.

Table 8.4   Second Most Significant Relationship

Sex of second significant partner

	Men	Women	Combined
Male	57.1%	33.3%	43.8%
Female	42.9	66.7	56.3
	(n = 14)	(n = 18)	

$x^2 = .976$ df = 1 p = .323.

Second partner's sexual preference

	Men	Women	Combined
Gay	28.6%	0	12.5%
Lesbian	0	5.6%	3.1
Bisexual	57.1	72.2	65.6
Straight	14.3	22.2	18.8
	(n = 14)	(n =18)	

Collapsed top two rows. $x^2 = 3.308$ df = 2 p = .191.

Uses "primary" and "secondary" labels for second relationship

	Men	Women	Combined
Yes	50.0%	50.0%	50.0%
No	50.0	50.0	50.0
	(n = 14)	(n = 18)	

$x^2 = 0.000$ df = 1 p = 1.000.

Label for second significant relationship

	Men	Women	Combined
Primary	14.3%	11.1%	12.5%
Secondary	57.1	88.9	75.0
Can't tell	28.6	0	12.5
	(n = 7)	(n = 9)	

Collapsed bottom two rows. Fisher's exact p = 1.000.

Location where respondent met second significant partner

	Men	Women	Combined
Friends, conventional parties, dinners, or get-togethers	38.5%	25.0%	31.0%
School, work, recreation group	15.4	31.3	24.1
Bars, sex related clubs, parties or get-togethers	7.7	12.5	10.3
Rap, support, personal growth or workshop groups	38.5	31.3	34.5
	(n = 13)	(n = 16)	

Collapsed top two rows as more conventional and bottom two rows as less conventional.
$x^2 = .000$ df = 1 p = 1.000.

Residential arrangement (second significant partner)

	Men	Women	Combined
Lives with second partner	7.1%	5.6%	6.3%
Does not live with second partner	92.9	94.4	93.8
	(n = 14)	(n = 18)	

Fisher's exact p = 1.000.

Frequency of visits (if not living together)

	Men	Women	Combined
< 1 month	15.4%	12.5%	13.8%
1-3 month	38.5	31.3	34.5
1-3 week	46.2	37.5	41.4
4 + week	0	18.8	10.3
	(n = 13)	(n = 16)	

Collapsed top two rows and bottom two rows. $x^2 = .028$ df = 1 p = .867.

Frequency of sexual activity with second partner

	Men	Women	Combined
Never	14.3%	16.7%	15.6%
< 1 month	21.4	27.8	25.0
1-3 month	35.7	22.2	28.1
1-3 week	28.6	33.3	31.3
4 + week	0	0	0
	(n = 14)	(n = 18)	

Collapsed top two rows and bottom three rows. $x^2 = .019$ df = 1 p = .892.

Respondent's evaluation of sex life with second partner

	Men	Women	Combined
Positive	69.2%	52.9%	60.0%
Okay	30.8	29.4	30.0
Negative	0	17.6	10.0
	(n = 13)	(n = 17)	

Collapsed bottom two rows. $x^2 = .277$ df=1 p=.599.

Length of relationship with second partner

	Men	Women	Combined
< 6 months	28.6%	35.3%	32.3%
1 year	7.1	5.9	6.5
2-4 years	42.9	29.4	35.5
5-9 years	14.3	29.4	22.6
10-19 years	7.1	0	3.2
	(n = 14)	(n = 17)	

Collapsed top two rows and bottom three rows. $x^2 = .000$ df = 1 p = 1.000.

Occurrence of separations or break-ups with second partner

	Men	Women	Combined
Yes	50.0%	38.9%	43.8%
No	50.0	61.1	56.3
	(n = 14)	(n = 18)	

$x^2 = .073$ df = 1 p = .788.

Length of time respondent expects relationship to last with second partner

	Men	Women	Combined
6 months - < year	14.3%	22.2%	18.8%
1-2 years	7.1	11.1	9.4
3-5 years	7.1	5.6	6.3
6-10 years	7.1	0	3.1
> 10 years	64.3	61.1	62.5
	(n = 14)	(n = 18)	

Collapsed top four rows. $x^2 = .000$ df = 1 p = 1.000.

How upset if second relationship broke up

	Men	Women	Combined
Not at all	0	5.6%	3.1%
A little	21.4%	5.6	12.5
Moderately	35.7	33.3	34.4
Very much	42.9	55.6	50.0
	(n = 14)	(n = 18)	

Collapsed top three rows. $x^2 = .127$ df = 1 p = .722.

Second relationship had effect on most significant relationship

	Men	Women	Combined
Yes	64.3%	55.6%	59.4%
No	35.7	44.4	40.6
	(n = 14)	(n = 18)	

$x^2 = .019$ df = 1 p = .892.

Effect of second relationship on most significant relationship

	Men	Women	Combined
Positive effect	88.9%	60.0%	73.7%
Negative effect	11.1	40.0	26.3
	(n = 9)	(n = 10)	

Fisher's exact p = .303.

Table 8.5   Other Significant Partners

Total sexual frequency with other significant partner(s)

	Men	Women	Combined
Never	0	14.3%	9.1%
< 1 month	0	28.6	18.2
1-3 month	75.0%	57.1	63.6
1-3 week	25.0	0	9.1
	(n = 4)	(n = 7)	

Collapsed top two rows and bottom two rows. Fisher's exact p = .236.

Sexual activity main reason for getting together with other partner(s)

	Men	Women	Combined
Yes	25.0%	33.3%	30.0%
No	75.0	66.7	70.0
	(n = 4)	(n = 6)	

Fisher's exact p = 1.000.

Sex of other significant partner

	Men	Women	Combined
Male	57.0%	55.0%	56.0%
Female	43.0	45.0	44.0
	(n = 4)	(n = 7)	

Fisher's exact p = 1.000.

Other significant partners had effect on most significant relationship

	Men	Women	Combined
Yes	50.0%	57.1%	54.5%
No	50.0	42.9	45.5
	(n = 4)	(n = 7)	

Fisher's exact p = 1.000.

Other significant partners had effect on second most significant relationship

	Men	Women	Combined
Yes	25.0%	28.6%	27.3%
No	75.0	71.4	72.7
	(n = 4)	(n = 7)	

Fisher's exact p = 1.000.

Effect of other significant relationships on first and/or second significant relationships

	Men	Women	Combined
Positive	66.7%	100.0%	85.7%
Negative	33.3	0	14.3
	(n = 3)	(n = 4)	

Fisher's exact p = .429.

Table 8.6   Casual and Anonymous Partners

Frequency of casual sex (for those with a significant partner)

	Men	Women	Combined
Never	24.2%	23.3%	23.8%
< 1 month	30.3	50.0	39.7
1-3 month	36.4	13.3	25.4
1-3 week	9.1	10.0	9.5
4 + week	0	3.3	1.6
	(n=33)	(n=30)	

Collapsed top two rows and bottom three rows. x^2 = 1.651 df = 1 p = .199.

Sex of casual partners (for those with a significant partner)

	Men	Women	Combined
All women	4.0%	4.3%	4.2%
More women	24.0	21.7	22.9
Equal men and women	16.0	13.0	14.6
More men	24.0	39.1	31.3
All men	32.0	21.7	27.1
	(n=25)	(n=23)	

$x^2 = .002$ df = 1 p = .961.

Location where respondent primarily met casual partners (for those with a significant partner)

	Men	Women	Combined
Friends, conventional parties, dinners, or get-togethers	16.0%	27.3%	21.3%
School, work, recreation group	24.0	36.4	29.8
Bars, sex related clubs, parties, or get-togethers	32.0	13.6	23.4
Rap, support, personal growth, or workshop groups	28.0	22.7	25.5
	(n=25)	(n=22)	

Collapsed top two rows as more conventional and bottom two rows as less conventional.
$x^2 = 1.756$ df = 1 p = .185.

Casual sex had effect on most significant relationship (for those with a significant partner)

	Men	Women	Combined
Yes	45.5%	38.1%	41.9%
No	54.5	61.9	58.1
	(n = 22)	(n = 21)	

$x^2 = .032$ df = 1 p = .857.

Casual sex had effect on second significant relationship (for those with a second significant partner)

	Men	Women	Combined
Yes	8.3%	0	4.0%
No	91.7	100.0%	96.0
	(n = 12)	(n = 13)	

Fisher's exact p = .480.

Effects on first and/or second significant relationship(s) (for those with two significant partners)

	Men	Women	Combined
Positive	66.7%	62.5%	64.7%
Negative	33.3	37.5	35.3
	(n = 9)	(n = 8)	

Fisher's exact p = 1.000.

Frequency of anonymous sex (for those with a significant partner)

	Men	Women	Combined
Never	42.4%	66.7%	54.0%
< 1 month	33.3	26.7	30.2
1-3 month	18.2	3.3	11.1
1-3 week	6.1	0	3.2
4 + week	0	3.3	1.6
	(n = 33)	(n = 30)	

Collapsed top two rows and bottom two rows. $x^2 = 2.438$ df = 1 p = .118.

Sex of anonymous partners (for those with a significant partner)

	Men	Women	Combined
All women	5.3%	10.0%	6.9%
More women	5.0	10.0	6.9
Equal men and women	0	30.0	10.3
More men	26.3	20.0	24.1
All men	63.2	30.0	51.7
	(n = 19)	(n = 10)	

Collapsed top three rows and bottom two rows. Fisher's exact p = .030.

Location where respondent primarily met anonymous partners (for those with a significant partner):

	Men	Women	Combined
Friends, conventional parties, dinners or get-togethers	11.1%	20.0%	14.3%
School, work, recreation group	11.1	10.0	10.7
Bars, sex related clubs, parties or get-togethers	77.8	60.0	71.4
Rap, support, personal growth or workshop groups	0	10.0	3.6
	(n = 18)	(n = 10)	

Collapsed top two rows as more conventional and bottom two rows as less conventional. Fisher's exact p=.310.

Anonymous sex had effect on most significant relationship (for those with a significant partner)

	Men	Women	Combined
Yes	28.6%	30.0%	29.2%
No	71.4	70.0	70.8
	(n = 14)	(n = 10)	

Fisher's exact p = 1.000.

Anonymous sex had effect on second significant relationship (for those with a second significant partner)

	Men	Women	Combined
Yes	14.3%	25.0%	18.2%
No	85.7	75.0	81.8
	(n = 7)	(n = 4)	

Fisher's exact p = 1.000.

Effects on first and/or second significant relationship(s) (for those with two significant partners)

	Men	Women	Combined
Positive	60.0%	66.7%	62.5%
Negative	40.0	33.3	37.5
	(n = 5)	(n = 3)	

Fisher's exact p = 1.000.

CHAPTER 9: MARRIAGE

Table 9.1   Marital History

Number of times married

	Men	Women	Combined
Never	50.0%	46.3%	48.3%
Once	37.5	39.0	38.2
Twice	10.4	9.8	10.1
Three or more times	2.1	4.9	3.4
	(n = 48)	(n = 41)	

Collapsed bottom three rows. $x^2 = .017$ df = 1 p = .895.

Current marital status

	Men	Women	Combined
Married	20.4%	18.2%	19.4%
Unmarried	79.6	81.8	80.6
	(n = 49)	(n = 44)	

$x^2 = .000$ df = 1 p = .993.

Length of current marriage

	Men	Women	Combined
Less than 4 years	10.0%	50.0%	27.8%
5-9 years	40.0	37.5	38.9
10-19 years	40.0	12.5	27.8
20 + years	10.0	0.0	5.6
	(n = 10)	(n = 8)	

Collapsed top two rows and bottom two rows. Fisher's exact p = .152.

Residence of spouse

	Men	Women	Combined
Lives with spouse	80.0%	37.5%	61.1%
Does not live with spouse	20.0	62.5	38.9
	(n = 10)	(n = 8)	

Fisher's exact p = .145.

Spouse's sexual preference

	Men	Women	Combined
Gay/Lesbian	10.0%	12.5%	11.1%
Bisexual	20.0	50.0	33.3
Straight	70.0	37.5	55.6
	(n = 10)	(n = 8)	

N.B.: Doesn't meet minimal expected frequencies for a x^2 test. There is also no clear basis for collapsing into a 2x2 table.

Open sexual relationship

	Men	Women	Combined
Yes	71.4%	83.3%	76.9%
No	28.6	16.7	23.1
	(n = 7)	(n = 6)	

Fisher's exact p = 1.000.

Labels spouse primary or secondary

	Men	Women	Combined
Primary	87.5%	100%	92.3%
Neither primary nor secondary	12.5	0	7.7
	(n = 8)	(n = 5)	

x^2=.000 df=1 p=1.000

Happiness of marriage

	Men	Women	Combined
Very unhappy	0	12.5%	5.6%
Moderately unhappy	10.0%	12.5	11.1
In between	0	12.5	5.6
Moderately happy	30.0	12.5	22.2
Very happy	60.0	50.0	55.6
	(n = 10)	(n = 8)	

Collapsed top four rows. Fisher's exact p = 1.000.

Table 9.2   Disclosure to Spouse

Disclosed bisexuality before marriage

	Men	Women	Combined
Yes	20.0%	50.0%	33.3%
No	80.0	50.0	66.7
	(n = 10)	(n = 8)	

Fisher's exact p = .321.

Thought self bisexual at that time

	Men	Women	Combined
Yes	25.0%	25.0%	25.0%
No	75.0	75.0	75.0
	(n = 8)	(n = 4)	

Fisher's exact p = 1.000.

Table 9.3   Marital dissolution

Bisexuality contributed to marital breakup

	Men	Women	Combined
Yes	50.0%	20.0%	34.5%
No	50.0	80.0	65.5
	(n = 14)	(n = 15)	

Fisher's exact p = .095.

Table 9.4   Desire for Marriage

Wants to remarry/get married

	Men	Women	Combined
Yes	34.2%	22.9%	28.8%
Don't know	36.8	28.6	32.9
No	28.9	48.6	38.4
	(n = 38)	(n = 35)	

$x^2 = 3.044$ df = 2 p = .218.

Bisexuality would affect a future marriage

	Men	Women	Combined
Yes	88.9%	83.3%	86.7%
No	11.1	16.7	13.3
	(n = 27)	(n = 18)	

$x^2 = .008$ df = 1   p = .933.

Wants to marry for love

	Men	Women	Combined
Not at all	0	3.8%	1.6%
A little	2.6%	0	1.6
Somewhat	5.3	0	3.1
Much	15.8	7.7	12.5
Very much	76.3	88.5	81.3
	(n = 38)	(n = 26)	

Collapsed first four rows. $x^2 = .804$ df = 1 p = .370.

Wants to marry for heterosexual sex

	Men	Women	Combined
Not at all	7.9%	42.3%	21.9%
A little	2.6	3.8	3.1
Somewhat	18.4	30.8	23.4
Much	36.8	3.8	23.4
Very much	34.2	19.2	28.1
	(n = 38)	(n = 26)	

Collapsed last four rows. $x^2 = 8.779$ df = 1   p = .005.

Wants to marry because desires children

	Men	Women	Combined
Not at all	31.6%	46.2%	37.5%
A little	18.4	3.8	12.5
Somewhat	7.9	7.7	7.8
Much	28.9	26.9	28.1
Very much	13.2	15.4	14.1
	(n = 38)	(n = 26)	

Collapsed last four rows. x^2 = .846 df = 1 p = .358.

Wants to marry to avoid being considered homosexual or bisexual

	Men	Women	Combined
Not at all	68.4%	73.1%	70.3%
A little	18.4	11.5	15.6
Somewhat	13.2	11.5	12.5
Much	0	3.8	1.6
Very much	0	0	0
	(n = 38)	(n = 26)	

Collapsed last four rows. x^2 = .015 df = 1 p = .903.

Wants to marry because of loneliness

	Men	Women	Combined
Not at all	15.8%	30.8%	21.9%
A little	10.5	15.4	12.5
Somewhat	42.1	19.2	32.8
Much	18.4	23.1	20.3
Very much	13.2	11.5	12.5
	(n = 38)	(n = 26)	

Collapsed last four rows. x^2 = 1.245 df = 1 p = .264.

Wants to marry because of family pressure

	Men	Women	Combined
Not at all	50.0%	69.2%	57.8%
A little	31.6	3.8	20.3
Somewhat	13.2	19.2	15.6
Much	5.3	7.7	6.3
Very much	0	0	0
	(n = 38)	(n = 26)	

Collapsed last four rows. x^2 = 1.619 df = 1 p = .203.

Wants to marry because of societal pressure

	Men	Women	Combined
Not at all	50.0%	57.7%	53.1%
A little	34.2	15.4	26.6
Somewhat	13.2	15.4	14.1
Much	2.6	11.5	6.3
Very much	0	0	0
	(n = 38)	(n = 26)	

Collapsed last four rows. x^2 = .123 df = 1 p = .726.

Table 9.5 Children

Has children

	Men	Women	Combined
Yes	32.7%	34.9%	33.7%
No	67.3	65.1	66.3
	(n = 49)	(n = 43)	

x^2 = .000 df = 1 p = .996.

Number of children

	Men	Women	Combined
One	43.8%	33.3%	38.7%
Two	31.3	26.7	29.0
Three or more	25.0	40.0	32.3
	(n = 16)	(n = 15)	

Collapsed last two rows. x^2 = .051 df = 1 p = .821.

Number of children thirteen years of age or older

	Men	Women	Combined
None	31.3%	20.0%	25.8%
One	18.8	13.3	16.1
Two	31.3	40.0	35.5
Three or more	18.8	26.7	22.6
	(n = 16)	(n = 15)	

Collapsed top two rows and bottom two rows. x^2 = .331 df = 1 p = .565.

Children (thirteen years old or older) who know or suspect respondent is bisexual

	Men	Women	Combined
None	45.5%	25.0%	34.8%
One	27.3	16.7	21.7
Two	18.2	50.0	34.8
Three or more	9.1	8.3	8.7
	(n = 11)	(n = 12)	

Collapsed top two rows and bottom two rows. Fisher's exact p = .214.

Effect on relationship of children's knowing

	Men	Women	Combined
Positive effect	0	44.4%	26.7%
No effect	66.7%	22.2	40.0
Negative effect[a]	33.3	33.3	33.3
	(n = 6)	(n = 9)	

Collapsed top two rows. Fisher's exact p = 1.000.
[a]One of the male respondents said there was a positive as well as a negative effect.

CHAPTER 10: JEALOUSY

Table 10.1   Monogamy in the Primary Relationship

Respondent is monogamous or nonmonogamous

	Men	Women	Combined
Monogamous	12.1%	23.3%	17.5%
Nonmonogamous	87.9	76.7	82.5
	(n = 33)	(n = 30)	

$x^2 = .703$ df = 1 p = .402.

"Open" vs. "closed" relationship

	Men	Women	Combined
Open	81.8%	95.0%	88.1%
Closed	18.2	5.0	11.9
	(n = 22)	(n = 20)	

$x^2 = .706$ df = 1 p = .401.

Table 10.2   Jealousy in the Primary Relationship

Perception of extent of partner's jealousy

	Men	Women	Combined
Not at all	13.8%	34.8%	23.1%
A little	51.7	39.1	46.2
Somewhat	20.7	21.7	21.2
Quite a bit	13.8	0	7.7
Extremely	0	4.3	1.9
	(n = 29)	(n = 23)	

Collapsed top two rows and bottom three rows. $x^2 = .122$ df = 1 p = .727.

Partner more jealous of opposite sex or same sex

	Men	Women	Combined
Opposite sex	54.2%	60.0%	56.4%
Same sex	25.0	13.3	20.5
No difference	20.8	26.7	23.1
	(n = 24)	(n = 15)	

Collapsed bottom two rows. $x^2 = .001$ df = 1 p = .977.

Spouse/partner is monogamous or nonmonogamous

	Men	Women	Combined
Monogamous	48.5%	30.0%	39.7%
Nonmonogamous	51.5	70.0	60.3
	(n = 33)	(n = 30)	

x^2 = 1.537 df = 1 p = .215.

Belief about extent of partner's outside sex

	Men	Women	Combined
Not sure	20.0%	10.0%	14.3%
Occasional	60.0	60.0	60.0
Moderate	13.3	10.0	11.4
Fairly extensive	6.7	10.0	8.6
Very extensive	0	10.0	5.7
	(n = 15)	(n = 20)	

Collapsed top two rows and bottom three rows. Fisher's exact p = .700.

Nature of partner's outside sexual involvements

	Men	Women	Combined
Significant relationships	21.4%	31.3%	26.7%
Significant to casual	28.6	25.0	26.7
Casual	28.6	31.3	30.0
Casual to anonymous	21.4	12.5	16.7
Anonymous	0	0	0
	(n = 14)	(n = 16)	

Collapsed top two rows and bottom three rows. x^2 = .000 df = 1 p = 1.000.

Extent of respondent's jealousy

	Men	Women	Combined
Not at all	50.0%	42.9%	45.9%
A little	37.5	28.6	32.4
Somewhat	12.5	14.3	13.5
Quite a bit	0	4.8	2.7
Extremely	0	9.5	5.4
	(n = 16)	(n = 21)	

Collapsed bottom four rows. x^2 = .010 df = 1 p = .921.

Table 10.3   Ground rules

Ground rules about outside activity

	Men	Women	Combined
Yes	82.1%	73.1%	77.8%
No	17.9	26.9	22.2
	(n = 28)	(n = 26)	

x^2 = .224 df = 1 p = .636.

Ground rules about eligible outside sexual partners

	Men	Women	Combined
Yes	39.1%	47.4%	42.9%
No	60.9	52.6	57.1
	(n = 23)	(n = 19)	

$x^2 = .050$ df = 1   p = .823.

Ground rules about number of outside sexual partners

	Men	Women	Combined
Yes	13.0%	5.3%	9.5%
No	87.0	94.7	90.5
	(n = 23)	(n = 19)	

$x^2 = .107$ df = 1   p = .750.

Ground rules about closeness of relationship with other partners

	Men	Women	Combined
Yes	21.7%	21.1%	21.4%
No	78.3	78.9	78.6
	(n = 23)	(n = 19)	

$x^2 = .012$ df = 1   p = .918.

Ground rules about scheduling time

	Men	Women	Combined
Yes	47.8%	57.9%	52.4%
No	52.2	42.1	47.6
	(n = 23)	(n = 19)	

$x^2 = .116$ df = 1   p = .740.

Spouse/partner likes to know outside partners

	Men	Women	Combined
Yes	52.2%	84.2%	66.7%
No	47.8	15.8	33.3
	(n = 23)	(n = 19)	

$x^2 = 3.472$ df = 1   p = .062.

Ground rules same for respondent and spouse/partner

	Men	Women	Combined
All same	71.4%	84.2%	77.5%
Mixed	14.3	15.8	15.0
All different	14.3	0	7.5
	(n = 21)	(n = 19)	

Collapsed bottom two rows. $x^2 = .345$ df = 1   p = .638.

Ground rules always been the same

	Men	Women	Combined
Yes	56.5%	52.6%	54.8%
No	43.5	47.4	45.2
	(n = 23)	(n = 19)	

x^2 = .636 df = 1 p = .801.

Table 10.4   Second Significant Relationship

Engages in sexual activity outside of both significant relationships

	Men	Women	Combined
Yes	85.7%	66.7%	75.0%
No	14.3	33.3	25.0
	(n = 14)	(n = 18)	

Fisher's exact p = .412.

Perception of extent of second partner's jealousy

	Men	Women	Combined
Not at all	50.0%	66.7%	58.6%
A little	28.6	6.7	17.2
Somewhat	21.4	13.3	17.2
Quite a bit	0	6.7	3.4
Extremely	0	6.7	3.4
	(n = 14)	(n = 15)	

Collapsed bottom four rows. x^2 = .284 df = 1 p = .594.

Second partner more jealous of males or females

	Men	Women	Combined
Male	0	0	0
Female	42.9%	40.0%	41.7%
No difference	57.1	60.0	58.3
	(n = 7)	(n = 5)	

Fisher's exact p = 1.000.

Second partner is monogamous or nonmonogamous

	Men	Women	Combined
Monogamous	7.1%	11.1%	9.4%
Nonmonogamous	92.9	88.9	90.6
	(n = 14)	(n = 18)	

Fisher's exact p = 1.000.

Belief about extent of second partner's outside sex

	Men	Women	Combined
Not sure	9.1%	7.1%	8.0%
Occasional	63.6	28.6	44.0
Moderate	18.2	21.4	20.0
Fairly extensive	9.1	14.3	12.0
Very extensive	0	28.6	16.0
	(n = 11)	(n = 14)	

Collapsed top two rows and bottom three rows. $x^2 = 2.061$ df = 1 p = .151.

Extent of respondent's jealousy

	Men	Women	Combined
Not at all	66.7%	43.8%	53.6%
A little	25.0	31.3	28.6
Somewhat	8.3	6.3	7.1
Quite a bit	0	18.8	10.7
	(n = 12)	(n = 16)	

Collapsed bottom three rows. $x^2 = .673$ df = 1 p = .412.

Ground rules regarding outside activity

	Men	Women	Combined
Yes	28.6%	33.3%	31.3%
No	71.4	66.7	68.8
	(n = 14)	(n = 18)	

Fisher's exact p = 1.000.

CHAPTER 11: BEING OUT

Table 11.1   Social Conflicts

Problems with homosexual men

	Men	Women	Combined
Yes	60.4%	44.2%	52.7%
No	39.6	55.8	47.3
	(n = 48)	(n = 43)	

$x^2 = 1.790$ df = 1 p = .181.

Problems with homosexual women

	Men	Women	Combined
Yes	37.5%	76.7%	56.0%
No	62.5	23.3	44.0
	(n = 48)	(n = 43)	

$x^2 = 12.632$ df = 1 p = .000.

Problems with heterosexual men

	Men	Women	Combined
Yes	60.4%	59.5%	60.0%
No	39.6	40.5	40.0
	(n = 48)	(n = 42)	

$x^2 = .000$ df = 1   p = 1.000.

Problems with heterosexual women

	Men	Women	Combined
Yes	60.4%	58.5%	59.6%
No	39.6	41.5	40.4
	(n = 48)	(n = 41)	

$x^2 = .000$ df = 1   p = 1.000.

Table 11.2   Disclosure of Bisexuality

Age when first disclosed

	Men	Women	Combined
14 years	4.1%	2.3%	3.3%
15-19 years	14.3	11.6	13.0
20-24 years	28.6	41.9	34.8
25-29 years	24.5	20.9	22.8
30-39 years	20.4	11.6	16.3
40-49 years	6.1	7.0	6.5
50 + years	2.0	4.7	3.3
	(n = 49)	(n = 43)	

Collapsed top three rows and bottom four rows. $x^2 = .723$ df = 1 p = .395.

Period of dual attraction before first disclosure

	Men	Women	Combined
< 6 months	14.6%	30.2%	22.0%
1-4 years	31.3	30.2	30.8
5-9 years	27.1	9.3	18.7
10-14 years	10.4	11.6	11.0
15-19 years	8.3	11.6	9.9
20-29 years	4.2	4.7	4.4
30 + years	4.2	2.3	3.3
	(n = 48)	(n = 43)	

Collapsed top two rows and bottom five rows. $x^2 = 1.405$ df = 1 p = .236.

Person first disclosed to

	Men	Women	Combined
Spouse	16.3%	7.7%	12.2%
Mother	2.3	7.7	4.9
Sister	0	5.1	2.4
Other relative	0	2.6	1.2
Female partner	32.6	12.8	23.2
Male partner	7.0	15.4	11.0
Homosexual friend	9.3	2.6	6.1
Heterosexual friend male sex	9.3	17.9	13.4
Heterosexual friend female sex	9.3	5.1	7.3
Support group	7.0	12.8	9.8
Therapist	7.0	2.6	4.9
Work associate	0	7.7	3.7
	(n = 43)	(n = 39)	

Doesn't meet minimal expected frequencies for a x^2 test.

Sister knows or suspects

	Men	Women	Combined
Definitely knows	54.1%	71.0%	61.8%
Suspects	13.5	3.2	8.8
Doesn't know or suspect	32.4	25.8	29.4
	(n = 37)	(n = 31)	

Collapsed bottom two rows. x^2 = .3587 df = 1 p = .549.

Brother knows or suspects

	Men	Women	Combined
Definitely knows	33.3%	69.2%	48.4%
Suspects	22.2	7.7	16.1
Doesn't know or suspect	44.4	23.1	35.5
	(n = 36)	(n = 26)	

x^2 = 8.160 df = 2 p = .017.

Mother knows or suspects

	Men	Women	Combined
Definitely knows	38.3%	42.9%	40.4%
Suspects	23.4	16.7	20.2
Doesn't know or suspect	38.3	40.5	39.3
	(n = 47)	(n = 42)	

x^2 = .644 df = 2 p = .725.

Father knows or suspects

	Men	Women	Combined
Definitely knows	34.2%	30.6%	32.4%
Suspects	18.4	13.9	16.2
Doesn't know or suspect	47.4	55.6	51.4
	(n = 38)	(n = 36)	

$x^2 = .533$ df = 2   p = .758.

Most other relatives know or suspect

	Men	Women	Combined
Definitely know	21.7%	30.0	25.6%
Suspect	17.4	17.5	17.4
Don't know or suspect	60.9	52.5	57.0
	(n = 46)	(n = 40)	

$x^2 = .833$ df = 2   p = .659.

Best heterosexual friend knows or suspects

	Men	Women	Combined
Definitely knows	72.9%	95.0%	83.0%
Suspects	10.4	5.0	8.0
Doesn't know or suspect	16.7	0	9.1
	(n = 48)	(n = 40)	

Collapsed bottom two rows. $x^2 = 6.044$ df = 1   p = .014.

Most heterosexual friends know or suspect

	Men	Women	Combined
Definitely know	52.1%	65.9%	58.4%
Suspect	18.8	22.0	20.2
Don't know or suspect	29.2	12.2	21.3
	(n = 48)	(n = 41)	

$x^2 = 3.965$ df = 2   p = .138.

Best homosexual friend knows or suspects

	Men	Women	Combined
Definitely knows	90.7%	95.1%	92.9%
Suspects	4.7	4.9	4.8
Doesn't know or suspect	4.7	0	2.4
	(n = 43)	(n = 41)	

Collapsed bottom two rows. $x^2 = .132$ df = 1   p = .716.

Most homosexual friends know or suspect

	Men	Women	Combined
Definitely know	72.1%	71.4%	71.8%
Suspect	25.6	21.4	23.5
Don't know or suspect	2.3	7.1	4.7
	(n = 43)	(n = 42)	

Collapsed bottom two rows. $x^2 = .000$ df = 1   p = 1.000.

Boss knows or suspects

	Men	Women	Combined
Definitely knows	38.2%	48.6%	43.7%
Suspects	17.6	13.5	15.5
Doesn't know or suspect	44.1	37.8	40.8
	(n = 34)	(n = 37)	

x^2 = .809 df = 2 p = .667.

Work associates know or suspect

	Men	Women	Combined
Definitely know	41.0%	56.4%	48.7%
Suspect	17.9	20.5	19.2
Don't know or suspect	41.0	23.1	32.1
	(n = 39)	(n = 39)	

x^2 = 3.005 df = 2 p = .223.

Neighbors know or suspect

	Men	Women	Combined
Definitely know	27.5%	18.9%	23.4%
Suspect	15.0	37.8	26.0
Don't know or suspect	57.5	43.2	50.6
	(n = 40)	(n = 37)	

x^2 = 5.334 df = 2 p = .069.

Table 11.3   Putative and Anticipated Reactions

Sister's reaction

	Men	Women	Combined
Rejecting	5.4%	3.2%	4.4%
Intolerant	10.8	12.9	11.8
Tolerant	27.0	16.1	22.1
Understanding	24.3	22.6	23.5
Accepting	32.4	45.2	38.2
	(n = 37)	(n = 31)	

Collapsed top two rows and bottom three rows. x^2 = .000   df = 1   p = 1.000.

Brother's reaction

	Men	Women	Combined
Rejecting	13.9%	3.8%	9.7%
Intolerant	25.0	23.1	24.2
Tolerant	36.1	19.2	29.0
Understanding	11.1	11.5	11.3
Accepting	13.9	42.3	25.8
	(n = 36)	(n = 26)	

Collapsed top two rows and bottom three rows. x^2 = .505   df = 1   p = .477.

Mother's reaction

	Men	Women	Combined
Rejecting	6.5%	14.3%	10.2%
Intolerant	26.1	21.4	23.9
Tolerant	34.8	28.6	31.8
Understanding	13.0	21.4	17.0
Accepting	19.6	14.3	17.0
	(n = 46)	(n = 42)	

$x^2 = 3.049$ df = 4 p = .550.

Father's reaction

	Men	Women	Combined
Rejecting	13.9%	22.9%	18.3%
Intolerant	38.9	25.7	32.4
Tolerant	36.1	34.3	35.2
Understanding	8.3	2.9	5.6
Accepting	2.8	14.3	8.5
	(n = 36)	(n = 35)	

Collapsed top two rows and bottom three rows. $x^2 = .014$ df = 1 p = .907.

Most other relatives' reaction

	Men	Women	Combined
Rejecting	13.3%	19.5%	16.3%
Intolerant	31.1	39.0	34.9
Tolerant	37.8	22.0	30.2
Understanding	8.9	7.3	8.1
Accepting	8.9	12.2	10.5
	(n = 45)	(n = 41)	

Collapsed top two rows and bottom three rows. $x^2 = 1.188$ df = 1 p = .276.

Best heterosexual friend's reaction

	Men	Women	Combined
Rejecting	2.1%	0	1.1%
Intolerant	4.2	0	2.3
Tolerant	20.8	20.0%	20.5
Understanding	10.4	7.5	9.1
Accepting	62.5	72.5	67.0
	(n = 48)	(n = 40)	

Collapsed top four rows. $x^2 = .587$ df = 1 p = .444.

Most heterosexual friends' reactions

	Men	Women	Combined
Rejecting	6.3%	2.4%	4.5%
Intolerant	8.3	7.3	7.9
Tolerant	33.3	24.4	29.2
Understanding	22.9	19.5	21.3
Accepting	29.2	46.3	37.1
	(n = 48)	(n = 41)	

Collapsed top four rows. $x^2 = 2.108$ df = 1   p = .147.

Best homosexual friend's reaction

	Men	Women	Combined
Rejecting	4.7%	0	2.4%
Intolerant	4.7	0	2.4
Tolerant	14.0	4.9%	9.5
Understanding	7.0	14.6	10.7
Accepting	69.8	80.5	75.0
	(n = 43)	(n = 41)	

Collapsed top four rows. $x^2 = .778$ df = 1   p = .378.

Most other homosexual friends' reactions

	Men	Women	Combined
Rejecting	0	0	0
Intolerant	14.0%	7.3%	10.7%
Tolerant	14.0	14.6	14.3
Understanding	18.6	24.4	21.4
Accepting	53.5	53.7	53.6
	(n = 43)	(n = 41)	

Collapsed top four rows. $x^2 = .000$ df = 1   p = 1.000.

Employer's Reaction

	Men	Women	Combined
Rejecting	20.6%	13.9%	17.1%
Intolerant	2.9	11.1	7.1
Tolerant	23.5	22.2	22.9
Understanding	11.8	11.1	11.4
Accepting	41.2	41.7	41.4
	(n = 34)	(n = 36)	

Collapsed top four rows. $x^2 = .000$ df = 1   p = 1.000.

Most work associates' reactions

	Men	Women	Combined
Rejecting	10.3%	5.1%	7.7%
Intolerant	10.3	15.4	12.8
Tolerant	33.3	28.2	30.8
Understanding	2.6	23.1	12.8
Accepting	43.6	28.2	35.9
	(n = 39)	(n = 39)	

Collapsed top four rows. $x^2 = 1.393$ df = 1 p = .238.

Most neighbors' reactions

	Men	Women	Combined
Rejecting	30.0%	34.3%	32.0%
Intolerant	17.5	25.7	21.3
Tolerant	32.5	34.3	33.3
Understanding	5.0	0	2.7
Accepting	15.0	5.7	10.7
	(n = 40)	(n = 35)	

Collapsed top two rows and bottom three rows. $x^2 = .723$ df = 1 p = .395.

Heterosexuals' reactions (in general)

	Men	Women	Combined
Rejecting	20.4%	7.1%	14.3%
Intolerant	28.6	23.8	26.4
Tolerant	40.8	50.0	45.1
Understanding	4.1	16.7	9.9
Accepting	6.1	2.4	4.4
	(n = 49)	(n = 42)	

Collapsed top two rows and bottom three rows. $x^2 = 2.345$ df = 1 p = .126.

Homosexuals' reactions (in general)

	Men	Women	Combined
Rejecting	2.0%	2.4%	2.2%
Intolerant	20.4	16.7	18.7
Tolerant	32.7	42.9	37.4
Understanding	24.5	23.8	24.2
Accepting	20.4	14.3	17.6
	(n = 49)	(n = 42)	

$x^2 = .019$ df = 1 p = .889.

Table 11.4   Discretion Versus Disclosure

Would like someone to know who does not

	Men	Women	Combined
Yes	46.9%	43.2%	45.2%
No	53.1	56.8	54.8
	(n = 49)	(n = 44)	

$x^2 = .024$ df = 1   p = .877.

Wishes someone who knows did not

	Men	Women	Combined
Yes	16.3%	25.0%	20.4%
No	83.7	75.0	79.6
	(n = 49)	(n = 44)	

$x^2 = .606$ df = 1   p = .436.

Table 11.5   Regrets

Regrets being bisexual

	Men	Women	Combined
Not at all	67.3%	79.5%	73.1%
A little	22.4	9.1	16.1
Somewhat	10.2	11.4	10.8
A lot	0	0	0
	(n = 49)	(n = 44)	

$x^2 = 3.187$ df = 2   p = .203.

Appendix B: Tables for the 1984–85 Mailed Questionnaire Study

The following abbreviations are used in the tables for the mailed questionnaire study:

BIM   bisexual men

BIW   bisexual women

HTM   heterosexual men

HTW   heterosexual women

HM   homosexual men

HW   homosexual women

OS   opposite-sex partners

SS   same-sex partners

As our primary interest in this part of the study is the difference between sexual preference groups, when there is a significant difference between men and women, we only provide the probability level. We do this to avoid excessive detail. In Chapters 13 and 14, however, we provide the relevant F, t, and chi-square values. Also, for simplicity, we sometimes provide only one row of percentages for a complete table; i.e., the residual row is omitted.

CHAPTER 12: SURVEYING THE SEXUAL UNDERGROUND

Table 12.1   Demographic and Background Data, 1984 and 1985 Questionnaire Sample

Age

	Men			Combined
	Heterosexual	Bisexual	Homosexual	
18-24	2.4%	5.2%	7.6%	5.6%
25-29	4.8	5.2	12.9	8.8
30-34	10.8	19.8	19.4	17.5
35-39	25.0	19.8	19.4	20.7
40-44	15.5	14.7	13.4	14.2
45-49	8.3	10.3	10.8	10.1
50+	33.3	25.0	16.7	22.8
	(n = 84)	(n = 116)	(n = 186)	

$x^2 = 21.411$ df = 12   p = .046.

	Women			Combined
	Heterosexual	Bisexual	Homosexual	
20-24	3.8%	2.1%	8.5%	4.8%
25-29	7.7	26.0	22.3	18.4
30-34	20.2	22.9	36.2	26.2
35-39	27.9	21.9	16.0	22.1
40-44	10.6	14.6	11.7	12.2
45-49	12.5	6.3	4.3	7.8
50+	17.3	6.3	1.1	8.5
	(n=104)	(n=96)	(n=94)	

$x^2 = 44.855$ df = 12   p = .000.
BIM vs. BIW: p = .000;   HM vs. HW: p = .000.

Education

	Men			
	Heterosexual	Bisexual	Homosexual	Combined
9-11 yrs	0	0	0.5%	0.3%
High school graduate	2.4%	0	4.4	2.7
Some college	10.6	19.3%	23.6	19.4
College graduate	38.8	33.9	28.6	32.4
Masters	20.0	25.7	23.6	23.4
Professional degree (MD, PhD, LLD, EdD)	28.2	21.1	19.2	21.8
	(n = 85)	(n = 109)	(n = 182)	

Collapsed top three rows. x^2 = 11.821   df = 6   p = .070.

	Women			
	Heterosexual	Bisexual	Homosexual	Combined
9-11 yrs.	1.0%	0	0	0.3%
High school graduate	2.9	0	1.1%	1.4
Some college	11.8	28.0%	18.1	19.0
College graduate	33.3	36.6	34.0	34.6
Masters	27.5	24.7	34.0	28.7
Professional degree (MD, PhD, LLD, EdD)	23.5	10.8	12.8	15.9
	(n = 102)	(n = 93)	(n = 94)	

Collapsed top three rows. x^2 = 11.200   df = 6   p = .086.

Income level

	Men			
	Heterosexual	Bisexual	Homosexual	Combined
< $5,000	2.5%	5.8%	3.3%	3.8%
5,000-9,999	2.5	9.6	8.3	7.4
10,000-14,999	7.5	5.8	12.2	9.3
15,000-19,999	10.0	11.5	11.6	11.2
20,000-24,999	5.0	14.4	17.7	14.0
25,000-29,999	10.0	11.5	8.8	9.9
30,000-39,999	20.0	11.5	21.0	18.1
40,000-49,999	13.8	12.5	5.5	9.3
50,000-59,999	6.3	3.8	4.4	4.7
60,000-69,999	8.8	2.9	3.3	4.4
70,000+	13.8	10.6	3.9	7.9
	(n = 80)	(n = 104)	(n = 181)	

$x^2 = 37.520$ df = 20 p = .010.

	Women			
	Heterosexual	Bisexual	Homosexual	Combined
< $5,000	7.4%	11.7%	5.7%	8.3%
5,000-9,999	2.1	11.7	11.4	8.3
10,000-14,999	10.6	18.1	18.2	15.6
15,000-19,999	8.5	16.0	15.9	13.4
20,000,24,999	20.2	11.7	25.0	18.8
25,000-29,999	16.0	14.9	8.0	13.0
30,000-39,999	22.3	10.6	11.4	14.9
40,000-49,999	6.4	2.1	2.3	3.6
50+	6.4	3.2	2.3	4.0
	(n = 94)	(n = 94)	(n = 88)	

$x^2 = 31.063$ df = 16 p = .014.
HTM vs. HTW: p = .000; BIM vs. BIW: p = .000; HM vs. HW: p = .045.

Race/Ethnicity

	Men			
	Heterosexual	Bisexual	Homosexual	Combined
White	95.3%	97.4%	90.8%	93.8%
Black	1.2	0.9	1.1	1.0
Asian	0	0.9	2.7	1.6
Hispanic	1.2	0	3.8	2.1
Other	2.4	0.9	1.6	1.6
	(n=85)	(n=115)	(n=185)	

Doesn't meet minimal expected frequencies for a x^2 test.

	Women			
	Heterosexual	Bisexual	Homosexual	Combined
White	92.4%	93.8%	90.4%	92.2%
Black	1.0	0	3.2	1.4
Asian	1.9	0	0	0.7
Hispanic	1.0	2.1	5.3	2.7
Other	3.8	4.2	1.1	3.1
	(n = 105)	(n = 96)	(n = 94)	

Collapsed bottom four rows. x^2 = .735   df = 2   p = .693.

Religious Background

	Men			
	Heterosexual	Bisexual	Homosexual	Combined
Catholic	25.0%	17.9%	23.1%	22.0%
Protestant	47.6	38.4	50.0	46.0
Jewish	13.1	22.3	10.4	14.6
Other	14.3	21.4	16.5	17.5
	(n = 84)	(n = 112)	(n = 182)	

x^2 = 11.883   df = 6   p = .065.

	Women			
	Heterosexual	Bisexual	Homosexual	Combined
Catholic	15.8%	23.7%	29.3%	22.7%
Protestant	43.6	28.0	30.4	34.3
Jewish	25.7	26.9	29.3	27.3
Other	14.9	21.5	10.9	15.7
	(n = 101)	(n = 93)	(n = 92)	

x^2 = 11.579   df = 6   p = .072.
HM vs. HW: p = .000.

Religiosity

	Men			
	Heterosexual	Bisexual	Homosexual	Combined
Not at all	65.9%	61.2%	56.6%	60.1%
A little	15.3	19.0	20.9	19.1
Moderate	8.2	9.5	11.0	9.9
Strong	5.9	6.0	5.5	5.7
Very strong	4.7	4.3	6.0	5.2
	(n = 85)	(n = 116)	(n = 182)	

x^2 = 2.797   df = 8   p = .947.

	Women			
	Heterosexual	Bisexual	Homosexual	Combined
Not at all	53.3%	60.4%	61.7%	58.3%
A little	21.9	17.7	18.1	19.3
Moderate	16.2	13.5	16.0	15.3
Strong	2.9	6.3	4.3	4.4
Very strong	5.7	2.1	0	2.7
	(n = 105)	(n = 96)	(n = 94)	

$x^2 = 3.149$ df = 6   p = .790.

Acceptance of feminist ideology

	Men			
	Heterosexual	Bisexual	Homosexual	Combine
Not at all	15.9%	8.6%	14.5%	12.9%
A little	14.5	14.3	11.9	13.2
Moderate	37.7	34.3	34.6	35.1
Strong	31.9	34.3	34.0	33.6
Radical	0	8.6	5.0	5.1
	(n = 69)	(n = 105)	(n = 159)	

Collapsed bottom two rows. $x^2 = 4.161$ df = 6   p = .656.

	Women			
	Heterosexual	Bisexual	Homosexual	Combined
Not at all	3.9%	2.1%	0	2.1%
A little	17.5	7.4	3.2%	9.6
Moderate	38.8	26.3	19.1	28.4
Strong	38.8	51.6	60.6	50.0
Radical	1.0	12.6	17.0	9.9
	(n=103)	(n=95)	(n=94)	

Collapsed top two rows. $x^2 = 39.951$ df = 6   p = .000.
HTM vs. HTW: p = .049; BIM vs. BIW: p=.025; HM vs. HW: p=.000.

CHAPTER 13: THE DEVELOPMENT OF SEXUAL PREFERENCE

Table 13.1   Age of First Attractions, Behaviors, Self-Labeling, and Coming Out

Men

First heterosexual attraction	Heterosexual	Bisexual	Homosexual
Mean	10.2	12.8	14.5
S.D.	4.5	5.3	6.6
	(n = 82)	(n = 107)	(n = 52)

BIM vs. HTM vs. HM: F = 11.47 p = .000.
BIM vs. HTM: t = 3.55 p = .000.

First heterosexual experience			
Mean	16.7	15.9	17.7
S.D.	6.3	5.6	6.5
	(n = 84)	(n = 113)	(n = 108)

BIM vs. HTM vs. HM: F = 2.38 p = .094.
BIM vs. HTM: t = .93 p = .353.

First homosexual attraction			
Mean	21.9	17.1	11.5
S.D.	13.9	10.6	5.8
	(n = 25)	(n = 111)	(n = 182)

BIM vs. HM vs. HTM: F = 25.82 p = .000.
BIM vs. HM: t = 5.15 p = .000.

First homosexual experience			
Mean	17.7	17.2	14.7
S.D.	11.9	10.3	8.2
	(n = 40)	(n = 114)	(n = 184)

BIM vs. HM vs. HTM: F = 3.21 p = .042.
BIM vs. HM: t = 2.19 p = .030.

Labeled themselves bisexual/homosexual			
Mean		29.0	21.1
S.D.		10.9	9.4
		(n = 115)	(n = 182)

BIM vs. HM: t = .659 p = .000.

First "came out"			
Mean		29.2	23.6
S.D.		10.0	9.2
		(n = 94)	(n = 164)

BIM vs. HM: t = 17.041 p = .000.

Women

First heterosexual attraction	Heterosexual	Bisexual	Homosexual
Mean	10.4	10.9	14.3
S.D.	5.0	4.8	4.1
	(n = 103)	(n = 82)	(n = 54)

BIW vs. HTW vs. HW: F = 12.83   p = .000.
BIW vs. HTW: t = .75   p = .452.

First heterosexual experience			
Mean	14.9	15.1	16.4
S.D.	4.6	3.7	3.3
	(n = 97)	(n = 89)	(n = 73)

BIW vs. HTW vs. HW: F = 3.88   p = .022.
BIW vs. HTW: t = .31   p = .755.

First homosexual attraction			
Mean	23.6	18.5	16.4
S.D.	10.9	9.1	7.3
	(n = 54)	(n = 96)	(n = 94)

BIW vs. HW vs. HTW: F = 11.24 p =.000.
BIW vs. HW: t = 1.78   p = .077.

First homosexual experience			
Mean	23.0	23.5	20.5
S.D.	10.5	8.7	6.8
	(n = 48)	(n = 96)	(n = 94)

BIW vs. HW vs. HTW: F = 3.26   p = .040.
BIW vs. HW: t = 2.63   p = .009

Labeled themselves bisexual/homosexual			
Mean		27.0	22.5
S.D.		8.6	5.8
		(n = 94)	(n = 91)

BIW vs. HW: t = 4.21   p = .000.

First "came out"			
Mean		26.7	23.0
S.D.		7.8	5.7
		(n = 81)	(n = 86)

BIW vs. HW: t = 8.70   p = .000.

Table 13.2   Comparisons between Men and Women

		t	p
Heterosexual			
	First heterosexual attraction	.31	.760
	First heterosexual experience	2.24	.027
	"Come out" (unconventional heterosexuals)	3.081	.007
Homosexual			
	First homosexual attraction	5.70	.000
	First homosexual experience	6.23	.000
	Label self	1.46	.145
	"Come out"	.320	.572
Bisexual			
	First heterosexual attraction	2.44	.016
	First heterosexual experience	1.31	.192
	First homosexual attraction	1.03	.304
	First homosexual experience	4.69	.000
	Label self	1.48	.141
	"Come out"	3.36	.002

Table 13.3   Ever Felt Confused About Their Sexual Identity

Men[a]			Women[b]		
Heterosexual	Bisexual	Homosexual	Heterosexual	Bisexual	Homosexual
27.2%	68.7%	63.8%	28.1%	76.0%	67.0%
(n = 81)	(n = 115)	(n = 185)	(n = 96)	(n = 96)	(n = 94)

[a]x^2 = 39.393   df = 2   p = .000,   [b]x^2 = 50.707   df = 2   p = .000.

Table 13.4   Percent Who Believe They Are Not "In Transition" on Various Dimensions

Men

	Heterosexual	Bisexual	Homosexual
Sexual feelings	97.6%	84.3%	94.1%
	(n = 84)	(n = 115)	(n = 186)

x^2 = 13.769   df = 2   p = .001.

	Heterosexual	Bisexual	Homosexual
Sexual behaviors	97.6%	78.9%	93.5%
	(n = 84)	(n = 114)	(n = 184)

x^2 = 23.473   df = 2   p = .000.

	Heterosexual	Bisexual	Homosexual
Romantic feelings	96.4%	82.0%	94.5%
	(n = 84)	(n = 111)	(n = 183)

x^2 = 17.272   df = 2   p = .000.

	Heterosexual	Bisexual	Homosexual
Self-definition	95.2%	85.8%	91.8%
	(n = 84)	(n = 113)	(n = 182)

x^2 = 5.488   df = 2   p = .064.

Women

	Heterosexual	Bisexual	Homosexual
Sexual feelings	98.1%	82.6%	97.8%
	(n = 103)	(n = 92)	(n = 92)

$x^2 = 22.693$ df = 2 p = .000.

Sexual behaviors	99.0%	74.2%	95.6%
	(n = 102)	(n = 93)	(n = 91)

$x^2 = 37.738$ df = 2 p = .000.

Romantic feelings	96.1%	74.2%	97.8%
	(n = 102)	(n = 93)	(n = 91)

$x^2 = 34.587$ df = 2 p = .000.

Self-definition	94.2%	79.6%	95.6%
	(n = 103)	(n = 93)	(n = 91)

$x^2 = 16.257$ df = 2 p = .000.

Table 13.5   Attempts to Change Sexual Preference

Men

	Heterosexual	Bisexual	Homosexual
Has tried to change sexual preference	7.4%	27.8%	32.8%
	(n = 81)	(n = 115)	(n = 186)

$x^2 = 19.256$ df = 2 p = .000.

Direction of desired change			
Heterosexual	0	58.1%	75.4%
Bisexual	83.3%	38.7	13.1
Homosexual	16.7	3.2	11.5
	(n = 6)	(n = 31)	(n = 61)

$x^2 = 20.688$ df = 4 p = .000.

Sexual preference actually changed	0	46.7%	18.0%
	(n = 6)	(n = 30)	(n = 61)

$x^2 = 10.839$ df = 2 p = .004.

Women

	Heterosexual	Bisexual	Homosexual
Has tried to change sexual preference	6.2%	31.3%	39.4%
	(n = 97)	(n = 96)	(n = 94)

$x^2 = 30.275$ df = 2 p = .000.

Direction of desired change			
Heterosexual	0	43.3%	30.6%
Bisexual	66.7%	23.3	8.3
Homosexual	33.3	33.3	61.1
	(n = 6)	(n = 30)	(n = 36)

$x^2 = 15.324$ df = 4 p = .004.
BIM vs. BIW: p = .004; HM vs. HW: p = .000.

Sexual preference actually changed	33.3%	44.8%	58.3%
	(n = 6)	(n = 29)	(n = 36)

$x^2 = 1.963$ df = 2 p = .375.
HM vs. HW: p = .000

CHAPTER 14: DIMENSIONS OF SEXUAL PREFERENCE

Table 14.1   Kinsey Scale Scores for Heterosexuals, Homosexuals, and Bisexuals

Sexual feelings

	Men			Women		
	Heterosexuals	Bisexual	Homosexual	Heterosexual	Bisexual	Homosexual
0 Exclusively heterosexual	67.8%	1.7%	0	52.0%	1.1%	0
1 Mainly heterosexual & small degree homosexual	29.8	17.2	0	44.0	16.8	0
2 Mainly heterosexual & significant degree homosexual	2.4	23.3	0	3.0	30.5	1.0%
3 Equally hetero- sexual & homosexual	0	19.8	0	0	29.5	0
4 Mainly homosexual & significant degree heterosexual	0	27.6	2.4%	1.0	12.6	3.0
5 Mainly homosexual & small degree heterosexual	0	8.6	29.8	0	9.5	44.0
6 Exclusively homosexual	0	1.7	67.9	0	0	52.0
	(n = 84)	(n = 116)	(n = 182)	(n = 100)	(n = 95)	(n = 93)
Mean	.35	2.87	5.60	.54	2.64	5.29

HTM $\bar{X}$ vs. HTW $\bar{X}$: t = 2.23   p = .027;   HM $\bar{X}$ vs. HW $\bar{X}$: t = 3.03   p = .003;
BIM $\bar{X}$ vs. BIW $\bar{X}$: t = 1.27   p = .204.

Sexual behaviors

	Men			Women		
	Heterosexuals	Bisexual	Homosexual	Heterosexual	Bisexual	Homosexual
0 Exclusively heterosexual	91.3%	11.0%	0	88.3%	19.3%	0
1 Mainly heterosexual & small degree homosexual	8.8	24.8	0	11.7	37.3	0
2 Mainly heterosexual & significant degree homosexual	0	20.2	0	0	18.1	0
3 Equally hetero-sexual & homosexual	0	12.8	0	0	10.8	0
4 Mainly homosexual & significant degree heterosexual	0	9.2	0	0	4.8	0
5 Mainly homosexual & small degree heterosexual	0	13.8	8.8%	0	4.8	11.9%
6 Exclusively homosexual	0	8.3	91.3	0	4.8	88.1
	(n = 80)	(n = 109)	(n = 168)	(n = 94)	(n = 83)	(n = 84)
Mean	.09	2.59	5.85	.12	1.78	5.85

HTM $\bar{X}$ vs. HTW $\bar{X}$: t = .63 p = .527; HM $\bar{X}$ vs. HW $\bar{X}$: t = .00 p = 1.000;
BIM $\bar{X}$ vs. BIW $\bar{X}$: t = 3.14 p = .002.

Romantic feelings

		Men			Women	
	Heterosexuals	Bisexual	Homosexual	Heterosexual	Bisexual	Homosexual
0 Exclusively heterosexual	84.0%	23.4%	0	76.0%	14.9%	0
1 Mainly heterosexual & small degree homosexual	13.6	17.1	0	20.8	21.3	0
2 Mainly heterosexual & significant degree homosexual	2.5	16.2	0	2.1	23.4	0
3 Equally hetero-sexual & homosexual	0	19.8	0	1.0	19.1	1.0%
4 Mainly homosexual & significant degree heterosexual	0	9.9	2.5%	0	11.7	2.1
5 Mainly homosexual & small degree heterosexual	0	9.0	13.6	0	5.3	20.8
6 Exclusively homosexual	0	4.5	84.0	0	4.3	76.0
	(n = 81)	(n = 111)	(n = 181)	(n = 96)	(n = 94)	(n = 92)
Mean	.19	2.21	5.54	.28	2.24	5.70

HTM $\bar{X}$ vs. HTW $\bar{X}$: t = 1.27 p = .206; HM $\bar{X}$ vs. HW $\bar{X}$: t = 1.63 p = .104;
BIM $\bar{X}$ vs. BIW $\bar{X}$: t = .16 p = .876.

Table 14.2   Composite Kinsey Scale Profiles

Homosexuals and heterosexuals

	Men		Women	
	Heterosexual	Homosexual	Heterosexual	Homosexual
Pure heterosexual (000)	65.0%	0	47.9%	0
Somewhat mixed heterosexual (100 and beyond)	35.0	0	52.1	0
Somewhat mixed homosexual (566 and beyond)	0	41.7%	0	54.8%
Pure homosexual (666)	0	58.3	0	45.2
	(n = 80)	(n = 168)	(n = 94)	(n = 84)

HTM vs. HTW: p = .218;   HM vs. HW: p = .347.

Bisexuals

	Bisexual men	Bisexual women
Heterosexual leaning bisexual(0-2 all dimensions or average ≤ 2)	43.1%	52.4%
Mid bisexual (average between 2 and 4)	21.6	17.9
Pure bisexual (333)	6.9	3.6
Homosexual leaning bisexual (4-6 all dimensions or average ≥ 4)	17.6	13.1
Varied bisexual (2 dimensions at least 3 points apart)	10.8	13.1
	(n = 102)	(n = 84)

BIM vs. BIW: p = .578.

Table 14.3   Overlap Between Composite Types

Somewhat mixed heterosexual men

Combination[a]	n
*100	12
*010	1
*110	1
*210	1
001	1
*101	6
011	1
*111	3
102	1
202	1
Total	28

Heterosexual-leaning bisexual men

Combination[a]	n
*100	5
*010	2
*110	5
200	3
*210	7
220	2
*101	2
*111	5
211	3
221	5
212	2
222	2
321	1
Total	44

Overlap: 85.7% of the "somewhat mixed heterosexual men" shared the same profile as the "heterosexual leaning bisexual men."

Overlap: 59.1% of the "heterosexual leaning bisexual men" shared the same profile as the "somewhat mixed heterosexual men."

Asterisk denotes self-defined bisexual and heterosexual men who shared the same combined Kinsey scale profiles.

Somewhat mixed heterosexual women

Combination[b]	n
*100	21
*110	5
001	3
*101	10
*111	5
201	2
102	1
*112	1
203	1
Total	49

Heterosexual leaning bisexual women

Combination[b]	n
000	1
*100	7
*110	1
200	2
210	1
*101	1
*111	4
121	1
211	5
221	1
*112	1
212	7
222	5
231	1
311	2
321	1
312	3
Total	44

Overlap: 86.7% of the "somewhat mixed heterosexual women" shared the same profile as the "heterosexual leaning bisexual women."

Overlap: 31.8% of the "heterosexual leaning bisexual women" shared the same profile as the "somewhat mixed heterosexual women."

[b] Asterisk denotes self-defined bisexual and heterosexual women who shared the same combined Kinsey scale profiles.

Somewhat mixed homosexual men

Combination[e]	n
*566	20
*656	7
*655	0
*565	13
*555	8
664	4
564	4
554	3
504	1
*464	1
*454	1
*444	2
*354	1
563	1
543	1
*453	1
662	1
661	1
Total	70

Homosexual leaning bisexual men

Combination[e]	n
*444	3
*454	1
*464	1
*554	1
445	1
455	5
*555	1
*565	1
*655	1
456	1
*566	2
*656	1
*453	1
*354	1
435	1
Total	22

Overlap: 81.4% of the "somewhat mixed homosexual men" shared the same profile as the "homosexual leaning bisexual men."

Overlap: 63.6% of the "homosexual leaning bisexual men" shared the same profile as the "somewhat mixed homosexual men."

[e] Asterisk denotes self-defined bisexual and homosexual men who shared the same combined Kinsey scale profiles.

Somewhat mixed homosexual women

Combination[d]	n
656	1
*566	22
556	2
*466	3
456	1
*565	8
*555	4
455	1
365	1
364	1
324	1
363	1
Total	46

Homosexual leaning bisexual women

Combination[d]	n
444	2
464	1
554	1
*555	3
*565	1
446	1
*466	1
*566	1
Total	11

Overlap: 80.4% of the "somewhat mixed homosexual women" shared the same profile as the "homosexual learning bisexual women."

Overlap: 54.5% of the "homosexual leaning bisexual women" shared the same profile as the "somewhat mixed homosexual women."

[d] Asterisk denotes self-defined bisexual and homosexual women who shared the same combined Kinsey scale profiles.

CHAPTER 15: THE INSTABILITY OF SEXUAL PREFERENCE

Table 15.1   Short-run Three Year Change on the Kinsey Scale for Heterosexuals, Homosexuals, and Bisexuals

Sexual feelings

	Men[a]			Women[b]		
	Heterosexual	Bisexual	Homosexual	Heterosexual	Bisexual	Homosexual
2 or more scale points heterosexual direction	1.2%	6.9%	0	2.0%	6.7%	2.2%
1 scale point heterosexual direction	7.4	10.3	8.4%	11.3	16.7	11.0
No change	82.7	54.3	81.6	80.4	51.1	74.7
1 scale point homosexual direction	7.4	21.6	6.7	6.2	15.6	4.4
2 or more scale points homosexual direction	1.2	7.0	3.4	0	10.0	7.7
	(n = 81)	(n = 116)	(n = 179)	(n = 97)	(n = 90)	(n = 91)

Collapsed table: "change" vs. "no change." [a]x^2 = 29.929 df = 2 p = .000;
[b]x^2 = 20.360 df = 2 p = .000.

Sexual behaviors

	Men[a]			Women[b]		
	Heterosexual	Bisexual	Homosexual	Heterosexual	Bisexual	Homosexual
2 or more scale points heterosexual direction	1.3%	17.0%	0	1.1%	14.3%	0
1 scale point heterosexual direction	5.3	13.2	1.9%	14.8	16.9	7.6%
No change	88.0	38.7	85.0	83.0	40.3	70.9
1 scale point homosexual direction	5.3	17.9	8.1	1.1	20.8	12.7
2 or more scale points homosexual direction	0	13.2	5.0	0	7.8	8.9
	(n = 75)	(n = 106)	(n = 160)	(n = 88)	(n = 77)	(n = 79)

Collapsed table: "change" vs. "no change." [a]x^2 = 77.291 df = 2 p = .000;
[b]x^2 = 34.718 df = 2 p = .000. HM vs. HW: p = .016.

Romantic feelings

	Men[a]			Women[b]		
	Heterosexual	Bisexual	Homosexual	Heterosexual	Bisexual	Homosexual
2 or more scale points heterosexual direction	0	12.6%	0	2.2%	14.4%	1.1%
1 scale point heterosexual direction	6.3%	6.3	6.3%	6.5	20.0	5.6
No change	85.9	56.8	82.4	88.0	38.9	75.6
1 scale point homosexual direction	7.7	14.4	5.7	3.3	14.4	11.1
2 or more scale points homosexual direction	0	9.9	5.7	0	12.2	6.6
	(n = 78)	(n = 111)	(n = 176)	(n = 92)	(n = 90)	(n = 90)

Collapsed table: "change" vs. "no change." [a]x^2 = 28.575 df = 2 p = .000; [b]x^2 = 54.702 df = 2 p = .000. BIM vs. BIW: p = .017.

Table 15.2   Major Change in Sexual Feelings on the Kinsey Scale

Ever major change

	Men[a]			Women[b]		
	Heterosexual	Bisexual	Homosexual	Heterosexual	Bisexual	Homosexual
Major change	15.3%	56.0%	31.9%	18.6%	61.7%	66.3%
	(n = 85)	(n = 116)	(n = 182)	(n = 102)	(n = 94)	(n = 92)

[a]x^2 = 37.560 df = 2 p = .000; [b]x^2 = 54.682 df = 2 p = .000. HM vs HW: p = .000.

Magnitude and direction of change on Kinsey scale

	Men			Women		
	Heterosexual	Bisexual	Homosexual	Heterosexual	Bisexual	Homosexual
3 or more scale points heterosexual direction	3.5%	.9%	0	0	4.3%	0
2 scale points heterosexual direction	3.5	6.9	0	0	5.3	1.1%
2 scale points homosexual direction	3.5	32.8	22.5%	4.9%	46.8	54.3
3 or more scale points homosexual direction	1.2	17.2	15.9	3.9	30.9	50.0
	(n = 85)	(n = 116)	(n = 182)	(n = 102)	(n = 94)	(n = 92)

N.B.: x^2 is not calculated because cells are not mutually exclusive and data are based on partial n's.

Table 15.3   Age at Time of Major Change in Sexual Feelings

	Men[a]			Women[b]		
	Heterosexual	Bisexual	Homosexual	Heterosexual	Bisexual	Homosexual
≤ to 20 years	25.0	3.2	24.1	5.3	13.8	17.2
21 to 29 years	16.7	27.0	31.5	36.8	39.7	60.3
30 to 39 years	33.3	38.1	29.6	36.8	37.9	22.4
40 or more years	25.0	31.7	14.8	21.1	8.6	0
	(n=12)	(n=63)	(n=54)	(n=19)	(n=58)	(n=58)

[a]x^2 = 14.315 df = 4 p = .008; [b]x^2 = 11.466 df = 4 p = .023. HM vs HW: p = .009;
BIM vs. BIW: p = .006.

CHAPTER 16: SEXUAL PROFILES

Table 16.1   Sexual Partners

Median number (and incidence) of partners in past twelve months

Men

	Heterosexuals	Bisexuals	Homosexuals
Opposite-sex	2 (79.8%)	1 (81.7%)[a]	0 (9.9%)
Same-sex	0 (6.0)	3 (85.2)[b]	6 (93.0)
Total	2	5[c]	6
	(n = 83/84)	(n = 114/115)	(n = 181/185)

Median tests.

[a] BIM vs. HTM: x^2 = .0417 df = 1 p = .838.
[b] BIM vs. HM: x^2 = 7.7612 df = 1 p = .006.
[c] BIM vs. HTM: x^2 = 25.127 df = 1 p = .000; BIM vs. HM: x^2 = .003 df = 1 p = .959.

Women

	Heterosexuals	Bisexuals	Homosexuals
Opposite-sex	2 (78.6%)	2 (84.4%)[a]	0 (12.9%)
Same-sex	0 (9.8)	1 (66.3)[b]	1 (87.1)
Total	2	3[c]	1
	(n = 102/103)	(n = 95/96)	(n = 92/93)

Median tests.

[a] BIW vs. HTW: x^2 = .013 df = 1 p = .910.
[b] BIW vs. HW: x^2 = .745 df = 1 p = .388. BIM vs. BIW: p = .000; HM vs. HW: p = .000.
[c] BIW vs. HTW: x^2 = 9.287 df = 1 p = .002; BIW vs. HW: x^2 = 17.692 df = 1 p = .000;
BIM vs. BIW: p = .028; HM vs. HW: p = .000.

Median number (and incidence) of partners in lifetime

Men

	Heterosexuals	Bisexuals	Homosexuals
Opposite-sex	20 (100%)	20 (100%)[a]	2 (67.4%)
Same-sex	0 (48.8)	30 (97.4)[b]	100 (100.0)
Total	20	65[c]	102
	(n = 82/85)	(n = 114)	(n = 178/185)

Median tests.

[a] BIM vs. HTM: x^2 = .003 df = 1 p = .959.
[b] BIM vs. HM: x^2 = 13.292 df = 1 p = .000.
[c] BIM vs. HTM: x^2 = 27.423 df = 1 p = .000; BIM vs. HM: x^2 = 3.093 df = 1 p = .079.

Women

	Heterosexuals	Bisexuals	Homosexuals
Opposite-sex	25 (100%)	30 (100%)[a]	10 (93.6%)
Same-sex	1 (55.9)	7 (100.0)[b]	10 (100.0)
Total	26	44[c]	22
	(n = 92-104)	(n = 96)	(n = 92-101)

Median tests.

[a] BIW vs. HTW: x^2 = 1.140 df = 1 p = .286; BIM vs. BIW: p = .007; HM vs. HW: p = .000.
[b] BIW vs. HW: x^2 = 3.513 df = 1 p = .061; BIM vs. BIW: p = .000; HM vs. HW: p = .000.
[c] BIW vs. HTW: x^2 = 2.622 df = 1 p = .105; BIW vs. HW: x^2 = 16.716 df = 1 p = .000.
  BIM vs. BIW: p = .051; HM vs. HW: p = .000.

Median number (and incidence) of unconventional sex partners in past 12 months

Men

	Heterosexuals	Bisexuals	Homosexuals
Sexual threesomes	0 (18.1%)	0 (43.7%)[a]	0 (35.0%)
Sex parties	0 (10.8)	0 (35.1)[b]	0 (21.7)
Casual partners			
Opposite-sex	0 (52.7)	0 (51.0)[c]	0 (1.6)
Same-sex	0 (2.4)	0 (70.6)[c]	0 (76.9)
Anonymous partners			
Opposite-sex	0 (22.9)	0 (19.8)[d]	0 (1.7)
Same-sex	0 (1.2)	1 (53.1)[d]	2 (64.1)
Paid partners			
Opposite-sex	0 (10.8)	0 (9.0)[e]	0
Same-sex	0	0 (2.7)[e]	0 (11.6)
	(n = 83/84)	(n = 111-115)	(n = 177-185)

Median tests.

[a] BIM vs. HTM: x^2 = 13.115 df = 1 p = .000; BIM vs. HM: x^2 = 1.909 df = 1 p = .167.
[b] BIM vs. HTM: x^2 = 13.773 df = 1 p = .000; BIM vs. HM: x^2 = 5.672 df = 1 p = .017.
[c] BIM vs. HTM: x^2 = .524 df = 1 p = .479; BIM vs. HM: x^2 = 6.450 df = 1 p = .011.
[d] BIM vs. HIM: x^2 = .116 df = 1 p = .733; BIM vs. HM: x^2 = 9.441 df = 1 p = .002.
[e] BIM vs. HTM: x^2 = .033 df = 1 p = .856; BIM vs. HM: x^2 = 6.188 df = 1 p = .013.

Women

	Heterosexuals	Bisexuals	Homosexuals
Sexual threesomes	0 (14.7%)	0 (47.3%)[f]	0 (8.7%)
Sex parties	0 (8.0)	0 (33.3)[g]	0 (9.8)
Casual partners			
Opposite-sex	0 (48.4)	0 (59.4)[h]	0 (6.7)
Same-sex	0 (4.8)	0 (46.3)[h]	0 (53.7)
Anonymous partners			
Opposite-sex	0 (20.6)	0 (29.3)[i]	0 (5.5)
Same-sex	0 (2.9)	0 (17.2)[i]	0 (9.8)
Paid partners			
Opposite-sex	0 (2.0)	0 (2.2)[j]	0 (1.1)
Same-sex	0	0 (0.0)[j]	0 (1.1)
	(n = 99-102)	(n = 92-96)	(n = 91-93)

Median tests.

[f] BIW vs. HTW: x^2 = 22.988 df = 1 p = .000; BIW vs. HW: x^2 = 32.245 df = 1 p = .000.
  HM vs. HW: p=.000.
[g] BIW vs. HTW: x^2 = 17.641 df = 1 p = .000; BIW vs. HW: x^2 = 13.779 df = 1 p = .000.
  HM vs. HW: p = .022.
[h] BIW vs. HTW: x^2 = .013 df = 1 p = .910; BIW vs. HW: x^2 = .745 df = 1 p = .388.
  BIM vs. BIW (w/ SS): p = 000; HM vs. HW: p = .000.
[i] BIW vs. HTW: x^2 = 1.551 df = 1 p = .213; BIW vs. HW: x^2 = 1.591 df = 1 p = .207.
  BIM vs. BIW (w/ SS): p = .000; HM vs. HW: p = .000.
[j] BIW vs. HTW: x^2 = .196 df = 1 p = .658; BIW vs. HW: x^2 = .000 df = 1 p = .991.
  BIM vs. BIW (w/ SS): p = .024; HM vs. HW: p = .008.

Table 16.2   Sexual Activity

Self-masturbation (past 12 months)

Men

	Heterosexual	Bisexual	Homosexual
Never	1.2%	.9%	1.1%
< 1 month	12.2	6.2	1.7
1-3 month	13.4	8.0	13.3
1-3 week	39.0	43.4	38.1
4 + week	34.1	41.6	45.9
	(n = 82)	(n = 113)	(n = 181)

Collapsed top two rows.   BIM vs. HTM: x^2 = 3.74 df = 3 p = .292;
BIM vs. HM: x^2 = 5.83 df = 3 p = .126.

Women

	Heterosexual	Bisexual	Homosexual
Never	3.9%	5.4%	2.2%
< 1 month	9.7	5.4	12.0
1-3 month	20.4	15.1	29.3
1-3 week	43.7	41.9	34.8
4 + week	22.3	32.3	21.7
	(n = 103)	(n = 93)	(n = 92)

Collapsed top two rows. BIW vs. HTW: x^2 = 3.14 df = 3 p = .381;
BIW vs. HW: x^2 = 7.98 df = 3 p = .048.   HM vs. HW: p = .000.

Orgasm in sleep (past 12 months)

Men

	Heterosexual	Bisexual	Homosexual
Never	83.3%	71.7%	81.2%
< 1 month	14.3	25.7	13.8
1-3 month	1.2	2.7	2.8
1-3 week	1.2	0	.6
4 + week	0	0	1.7
	(n = 84)	(n = 113)	(n = 181)

Collapsed bottom four rows. BIM vs. HTM: x^2 = 3.033 df = 1 p = .086;
BIM vs. HM: x^2 = 3.105 df = 1 p = .083.

Women

	Heterosexual	Bisexual	Homosexual
Never	60.6%	57.4%	61.1%
< 1 month	29.3	33.0	32.2
1-3 month	10.1	7.4	6.7
1-3 week	0	1.1	0
4 + week	0	1.1	0
	(n - 99)	(n = 94)	(n = 90)

Collapsed bottom four rows. BIW vs. HTW: x^2 = .090 df = 1 p = .770;
BIW vs. HW: x^2 = .126 df = 1 p = .728.   HTM vs. HTW: p = .001; HM vs. HW: p = .001.

Masturbate partner (past 12 months)

Men

	Heterosexual	Bisexual (with opposite-sex partner)	Homosexual	Bisexual (with same-sex partner)
Never	19.8%	21.8%	9.3%	24.8%
< 1 month	23.5	27.7	31.7	41.0
1-3 month	38.3	29.7	30.1	21.9
1-3 week	16.0	19.8	24.0	11.4
4 + week	2.5	1.0	4.9	1.0
	(n = 81)	(n = 101)	(n = 183)	(n = 105)

Collapsed bottom two rows. BIM vs. HTM: $x^2 = 1.510$ df = 3 p = .681;
BIM vs. HM: $x^2 = 23.067$ df = 4 p = .000.

Women

	Heterosexual	Bisexual (with opposite-sex partner)	Homosexual	Bisexual (with same-sex partner)
Never	14.4%	29.7%	11.1%	44.8%
< 1 month	36.5	25.3	23.3	29.9
1-3 month	26.9	23.1	31.1	14.9
1-3 week	16.3	18.7	30.0	10.3
4 + week	5.8	3.3	4.4	0
	(n = 104)	(n = 91)	(n = 90)	(n = 87)

BIW vs. HTW: $x^2 = 8.359$ df = 4 p = .079. Collapsed bottom two rows.
BIW vs. HW: $x^2 = 35.240$ df = 3 p = .000. BIM vs. BIW (w/ SS): p = .037.

Masturbated by partner (past 12 months)

Men

	Heterosexual	Bisexual (with opposite-sex partner)	Homosexual	Bisexual (with same-sex partner)
Never	31.0%	35.9%	10.3%	30.6%
< 1 month	41.7	33.0	35.1	38.7
1-3 month	19.0	22.3	29.7	19.8
1-3 week	8.3	7.8	21.1	9.0
4 + week	0	1.0	3.8	1.8
	(n = 84)	(n = 103)	(n = 185)	(n = 111)

Collapsed bottom two rows. BIM vs. HTM: $x^2 = 1.530$ df = 3 p = .677;
BIM vs. HM: $x^2 = 26.012$ df = 4 p = .000.

Women

	Heterosexual	Bisexual (with opposite-sex partner)	Homosexual	Bisexual (with same-sex partner)
Never	14.4%	20.7%	10.9%	44.3%
< 1 month	24.0	27.2	21.7	30.7
1-3 month	34.6	26.1	32.6	15.9
1-3 week	19.2	22.8	31.5	9.1
4 + week	7.7	3.3	3.3	0
	(n = 104)	(n = 92)	(n = 92)	(n = 88)

BIW vs. HTW: x^2 = 4.535 df = 4 p = .338.   Collapsed bottom two rows.
BIW vs. HW: x^2 = 38.380 df = 3 p = .000.   BIM vs. BIW (w/ OS): p = .006;   BIM vs. BIW: p = .006;
HTM vs. HTW: p = .000.

Self-masturbate in front of partner (past 12 months)

Men

	Heterosexual	Bisexual (with opposite-sex partner)	Homosexual	Bisexual (with same-sex partner)
Never	40.5%	37.4%	19.5%	35.4%
< 1 month	38.1	32.7	34.1	39.8
1-3 month	14.3	18.7	25.9	17.7
1-3 week	4.8	8.4	13.0	7.1
4 + week	2.4	2.8	7.6	0
	(n = 84)	(n = 107)	(n = 185)	(n = 113)

Collapsed bottom two rows.   BIM vs. HTM: x^2 = 1.890 df = 3 p = .601;
BIM vs. HM: x^2 = 17.950 df = 3 p = .000.

Women

	Heterosexual	Bisexual (with opposite-sex partner)	Homosexual	Bisexual (with same-sex partner)
Never	30.5%	39.4%	42.4%	60.7%
< 1 month	34.3	24.5	30.4	29.2
1-3 month	27.6	18.1	14.1	5.6
1-3 week	6.7	11.7	8.7	3.4
4 + week	1.0	6.4	4.3	1.1
	(n = 105)	(n = 94)	(n = 92)	(n = 89)

BIW vs. HTW: x^2 = 10.668 df = 4 p = .031;   BIW vs. HW: x^2 = 10.423 df = 4 p = .034.
BIM vs. BIW (w/ SS): p = .004;   HM vs. HW: p = .002.

Perform oral-genital sex (past 12 months)

Men

	Heterosexual	Bisexual (with opposite-sex partner)	Homosexual	Bisexual (with same-sex partner)
Never	8.2%	17.8%	7.0%	18.4%
< 1 month	17.6	17.8	22.7	44.7
1-3 month	42.4	29.9	31.4	18.4
1-3 week	25.9	23.4	31.9	16.7
4 + week	5.9	11.2	7.0	1.8
	(n = 85)	(n = 107)	(n = 185)	(n = 114)

BIM vs. HTM: x^2 = 7.096 df = 4 p = .131; BIM vs. HM: x^2 = 34.313 df = 4 p = .000.

Women

	Heterosexual	Bisexual (with opposite-sex partner)	Homosexual	Bisexual (with same-sex partner)
Never	6.7%	14.9%	10.9%	34.8%
< 1 month	19.0	18.1	26.1	37.1
1-3 month	27.6	19.1	29.3	15.7
1-3 week	36.2	29.8	26.1	9.0
4 + week	10.5	18.1	7.6	3.4
	(n = 105)	(n = 94)	(n = 92)	(n = 89)

BIW vs. HTW: x^2 = 7.429 df = 4 p = .115; BIW vs. HW: x^2 = 26.873 df = 4 p = .000.

Receive oral-genital sex (past 12 months)

Men

	Heterosexual	Bisexual (with opposite-sex partner)	Homosexual	Bisexual (with same-sex partner)
Never	14.1%	15.9%	5.9%	20.4%
< 1 month	24.7	25.2	23.2	40.7
1-3 month	32.9	30.8	34.6	20.4
1-3 week	22.4	16.8	29.7	16.8
4 + week	5.9	11.2	6.5	1.8
	(n = 85)	(n = 107)	(n = 185)	(n = 113)

BIM vs. HTM: x^2 = 2.500 df = 4 p = .645; BIM vs. HM: x^2 = 33.173 df = 4 p = .000.

Women

	Heterosexual	Bisexual (with opposite-sex partner)	Homosexual	Bisexual (with same-sex partner)
Never	10.5%	15.2%	9.8%	37.8%
< 1 month	23.8	21.7	30.4	36.7
1-3 month	31.4	21.7	27.2	16.7
1-3 week	26.7	26.1	27.2	7.8
4 + week	7.6	15.2	5.4	1.1
	(n = 105)	(n = 92)	(n = 92)	(n = 90)

BIW vs. HTW: x^2 = 5.246 df = 4 p = .263; BIW vs. HW: x^2 = 32.057 df = 4 p = .000.
BIM vs. BIW (w/ SS): p = .028.

Vaginal intercourse (past 12 months)

Men

	Heterosexual	Bisexual
Never	8.3%	17.9%
< 1 month	8.3	11.6
1-3 month	27.4	19.6
1-3 week	40.5	42.0
4 + week	15.5	8.9
	(n = 84)	(n = 112)

BIM vs. HTM: x^2 = 6.851 df = 4 p = .144.

Women

	Heterosexual	Bisexual
Never	3.9%	12.9%
< 1 month	9.7	17.2
1-3 month	26.2	17.2
1-3 week	41.7	32.3
4 + week	18.4	20.4
	(n = 103)	(n = 93)

BIW vs. HTW: x^2 = 10.246 df = 4 p = .036.

Perform finger-anal sex (past 12 months)

Men

	Heterosexual	Bisexual (with opposite-sex partner)	Homosexual	Bisexual (with same-sex partner)
Never	29.8%	45.5%	27.6%	42.5%
< 1 month	38.1	23.6	35.7	31.0
1-3 month	20.2	21.8	22.2	18.6
1-3 week	9.5	7.3	10.8	7.1
4 + week	2.4	1.8	3.8	.9
	(n = 84)	(n = 110)	(n = 185)	(n = 113)

Collapsed bottom two rows. BIM vs. HTM: x^2 = 6.780 df = 3 p = .083;
BIM vs. HM: x^2 = 9.135 df = 4 p = .058.

Women

	Heterosexual	Bisexual (with opposite-sex partner)	Homosexual	Bisexual (with same-sex partner)
Never	45.7%	37.6%	45.1%	67.8%
< 1 month	29.5	32.3	31.9	23.3
1-3 month	14.3	17.2	13.2	6.7
1-3 week	10.5	8.6	7.7	2.2
4 + week	0	4.3	2.2	0
	(n=105)	(n=93)	(n=91)	(n=90)

Collapsed bottom two rows. BIW vs. HTW: x^2 = 1.406 df = 3 p = .703;
BIW vs. HW: x^2 = 11.649 df = 3 p = .009. BIM vs. BIW (w/ SS): p = .003;
HM vs. HW: p = .050.

Receive finger-anal sex (past 12 months)

Men

	Heterosexual	Bisexual (with opposite-sex partner)	Homosexual	Bisexual (with same-sex partner)
Never	42.9%	49.5%	29.9%	42.0%
< 1 month	42.9	24.8	32.1	33.0
1-3 month	9.5	19.3	25.0	18.8
1-3 week	2.4	5.5	9.8	5.4
4 + week	2.4	.9	3.3	.9
	(n=84)	(n=109)	(n=184)	(n=112)

Collapsed bottom two rows. BIM vs. HTM: x^2 = 8.430 df = 3 p = .041;
BIM vs. HM: x^2 = 7.824 df = 4 p = .098.

Women

Women	Heterosexual	Bisexual (with opposite-sex partner)	Homosexual	Bisexual (with same-sex partner)
Never	38.1%	30.9%	51.1%	69.7%
< 1 month	35.2	33.0	22.8	20.2
1-3 month	19.0	21.3	16.3	9.0
1-3 week	6.7	9.6	8.7	1.1
4 + week	1.0	5.3	1.1	0
	(n = 105)	(n = 94)	(n = 92)	(n = 89)

BIW vs. HTW: x^2 = 4.845 df = 4 p = .304. Collapsed bottom two rows.
BIW vs. HW: x^2 = 10.787 df = 3 p = .014. BIM vs. BIW (w/ OS): p = .035;
BIM vs. BIW (w/ SS): p = .001; HM vs. HW: p = .013.

Perform oral-anal sex (past 12 months)

Men

	Heterosexual	Bisexual (with opposite-sex partner)	Homosexual	Bisexual (with same-sex partner)
Never	65.1%	64.2%	60.0%	77.0%
< 1 month	25.3	19.3	21.1	15.9
1-3 month	6.0	11.9	11.4	4.4
1-3 week	3.6	1.8	6.5	1.8
4 + week	0	2.8	1.1	.9
	(n = 83)	(n = 109)	(n = 185)	(n = 113)

Collapsed bottom two rows. BIM vs. HTM: $x^2 = 2.646$ df = 3 p = .457;
BIM vs. HM: $x^2 = 10.860$ df = 3 p = .013.

Women

	Heterosexual	Bisexual (with opposite-sex partner)	Homosexual	Bisexual (with same-sex partner)
Never	72.4%	63.4%	81.5%	77.8%
< 1 month	19.0	16.1	10.9	15.6
1-3 month	2.9	10.8	4.3	4.4
1-3 week	5.7	7.5	2.2	0
4 + week	0	2.2	1.1	2.2
	(n = 105)	(n = 93)	(n = 92)	(n = 90)

Collapsed bottom two rows. BIW vs. HTW: $x^2 = 6.507$ df = 3 p = .092;
BIW vs. HW: $x^2 = 1.014$ df = 3 p = .797.   HM vs. HW: p = .007.

Receive oral-anal sex (past 12 months)

Men

	Heterosexual	Bisexual (with opposite-sex partner)	Homosexual	Bisexual (with same-sex partner)
Never	66.7%	69.7%	48.9%	71.4%
< 1 month	25.0	18.3	31.5	20.5
1-3 month	4.8	9.2	14.7	6.3
1-3 week	3.6	1.8	4.3	.9
4 + week	0	.9	.5	.9
	(n = 84)	(n = 109)	(n = 184)	(n = 112)

Collapsed bottom three rows. BIM vs. HTM: $x^2 = 1.644$ df = 2 p = .450;
BIM vs. HM: $x^2 = 17.288$ df = 4 p = .002.

Women

	Heterosexual	Bisexual (with opposite-sex partner)	Homosexual	Bisexual (with same-sex partner)
Never	64.4%	62.8%	81.3%	79.5%
< 1 month	26.0	22.3	12.1	14.8
1-3 month	5.8	8.5	4.4	4.5
1-3 week	2.9	5.3	2.2	0
4 + week	1.0	1.1	0	1.1
	(n = 104)	(n = 94)	(n = 91)	(n = 88)

Collapsed bottom two rows. BIW vs. HTW: $x^2 = 1.438$ df = 3 p = .696. Collapsed bottom three rows. BIW vs. HW: $x^2 = .319$ df = 2 p = .855. HM vs. HW: p = .000.

Perform anal intercourse (past 12 months)

Men

	Heterosexual	Bisexual (with opposite-sex partner)	Homosexual	Bisexual (with same-sex partner)
Never	73.8%	66.7%	29.3%	46.4%
< 1 month	17.9	21.3	37.5	34.8
1-3 month	6.0	11.1	21.7	10.7
1-3 week	2.4	.9	10.3	7.1
4 + week	0	0	1.1	.9
	(n = 84)	(n = 108)	(n = 184)	(n = 112)

Collapsed bottom two rows. BIM vs. HTM: $x^2 = 1.249$ df = 2 p = .541; BIM vs. HM: $x^2 = 11.411$ df = 3 p = .010.

Receive anal intercourse (past 12 months)

Men

	Homosexual	Bisexual
Never	33.0%	54.9%
< 1 month	38.9	29.2
1-3 month	15.1	11.5
1-3 week	10.3	3.5
4 + week	2.7	.9
	(n = 185)	(n = 113)

BIM vs. HM: $x^2 = 16.437$ df = 4 p = .002.

Women

	Heterosexual	Bisexual
Never	59.0%	51.1%
< 1 month	32.4	34.0
1-3 month	7.6	11.7
1-3 week	1.0	3.2
4 + week	0	0
	(n = 105)	(n = 94)

Collapsed bottom two rows. BIW vs. HTW: $x^2 = 2.761$ df = 3 p = .430.

Non-genital activity such as kissing, hugging, etc. (past 12 months)

Men

	Heterosexual	Bisexual (with opposite-sex partner)	Homosexual	Bisexual (with same-sex partner)
Not at all	1.2%	3.6%	.5%	8.2%
A little	2.4	3.6	3.8	15.5
Somewhat	16.7	14.5	20.5	25.5
Great deal	79.8	78.2	75.1	50.9
	(n = 84)	(n = 110)	(n = 185)	(n = 110)

Collapsed bottom two rows. BIM vs. HTM: x^2 = 1.301 df = 2 p = .526;
BIM vs. HM: x^2 = 30.381 df = 3 p = .000.

Women

	Heterosexual	Bisexual (with opposite-sex partner)	Homosexual	Bisexual (with same-sex partner)
Not at all	3.8%	0	3.3%	1.2%
A little	1.9	8.7%	0	3.5
Somewhat	21.0	21.7	4.4	11.6
Great deal	73.3	69.6	92.3	83.7
	(n = 105)	(n = 92)	(n = 91)	(n = 86)

Collapsed top two rows. BIW vs. HTW: x^2 = .724 df = 2 p = .696.
Collapsed top three rows. BIW vs. HW: x^2 = 2.351 df = 1 p = .134.
Collapsed top two rows. BIM vs. BIW (w/ SS): p = .000.
Collapsed top three rows. HM vs. HW: p = .002.

Incidence of unconventional sexual behaviors with opposite-sex partners (past 12 months)

	Men		Women	
Behavior	Heterosexuals	Bisexuals	Heterosexuals	Bisexuals
Sadomasochism	4.8%[a]	16.5%	9.6%[f]	24.5%
Anal Fisting	1.2[b]	5.5	1.97[g]	9.7
Urination Play	11.9[c]	8.2	4.8[h]	10.9
Enema Play	1.2[d]	6.4	2.9[i]	7.4
Feces Play	1.2[e]	1.8	1.0[j]	2.1
	(n = 84)	(n = 109/110)	(n = 104/105)	(n = 92-94)

[a]HTM vs. BIM: x^2 = 7.097 df = 1 p = .008.
[b]HTM vs. BIM: x^2 = 2.870 df = 1 p = .092.
[c]HTM vs. BIM: x^2 = .740 df = 1 p = .390.
[d]HTM vs. BIM: x^2 = 3.769 df = 1 p = .052.
[e]HTM vs. BIM: x^2 = .132 df = 1 p = .716.

[f]HW vs. BIW: x^2 = 7.974 df = 1 p = .005.
[g]HW vs. BIW: x^2 = 6.024 df = 1 p = .014.
[h]HW vs. BIW: x^2 = 2.624 df = 1 p = .105.
[i]HW vs. BIW: x^2 = 2.229 df = 1 p = .135.
[j]HW vs. BIW: x^2 = .467 df = 1 p = .495.

Incidence of unconventional sexual behaviors with same-sex partners (past 12 months)

	Men		Women	
Behavior	Homosexuals	Bisexuals	Homosexuals	Bisexuals
Sadomasochism	16.8%	10.6%[a]	19.6%	17.6%[f]
Anal Fisting	11.5	6.2[b]	2.2	6.6[g]
Urination Play	12.5	4.4[c]	7.6	4.4[h]
Enema Play	11.4	7.1[d]	3.3	3.3[i]
Feces Play	3.3	0.9[e]	2.2	1.1[j]
	(n = 183/184)	(n = 113)	(n = 92)	(n= 90-94)

[a]HM vs. BIM: $x^2 = 2.275$ df = 1 p = .131. [f]HW vs. BIW: $x^2 = .119$ df = 1 p = .730.
[b]HM vs. BIM: $x^2 = 2.402$ df = 1 p = .121. [g]HW vs. BIW: $x^2 = 2.230$ df = 1 p = .135.
[c]HM vs. BIM: $x^2 = 5.912$ df = 1 p = .015. [h]HW vs. BIW: $x^2 = .813$ df = 1 p = .367.
[d]HM vs. BIM: $x^2 = 1.071$ df = 1 p = .300. [i]HW vs. BIW: $x^2 = .000$ df = 1 p = 1.000.
[e]HM vs. BIM: $x^2 = 1.977$ df = 1 p = .160. [j]HW vs. BIW: $x^2 = .345$ df = 1 p = .557.

HM vs. HW (AF): p = .016; HM vs. HW (EP): p = .043.

Total sexual outlet with a partner (past 12 months)

Men

	Heterosexual	Bisexual	Homosexual
Never	8.8%	2.7%	7.3%
< 1 month	6.3	12.7	20.2
1-3 month	31.3	19.1	29.8
1-3 week	40.0	47.3	31.5
4+ week	13.8	18.2	11.2
	(n = 80)	(n = 110)	(n = 178)

$x^2 = 23.002$ df = 8 p = .003.

Women

	Heterosexual	Bisexual	Homosexual
Never	7.8%	4.3%	8.0%
< 1 month	8.8	16.0	20.7
1-3 month	21.6	21.3	25.3
1-3 week	40.2	38.3	41.4
4+ week	21.6	20.2	4.6
	(n = 102)	(n = 94)	(n = 87)

$x^2 = 16.530$ df = 8 p = .038.

Table 16.3   Sexual Problems

Difficulty finding suitable partner

Men

	Heterosexual	Bisexual (with opposite-sex partner)	Homosexual	Bisexual (with same-sex partner)
Not at all	43.4%	47.6%	28.0%	28.0%
A little	19.3	10.5	23.1	11.2
Somewhat	14.5	13.3	24.2	32.7
Great deal	22.9	28.6	24.7	28.0
	(n = 83)	(n = 105)	(n = 182)	(n = 107)

BIM vs. HTM: x^2 = 3.285 df = 3 p = .350; BIM vs. HM: x^2 = 7.518 df = 3 p = .057.

Women

	Heterosexual	Bisexual (with opposite-sex partner)	Homosexual	Bisexual (with same-sex partner)
Not at all	63.4%	49.5%	55.4%	22.7%
A little	8.9	11.8	9.8	15.9
Somewhat	13.9	17.2	10.9	21.6
Great deal	13.9	21.5	23.9	39.8
	(n = 101)	(n = 93)	(n = 92)	(n = 88)

BIW vs. HTW: x^2 = 4.027 df = 3 p = .259; BIW vs. HW: x^2 = 20.840 df = 3 p = .000.
HTM vs. HTW: p = .028; HM vs. HW: p = .000.

Hard feeling good about sexual experiences

Men

	Heterosexual	Bisexual (with opposite-sex partner)	Homosexual	Bisexual (with same-sex partner)
Not at all	63.5%	78.9%	58.7%	79.1%
A little	21.2	12.8	30.4	10.0
Somewhat	12.9	4.6	7.6	5.5
Great deal	2.4	3.7	3.3	5.5
	(n = 85)	(n = 109)	(n = 184)	(n = 110)

Collapsed bottom two rows. BIM vs. HTM: x^2 = 5.659 df = 2 p = .062;
BIM vs. HM: x^2 = 18.214 df = 3 p = .000.

Women

	Heterosexual	Bisexual (with opposite-sex partner)	Homosexual	Bisexual (with same-sex partner)
Not at all	63.7%	66.0%	64.8%	62.5%
A little	28.4	13.8	23.1	18.2
Somewhat	5.9	9.6	9.9	13.6
Great deal	2.0	10.6	2.2	5.7
	(n = 102)	(n = 94)	(n = 91)	(n = 88)

BIW vs. HTW: x^2 = 11.778 df = 3 p = .009; BIW vs. HW: x^2 = 2.679 df = 3 p = .452.
BIM vs. BIW (w/ SS): p = .049.

Do not feel sexually adequate with partner

Men

	Heterosexual	Bisexual (with opposite-sex partner)	Homosexual	Bisexual (with same-sex partner)
Not at all	52.9%	68.8%	48.1%	76.6%
A little	27.1	13.8	31.1	16.8
Somewhat	15.3	10.1	15.8	3.7
Great deal	4.7	7.3	4.9	2.8
	(n = 85)	(n = 109)	(n = 183)	(n = 107)

BIM vs. HTM: x^2 = 7.826 df = 3 p = .050. Collapsed bottom two rows.
BIM vs. HM: x^2 = 23.544 df = 2 p = .000.

Women

	Heterosexual	Bisexual (with opposite-sex partner)	Homosexual	Bisexual (with same-sex partner)
Not at all	65.4%	66.0%	49.5%	47.8%
A little	24.0	18.1	37.4	28.9
Somewhat	8.7	9.6	4.9	15.6
Great deal	1.9	6.4	3.3	7.8
	(n = 104)	(n = 94)	(n = 91)	(n = 90)

BIW vs. HTW: x^2 = 3.398 df = 3 p = .334. Collapsed bottom two rows.
BIW vs. HW: x^2 = 3.558 df = 2 p = .176. BIM vs. BIW (w/ SS): p = .000.

Becoming and staying sexually aroused

Men

	Heterosexual	Bisexual (with opposite-sex partner)	Homosexual	Bisexual (with same-sex partner)
Not at all	61.2%	69.9%	67.4%	75.5%
A little	23.5	16.5	17.9	15.1
Somewhat	8.2	8.7	10.3	3.8
Great deal	7.1	4.9	4.3	5.7
	(n = 85)	(n = 103)	(n = 184)	(n = 106)

BIM vs. HTM: x^2 = 2.099 df = 3 p = .552; BIM vs. HM: x^2 = 5.265 df = 3 p = .153.

Women

	Heterosexual	Bisexual (with opposite-sex partner)	Homosexual	Bisexual (with same-sex partner)
Not at all	48.5%	47.3%	50.0%	53.5%
A little	36.9	28.6	34.8	31.4
Somewhat	9.7	18.7	12.0	11.6
Great deal	4.9	5.5	3.3	3.5
	(n = 103)	(n = 91)	(n = 92)	(n = 86)

BIW vs. HTW: x^2 = 3.884 df = 3 p = .274. Collapsed bottom two rows.
BIW vs. HW: x^2 = .260 df = 2 p = .879. BIM vs. BIW (w/ OS): p = .013;
BIM vs. BIW (w/ SS): p = .008; HM vs. HW: p = .014.

Difficulty telling sex needs to partner

Men

	Heterosexual	Bisexual (with opposite-sex partner)	Homosexual	Bisexual (with same-sex partner)
Not at all	48.2%	52.8%	44.8%	62.4%
A little	33.7	24.1	35.5	22.0
Somewhat	16.9	16.7	12.6	12.8
Great deal	1.2	6.5	7.1	2.8
	(n = 83)	(n = 108)	(n = 183)	(n = 109)

BIM vs. HTM: x^2 = 5.351 df = 3 p = .147; BIM vs. HM: x^2 = 10.555 df = 3 p = .015.

Women

	Heterosexual	Bisexual (with opposite-sex partner)	Homosexual	Bisexual (with same-sex partner)
Not at all	47.5%	42.6	46.7%	42.5%
A little	36.6	33.0	32.6	31.0
Somewhat	12.9	13.8	18.5	16.1
Great deal	3.0	10.6	2.2	10.3
	(n = 101)	(n = 94)	(n = 92)	(n = 87)

BIW vs. HTW: x^2 = 4.984 df = 3 p = .173; BIW vs. HW: x^2 = 5.578 df = 3 p = .134.
BIM vs. BIW (w/ SS): p = .017.

Lack of orgasm on your part

Men

	Heterosexual	Bisexual (with opposite-sex partner)	Homosexual	Bisexual (with same-sex partner)
Not at all	65.5%	78.9%	66.5%	77.6%
A little	22.6	13.8	21.4	13.1
Somewhat	7.1	4.6	9.9	3.7
Great deal	4.8	2.8	2.2	5.6
	(n = 84)	(n = 109)	(n = 182)	(n = 107)

Collapsed bottom two rows. BIM vs. HTM: x^2 = 4.346 df = 2 p = .118;
BIM vs. HM: x^2 = 4.188 df = 2 p = .130.

Women

	Heterosexual	Bisexual (with opposite-sex partner)	Homosexual	Bisexual (with same-sex partner)
Not at all	51.0%	48.9%	57.1%	55.3%
A little	28.4	24.5	28.6	23.5
Somewhat	5.9	18.1	7.7	10.6
Great deal	14.7	8.5	6.6	10.6
	(n = 102)	(n = 94)	(n = 91)	(n = 85)

BIW vs. HTW: x^2 = 8.382 df = 3 p = .039; BIW vs. HW: x^2 = 1.688 df = 3 p = .640.
BIM vs. BIW (w/ OS): p = .000;   BIM vs. BIW (w/ SS): p = .011.

Lack of orgasm by partner

Men

	Heterosexual	Bisexual (with opposite-sex partner)	Homosexual	Bisexual (with same-sex partner)
Not at all	53.1%	59.4%	70.9%	80.0%
A little	25.9	28.3	24.0	14.3
Somewhat	14.8	9.4	3.9	3.8
Great deal	6.2	2.8	1.1	1.9
	(n = 81)	(n = 106)	(n = 179)	(n = 105)

BIM vs. HTM: x^2 = 2.728 df = 3 p = .435; BIM vs. HM: x^2 = 3.856 df = 2 p = .154.

Women

	Heterosexual	Bisexual (with opposite-sex partner)	Homosexual	Bisexual (with same-sex partner)
Not at all	81.6%	82.8%	64.8%	73.8%
A little	16.5	14.0	25.3	20.2
Somewhat	1.9	1.1	7.7	3.6
Great deal	0	2.2	2.2	2.4
	(n = 103)	(n = 93)	(n = 91)	(n = 84)

Collapsed bottom three rows. BIW vs. HTW: x^2 = .002 df = 1 p = .935. Collapsed bottom two rows.
BIW vs. HW. x^2 = 1.840 df = 2 p = .412.   BIM vs. BIW (w/ OS): p = .001;
HTM vs. HTW: p = .000.

Reach orgasm too fast

Men

	Heterosexual	Bisexual (with opposite-sex partner)	Homosexual	Bisexual (with same-sex partner)
Not at all	61.0%	66.5%	73.2%	82.5%
A little	19.5	19.6	18.4	11.7
Somewhat	18.3	11.2	6.7	3.9
Great deal	1.2	2.8	1.7	1.9
	(n = 82)	(n = 107)	(n = 179)	(n = 103)

Collapsed bottom two rows. BIM vs. HTM: x^2 = 1.065 df = 2 p = .596;
BIM vs. HM: x^2 = 3.208 df = 2 p = .202.

Women

	Heterosexual	Bisexual (with opposite-sex partner)	Homosexual	Bisexual (with same-sex partner)
Not at all	92.1%	94.6%	88.9%	93.8%
A little	5.0	2.2	5.6	4.9
Somewhat	3.0	2.2	5.6	1.2
Great deal	0	1.1	0	0
	(n = 101)	(n = 92)	(n = 90)	(n = 81)

Collapsed bottom three rows. BIW vs. HTW: x^2 = .161 df = 1 p = .693;
BIW vs. HW: x^2 = .755 df = 1 p = .403.   BIM vs. BIW (w/ OS): p = .000;
BIM vs. BIW (w/ SS): p = .041;   HTM vs. HTW: p = .000;   HM vs. HW: p = .009.

Partner reaches orgasm too fast

Men

	Heterosexual	Bisexual (with opposite-sex partner)	Homosexual	Bisexual (with same-sex partner)
Not at all	93.8%	93.1%	75.8%	85.7%
A little	6.3	4.9	15.7	8.2
Somewhat	0	1.0	5.6	4.1
Great deal	0	1.0	2.8	2.0
	(n = 80)	(n = 102)	(n = 178)	(n = 98)

Collapsed bottom three rows. BIM vs. HTM: $x^2 = .018$ df = 1 p = .895. Collapsed bottom two rows. BIM vs. HM: $x^2 = 3.991$ df = 2 p = .144.

Women

	Heterosexual	Bisexual (with opposite-sex partner)	Homosexual	Bisexual (with same-sex partner)
Not at all	62.1%	63.0%	87.6%	93.8%
A little	24.3	19.6	10.1	4.9
Somewhat	7.8	10.9	2.2	1.2
Great deal	5.8	6.5	0	0
	(n = 103)	(n = 92)	(n = 89)	(n = 81)

BIW vs. HTW: $x^2 = 1.042$ df = 3 p = .791. Collapsed bottom three rows.
BIW vs. HW: $x^2 = 1.247$ df = 1 p = .269. BIM vs. BIW (w/ OS): p = .000;
HTM vs. HTW: p = .000; HM vs. HW: p = .051.

Reach orgasm too slowly

Men

	Heterosexual	Bisexual (with opposite-sex partner)	Homosexual	Bisexual (with same-sex partner)
Not at all	71,3%	78.4%	70.0%	80.8%
A little	21.3	14.7	17.8	8.1
Somewhat	1.3	2.9	8.3	5.1
Great deal	6.3	3.9	3.9	6.1
	(n = 80)	(n = 102)	(n = 180)	(n = 99)

Collapsed bottom two rows. BIM vs. HTM: $x^2 = 1.422$ df = 2 p = .493;
BIM vs. HM: $x^2 = 7.230$ df = 3 p = .065.

Women

	Heterosexual	Bisexual (with opposite-sex partner)	Homosexual	Bisexual (with same-sex partner)
Not at all	46.5%	54.8%	63.7%	58.8%
A little	23.8	22.6	18.7	25.0
Somewhat	16.8	11.8	11.0	8.8
Great deal	12.9	10.8	6.6	7.5
	(n = 101)	(n = 93)	(n = 91)	(n = 80)

BIW vs. HTW: $x^2 = 1.722$ df = 3 p = .632; BIW vs. HW: $x^2 = 1.222$ df = 3 p = .748.
BIM vs. BIW (w/ OS): p = .000; BIM vs. BIW (w/ SS): p = .006; HTM vs. HTW: p = .000.

Partner reaches orgasm too slowly

Men

	Heterosexual	Bisexual (with opposite-sex partner)	Homosexual	Bisexual (with same-sex partner)
Not at all	58.8%	63.1%	70.2%	80.8%
A little	21.2	24.3	21.3	13.1
Somewhat	13.8	8.7	7.3	4.0
Great deal	6.3	3.9	1.1	2.0
	(n = 80)	(n = 103)	(n = 178)	(n = 99)

BIM vs. HTM: x^2 = 1.852 df = 3 p = .604. Collapsed bottom two rows.
BIM vs. HM: x^2 = 3.770 df = 2 p = .160.

Women

	Heterosexual	Bisexual (with opposite-sex partner)	Homosexual	Bisexual (with same-sex partner)
Not at all	79.6%	84.6%	76.7%	84.0%
A little	17.5	8.8	16.7	11.1
Somewhat	1.9	5.5	4.4	4.9
Great deal	1.0	1.1	2.2	0
	(n = 103)	(n = 91)	(n = 90)	(n = 81)

Collapsed bottom two rows. DIW vs. HTW. x^2 = 4.286 df = 2 p = .123;
BIW vs. HW: x^2 = 1.442 df = 2 p = .490. BIM vs. BIW (w/ OS): p = .005;
HTM vs. HTW: p = .001.

Table 16.4   Sexually Transmitted Diseases

Number of times contracted STDs in lifetime

Men

	Heterosexual	Bisexual	Homosexual
0	53.7%	33.0%	27.1%
1	26.8	13.9	16.0
2	13.4	18.3	16.6
3	3.7	10.4	10.5
4	1.2	9.6	3.9
5	0	3.5	7.2
6	0	2.6	4.4
7	0	0	1.1
8	0	1.7	2.2
10	0	1.7	3.9
12	0	1.7	.6
15	1.2	1.7	2.2
20	0	1.7	3.3
25	0	0	1.1
	(n = 82)	(n = 115)	(n = 181)

Median tests. BIM vs. HTM: x^2 = 21.219 df = 1 p = .000; BIM vs. HM: x^2 = .697 df = 1 p = .404. ·

Women

	Heterosexual	Bisexual	Homosexual
0	54.8%	38.9%	63.0%
1	21.2	13.7	17.4
2	10.6	16.8	6.5
3	5.8	3.2	5.4
4	1.0	5.3	4.3
5	0	5.3	1.1
6	1.9	4.2	0
7	0	1.1	0
8	0	2.1	0
10	1.9	7.4	0
12	1.0	1.1	0
13	1.0	0	0
20	0	1.1	1.1
25	1.0	0	0
30	0	0	1.1
	(n = 104)	(n = 95)	(n = 92)

Median tests. BIW vs. HTW: $x^2 = 10.850$ df = 1 p = .001; BIW vs. HW: $x^2 = 9.915$ df = 1 p = .002. HM vs. HW: p = .000.

STDs led to cutback in sex

Men

	Heterosexual	Bisexual (with opposite-sex partner)	Homosexual	Bisexual (with same-sex partner)
Not at all	67.5%	51.4%	19.8%	17.1%
A little	13.7	15.0	11.5	9.0
Somewhat	12.5	19.6	29.1	22.5
Great deal	6.3	14.0	39.6	51.4
	(n = 80)	(n = 107)	(n = 182)	(n = 111)

BIM vs. HTM: $x^2 = 6.250$ df = 3 p = .100; BIM vs. HM: $x^2 = 3.981$ df = 3 p = .264.

Women

	Heterosexual	Bisexual (with opposite-sex partner)	Homosexual	Bisexual (with same-sex partner)
Not at all	49.0%	48.4%	81.4%	66.3%
A little	12.5	9.7	11.6	14.1
Somewhat	22.9	19.4	5.8	12.0
Great deal	15.6	22.6	1.2	7.6
	(n = 96)	(n = 93)	(n = 86)	(n = 92)

BIW vs. HTW: $x^2 = 1.831$ df = 3 p = .608; BIW vs. HW: $x^2 = 8.177$ df = 3 p = .042. BIM vs. BIW (w/ SS): p = .000; HTM vs. HTW: p = .030; HM vs. HW: p = .000.

CHAPTER 17: INTIMATE RELATIONSHIPS

Table 17.1   Total Number of Involved Relationships with Opposite-sex or Same-sex Partners

Opposite-sex relationships

	Men[a]		Women[b]	
	Heterosexual	Bisexual	Heterosexual	Bisexual
0	14.1%	25.7%	12.5%	27.1%
1	54.1	52.2	64.4	47.9
2 or more	31.8	22.1	23.1	25.0
	(n = 85)	(n = 113)	(n = 104)	(n = 96)

[a]BIM vs. HTM: x^2 = 18.930 df = 2 p = .000; [b]BIW vs. HTW: x^2 = 8.023 df = 2 p = .018.
HIM vs. HTW: p = .000.

Same-sex relationships

	Men[a]		Women[b]	
	Homosexual	Bisexual	Homosexual	Bisexual
0	31.2%	54.4%	23.4%	55.2%
1	36.6	26.3	69.1	21.9
2 or more	32.2	19.3	7.4	22.9
	(n = 186)	(n = 114)	(n = 94)	(n = 96)

[a]BIM vs. HM: x^2 = 16.113 df = 2 p = .000; [b]BIW vs. HW: x^2 = 44.946 df = 2 p = .000.
HM vs. HW: p = .000.

Table 17.2   Total number of involved relationships

	Men[a]			Women[b]		
	Heterosexual	Bisexual	Homosexual	Heterosexual	Bisexual	Homosexual
0	14.6%	15.9%	30.6%	10.9%	15.6%	19.6%
1	53.7	35.4	37.2	66.3	38.5	69.6
2 or more	31.7	48.7	32.2	22.8	45.8	10.9
	(n = 82)	(n = 113)	(n = 183)	(n = 101)	(n = 96)	(n = 92)

[a]x^2 = 19.829 df = 4 p = .001; [b]x^2 = 35.285 df = 4 p = .000.
HM vs. HW: p = .000.

Table 17.3   Basic Structure of Involved Relationships

Among bisexuals

	Males	Females
Uncoupled	15.9%	15.6%
Single relationship with opposite-sex	28.3	30.2
Single relationship with same-sex	7.1	8.3
Two relationships one with each sex	15.0	9.4
Two or more opposite-sex relationships	9.7	9.4
Two or more same-sex relationships	2.7	3.1
Two or more relationships with opposite-sex and one with same-sex	4.4	4.2
Two or more relationships with same-sex and one with opposite-sex	8.8	8.3
Two or more relationships with each sex	8.0	11.5
	(n = 113)	(n = 96)

$x^2 = 2.241$ df = 8 p = .971.

Among heterosexuals

	Men	Women
Uncoupled	14.6%	10.9%
Single relationship with opposite-sex	53.7	66.3
Single relationship with same-sex	0	0
Two relationships one with each sex	0	0
Two or more opposite-sex relationships	28.0	19.8
Two or more same-sex relationships	0	1.0
Two or more relationships with opposite-sex and one with same-sex	2.4	1.0
Two or more relationships with same-sex and one with opposite-sex	0	0
Two or more relationships with each sex	1.2	1.0
	(n = 82)	(n = 101)

Only used rows 1, 2, and 5 because of low expected frequencies in other cells.
$x^2 = 3.001$ df = 2 p = .226.

Among homosexuals

	Men	Women
Uncoupled	30.6%	19.6%
Single relationship with opposite-sex	1.1	2.2
Single relationship with same-sex	36.1	67.4
Two relationships one with each sex	.5	3.3
Two or more opposite-sex relationships	0	1.1
Two or more same-sex relationships	26.8	5.4
Two or more relationships with opposite-sex and one with same-sex	0	0
Two or more relationships with same-sex and one with opposite-sex	1.6	1.1
Two or more relationships with each sex	3.3	0
	(n = 183)	(n = 92)

Only used rows 1, 3, and 6 because of low expected frequencies in other cells.
$x^2 = 29.891$ df = 2 p = .000.

Table 17.4   Percent With One or Two or More Relationships Who Indicated They Were Monogamous or Nonmonogamous Outside of These Relationships

	One relationship					
	Men[a]			Women[b]		
	Heterosexual	Bisexual	Homosexual	Heterosexual	Bisexual	Homosexual
Monogamous	88.6%	55.0%	63.2%	85.1%	54.1%	87.5%
Non-monogamous	11.4	45.0	36.8	14.9	45.9	12.5
	(n=44)	(n=40)	(n=68)	(n=67)	(n=37)	(n=64)

[a]x^2 = 13.937 df = 2 p = .001; [b]x^2 = 16.199 df = 2 p = .000.
HM vs. HW: p = .001.

	Two or more relationships					
	Men[a]			Women[b]		
	Heterosexual	Bisexual	Homosexual	Heterosexual	Bisexual	Homosexual
Monogamous	38.5%	12.7%	33.9%	30.4%	22.7%	60.0%
Non-monogamous	61.5	87.3	66.1	69.6	77.3	40.0
	(n=26)	(n=55)	(n=59)	(n=23)	(n=44)	(n=10)

[a]x^2 = 9.560 df = 2 p = .008; [b]x^2 = 5.011 df = 2 p = .082.

Table 17.5   Percent Who Met One or More Opposite-Sex or Same-Sex Significant Partners at Each of Four Settings

Opposite-sex partners

	Men		Women	
	Heterosexual	Bisexual	Heterosexual	Bisexual
Friends and conventional parties	39.7%	36.9%	36.3%	37.1%
	(n = 73)	(n = 84)	(n = 91)	(n = 70)

BIM vs. HTM (w/ OS): x^2 = .039 df = 1 p = .843;
BIW vs. HTW (w/ OS): x^2 = .000 df = 1 p = 1.000.

	Men		Women	
Mainstream activities	52.1%	44.0%	54.9%	40.0%
	(n = 73)	(n = 84)	(n = 91)	(n = 70)

BIM vs. HTM (w/ OS): x^2 = .708 df = 1 p = .400;
BIW vs. HTW (w/ OS): x^2 = 2.965 df = 1 p = .085.

	Men		Women	
Underground sex scenes	8.2%	15.5%	9.9%	12.9%
	(n = 73)	(n = 84)	(n = 91)	(n = 70)

BIM vs. HTM (w/ OS): x^2 = 1.312 df = 1 p = .252;
BIW vs. HTW (w/ OS): x^2 = .116 df = 1 p = .734.

	Men		Women	
Underground support groups	8.2%	22.6%	4.4%	17.1%
	(n = 73)	(n = 84)	(n = 91)	(n = 70)

BIM vs. HTM (w/ OS): x^2 = 5.021 df = 1 p = .025;
BIW vs. HTW (w/ OS): x^2 = 5.829 df = 1 p = .016.

Same-sex partners

	Men		Women	
	Bisexual	Homosexual	Bisexual	Homosexual
Friends and conventional parties	23.1%	33.6%	32.6%	36.1%
	(n = 53)	(n = 128)	(n = 43)	(n = 72)

BIM vs. HM (w/ SS): x^2 = 1.464 df = 1 p = .226;
BIW vs. HW (w/ SS): x^2 = .034 df = 1 p = .853.

Mainstream activities	32.7%	28.1%	27.9%	40.3%
	(n = 52)	(n = 128)	(n = 43)	(n = 72)

BIM vs. HM (w/ SS): x^2 = .184 df = 1 p = .668;
BIW vs. HW (w/ SS): x^2 = 1.297 df = 1 p = .255.

Underground sex scenes	36.5%	35.2%	23.3%	5.6%
	(n = 52)	(n = 128)	(n = 43)	(n = 71)

BIM vs. HM (w/ SS): x^2 = .000 df = 1 p = .997;
BIW vs. HW (w/ SS): x^2 = 6.171 df = 1 p = .013.   HM vs. HW: p = .000.

Underground support groups	36.5%	13.3%	34.9%	19.4%
	(n = 52)	(n = 128)	(n = 43)	(n = 72)

BIM vs. HM (w/ SS): x^2 = 11.089 df = 1 p = .001;
BIW vs. HW (w/ SS): x^2 = 2.633 df = 1 p = .105.

Table 17.6   The Length of Longest Relationship

	Men[a]			Women[b]		
	Heterosexual	Bisexual	Homosexual	Heterosexual	Bisexual	Homosexual
Less than 2	4.7%	12.0%	18.5%	4.8%	7.3%	10.6%
2 to 4	23.5	22.4	32.6	21.9	23.9	40.4
5 to 9	20.0	28.4	27.2	24.8	37.5	37.3
10 or more	51.8	37.1	21.7	48.6	31.3	11.8
	(n = 85)	(n = 116)	(n = 184)	(n = 105)	(n = 96)	(n = 94)

[a]x^2 = 31.013 df = 6 p = .000; [b]x^2 = 36.095 df = 6 p = .000.
HM vs. HW: p = .025.

Table 17.7   Sex of Partner in Longest Relationship

	Men			Women		
	Heterosexual	Bisexual	Homosexual	Heterosexual	Bisexual	Homosexual
Opposite-sex	98.8%	84.3%	17.8%	98.1%	87.4%	28.6%
Same-sex	1.2	15.7	82.2	1.9	12.6	71.4
	(n = 84)	(n = 108)	(n = 174)	(n = 103)	(n = 95)	(n = 91)

BIM vs. HTM: x^2 = 10.124 df = 1 p = .003;   BIM vs. HM: x^2 = 117.160 df = 1 p = .000;
BIW vs. HTW: x^2 = .031 df = 1 p = .869;   BIW vs. HW: x^2 = 63.831 df = 1 p = .000.

Table 17.8   Currently in Longest Relationship

	Men[a]			Women[b]		
	Heterosexual	Bisexual	Homosexual	Heterosexual	Bisexual	Homosexual
Yes	33.7%	39.1%	29.6%	48.1	29.3	31.5
No	66.3	60.9	70.2	51.9	70.8	68.5
	(n = 83)	(n = 115)	(n = 179)	(n = 104)	(n = 96)	(n = 92)

[a]x^2 = 2.839 df = 2 p = .242; [b]x^2 = 9.117 df = 2 p = .010.   HTM vs. HTW: p = .047.

Table 17.9   Years Expect Current Relationship to Last

	Men[a]			Women[b]		
	Heterosexual	Bisexual	Homosexual	Heterosexual	Bisexual	Homosexual
Less than 1	18.5%	10.8%	11.2%	6.9%	4.2%	6.2%
1 to 4	18.5	16.1	22.4	11.5	25.4	27.7
5 to 10	3.1	18.3	12.9	10.3	11.3	9.2
More than 10	60.0	54.8	53.4	71.3	59.2	56.9
	(n = 65)	(n = 93)	(n = 116)	(n = 87)	(n = 71)	(n = 65)

[a]x^2 = 12.240 df = 6 p = .056; [b]x^2 = 8.454 df = 6 p = .207   HTM vs HTW: p = .028

Table 17.10   Extent to Which Sexual Preference or Practices Had a Negative Effect on Respondent's Ability to Sustain Long-term Relationships

	Men[a]			Women[b]		
	Heterosexual	Bisexual	Homosexual	Heterosexual	Bisexual	Homosexual
Not at all	69.1%	54.4%	53.6%	78.8%	78.7%	61.5%
A little	14.8	17.5	18.6	5.8	7.4	18.7
Some-what	8.6	21.9	20.8	11.5	7.4	17.6
A great deal	7.4	6.1	7.1	3.8	6.4	2.2
	(n = 81)	(n = 114)	(n = 183)	(n = 104)	(n = 94)	(n = 91)

Collapsed table: "Not at all" vs. "A little" through "A great deal."
[a]x^2 = 9.582 df = 6 p = .143; [b]x^2 = 9.484 df = 2 p = .009.   BIM vs. BIW: p = .001.

CHAPTER 18: MANAGING IDENTITIES

Table 18.1   Estimated Proportions of Persons Thought to Know or Suspect Respondent's
                Homosexuality, Bisexuality, or Unconventional Heterosexual Activity, and if Directly Told
                by Respondent

Family and friends who know or suspect

	Men[a]			Women[b]		
	Heterosexual	Bisexual	Homosexual	Heterosexual	Bisexual	Homosexual
Many/all	31.8%	51.3%	79.3%	46.6%	60.4%	92.1%
Some	22.7	16.5	12.3	10.0	15.4	3.4
A few	22.7	25.7	6.7	33.3	18.7	4.5
None	22.7	6.4	1.7	10.0	5.5	0
	(n = 22)	(n = 109)	(n = 179)	(n = 30)	(n = 91)	(n = 88)

Collapsed table: "Many" and "all" vs. "Some" through "None."
[a]$x^2 = 35.943$ df = 2 p = .000; [b]$x^2 = 36.276$ df = 2 p = .000.
HTM vs. HTW: p = .052.

Family and friends directly told

	Men[a]			Women[b]		
	Heterosexual	Bisexual	Homosexual	Heterosexual	Bisexual	Homosexual
Many/all	47.6%	65.4%	69.4%	55.2%	71.3%	82.8%
Some	9.5	9.6	10.6	10.3	6.9	9.9
A few	19.0	17.3	13.5	24.1	14.9	6.2
None	23.8	7.7	6.5	10.3	6.9	1.2
	(n = 21)	(n = 104)	(n = 170)	(n = 29)	(n = 87)	(n = 81)

[a]$x^2 = 3.869$ df = 2 p = .145; [b]$x^2 = 8.571$ df = 2 p = .014.   HM vs. HW: p = .022.

Workmates and acquaintances who know or suspect

	Men[a]			Women[b]		
	Heterosexual	Bisexual	Homosexual	Heterosexual	Bisexual	Homosexual
Many/all	13.6%	37.0%	69.7%	34.5%	30.8%	61.7%
Some	31.8	17.6	17.4	13.8	17.6	27.4
A few	31.8	27.8	10.1	17.2	33.0	8.3
None	22.7	17.6	2.8	34.5	18.7	2.4
	(n = 22)	(n = 108)	(n = 178)	(n = 29)	(n = 91)	(n = 84)

[a]$x^2 = 46.404$ df = 2 p = .000; [b]$x^2 = 18.632$ df = 2 p = .000.

Workmates and acquaintances directly told

	Men[a]			Women[b]		
	Heterosexual	Bisexual	Homosexual	Heterosexual	Bisexual	Homosexual
Many/all	36.3%	45.1%	46.7%	42.3%	47.6%	45.5%
Some	13.6	13.7	17.8	15.4	14.3	28.4
A few	27.3	18.6	21.7	11.5	16.7	18.2
None	22.7	22.5	13.9	30.8	21.4	8.0
	(n = 22)	(n = 102)	(n = 180)	(n = 26)	(n = 84)	(n = 88)

[a]$x^2 = .857$ df = 2 p = .651; [b]$x^2 = .242$ df = 2 p = .890.

Table 18.2   Estimated Percentage of Persons Thought to be Accepting or Rejecting

Family and traditional friends who were accepting

	Men[a]			Women[b]		
	Heterosexual	Bisexual	Homosexual	Heterosexual	Bisexual	Homosexual
Many/all	45.5%	52.7%	63.5%	37.9%	55.5%	65.5%
Some	22.7	24.5	23.0	37.9	17.8	21.1
A few	27.3	20.9	11.8	17.2	20.0	12.2
None	4.5	1.8	1.7	6.9	6.7	1.1
	(n = 22)	(n = 110)	(n = 178)	(n = 29)	(n = 90)	(n = 90)

$^a x^2 = 4.850$ df = 2 p = .088; $^b x^2 = 6.060$ df = 2 p = .049.

Family and traditional friends who were rejecting

	Men[a]			Women[b]		
	Heterosexual	Bisexual	Homosexual	Heterosexual	Bisexual	Homosexual
Many/all	28.6%	11.4%	8.7%	28.0%	17.3%	9.3%
Some	19.0	22.9	14.5	8.0	18.5	14.0
A few	38.1	41.0	46.8	48.0	43.2	57.0
None	14.3	24.8	30.1	16.0	21.0	19.8
	(n = 21)	(n = 105)	(n = 173)	(n = 25)	(n = 81)	(n = 86)

$^a x^2 = 5.901$ df = 2 p = .052; $^b x^2 = 5.800$ df = 2 p = .057.

Workmates and acquaintances who were accepting

	Men[a]			Women[b]		
	Heterosexual	Bisexual	Homosexual	Heterosexual	Bisexual	Homosexual
Many/all	36.4%	41.6%	65.2%	29.6%	39.8%	48.3%
Some	31.8	31.1	22.1	22.2	20.5	37.9
A few	31.8	23.6	12.7	25.9	35.2	12.6
None	0	3.8	0	22.2	4.5	1.1
	(n = 22)	(n = 106)	(n = 181)	(n = 27)	(n = 88)	(n = 87)

Collapsed table: "Many" and "all" vs. "A few" and "None."
$^a x^2 = 18.593$ df = 2 p = .000; $^b x^2 = 3.335$ df = 2 p = .189.   HM vs. HW: p = .008.

Workmates and acquaintances who were rejecting

	Men[a]			Women[b]		
	Heterosexual	Bisexual	Homosexual	Heterosexual	Bisexual	Homosexual
Many/all	27.2%	15.5%	9.6%	40.0%	26.8%	10.5%
Some	31.8	31.1	21.5	24.0	25.6	31.4
A few	36.4	41.7	53.7	32.0	36.6	47.7
None	4.5	11.7	15.3	4.0	11.0	10.5
	(n = 22)	(n = 103)	(n = 21)	(n = 25)	(n = 82)	(n = 86)

$^a x^2 = 5.670$ df = 2 p = .059; $^b x^2 = 12.949$ df = 2 p = .002.

Table 18.3   Socializing with Same-sex Heterosexuals, Bisexuals, and Homosexuals

Men socializing with men

	Heterosexual	Bisexual	Homosexual	x^2	df	p
Homosexual Men	31.0%	57.0%	95.1%	137.492	2	.000
	(n = 84)	(n = 114)	(n = 184)			
Bisexual Men	22.0	47.8	26.9	18.308	2	.000
	(n = 82)	(n = 113)	(n = 171)			
Heterosexual Men	85.9	68.1	56.4	24.573	2	.000
	(n = 85)	(n = 113)	(n = 179)			

Collapsed tables: "Great deal" and "Somewhat" vs. "A little" and "Not at all."

Women socializing with women

	Heterosexual	Bisexual	Homosexual	x^2	df	p
Homosexual Women	44.0%	50.5%	97.8%	89.520	2	.000
	(n = 100)	(n = 93)	(n = 93)			
Bisexual Women	37.1	55.3	16.1	31.485	2	.000
	(n = 97)	(n = 94)	(n = 87)			
Heterosexual Women	90.4	74.5	54.9	33.027	2	.000
	(n = 104)	(n = 94)	(n = 91)			

Collapsed tables: "Great deal" and "Somewhat" vs. "A little" and "Not at all."
HTM vs. HTW: p = .026; HM vs. HW: p = .047.

Table 18.4   Respondents Who Felt Uncomfortable in the Company of Homosexual, Bisexual, or Heterosexual Men or Women

	Men			x^2	df	p
	Heterosexual	Bisexual	Homosexual			
Homosexual men	44.7%	36.5%	17.9%[†]	1.004	1	.31
	(n = 85)	(n = 115)	(n = 184)			
Homosexual women	45.9	51.8	40.0	3.946	2	.146
	(n = 85)	(n = 114)	(n = 180)			
Bisexual men	36.5	16.5[†]	19.6	8.006	1	.007
	(n = 85)	(n = 115)	(n = 184)			
Bisexual women	29.4	19.1	22.0	2.974	2	.226
	(n = 85)	(n = 115)	(n = 182)			
Heterosexual men	28.2[†]	51.3	54.9	.236	1	.645
	(n = 85)	(n = 115)	(n = 184)			
Heterosexual women	25.9	32.2	35.0	2.250	2	.325
	(n = 85)	(n = 115)	(n = 183)			

	Women			x^2	df	p
	Heterosexual	Bisexual	Homosexual			
Homosexual men	25.5%	27.4%	31.9%	.999	2	.607
	(n = 102)	(n = 95)	(n = 91)			
Homosexual women	42.2	48.4	15.2[t]	15.645	1	.000
	(n = 102)	(n = 95)	(n = 92)			
Bisexual men	22.3[b]	24.2	56.0[d]	29.441	2	.000
	(n = 103)	(n = 95)	(n = 91)			
Bisexual women	27.5	16.8[t]	46.2[e]	6.230	1	.013
	(n = 102)	(n = 95)	(n = 91)			
Heterosexual men	18.4	54.7	76.3[f]	72.154	2	.000
	(n = 103)	(n = 95)	(n = 93)			
Heterosexual women	9.7[t]	37.9	44.1	.511	1	.483
	(n = 103)	(n = 95)	(n = 93)			

[t]Feelings toward members of the respondent's own sexual preference were not included in the x^2 tests because their small values were producing significant differences; this procedure decreases the degrees of freedom from 2 to 1.

[a]HTM vs. HTW: p = .006; [b]HTM vs. HTW: p = .033; [c]HM vs. HW: p = .011;
[d]HM vs. HW: p = .000; [e]HM vs. HW: p = .000; [f]HM vs. HW: p = .000.

Table 18.5   Respondents Who Felt Negative Toward Homosexual, Bisexual, or Heterosexual Men or Women

	Men			x^2	df	p
	Heterosexual	Bisexual	Homosexual			
Homosexual men	27.4%	27.2%	14.7[t]	0.017	1	.879
	(n = 84)	(n = 114)	(n = 184)			
Homosexual women	27.7	35.1	28.8	1.665	2	.435
	(n = 83)	(n = 114)	(n = 184)			
Bisexual men	19.0	1.7[t]	18.5	0.003	1	.955
	(n = 84)	(n = 115)	(n = 184)			
Bisexual women	13.1	6.1	15.2	6.318	2	.042
	(n = 84)	(n = 115)	(n = 184)			
Heterosexual men	15.5[t]	42.6	45.7	0.157	1	.696
	(n = 84)	(n = 115)	(n = 184)			
Heterosexual women	9.5	26.1	24.0	10.604	2	.005
	(n = 84)	(n = 115)	(n = 183)			

	Women			x^2	df	p
	Heterosexual	Bisexual	Homosexual			
Homosexual men	20.6%	17.0%	38.0%	12.590	2	.003
	(n = 102)	(n = 94)	(n = 92)			
Homosexual women	29.4	29.8	10.9[t]	0.029	1	.013
	(n = 102)	(n = 94)	(n = 92)			
Bisexual men	17.6	14.9[a]	41.3[c]	21.440	2	.000
	(n = 102)	(n = 94)	(n = 92)			
Bisexual women	15.7	5.3[t]	46.7[d]	3.524	1	.064
	(n = 102)	(n = 94)	(n = 92)			
Heterosexual men	21.4	51.6	82.6[e]	31.328	2	.000
	(n = 103)	(n = 95)	(n = 92)			
Heterosexual women	3.9[t]	25.3	44.6[f]	1.657	1	.198
	(n = 102)	(n = 95)	(n = 92)			

[t]Feelings toward members of the respondent's own sexual preference were not included in the x^2 tests because their small values were producing significant differences; this procedure decreases the degrees of freedom from 2 to 1.
[a]BIM vs. BIW: p =.000; [b]HM vs. HW: p = .000; [c]HM vs. HW: p = .000;
[d]HM vs. HW: p = .000; [e]HM vs. HW: p = .000; [f]HM vs. HW: p = .001.

Table 18.6 Social Isolation

Felt socially isolated

	Men[a]			Women[b]		
	Heterosexual	Bisexual	Homosexual	Heterosexual	Bisexual	Homosexual
Any amount	56.5%	61.2%	57.8%	52.4%	59.4%	59.6%
Not at all	43.5	38.8	42.2	47.6	40.6	40.4
	(n = 85)	(n = 116)	(n = 185)	(n = 105)	(n = 96)	(n = 94)

[a]x^2 = .529 df = 2 p = .768; [b]x^2 = 1.385 df = 2 p = .500.

Attributed social isolation to their sexuality

	Men[a]			Women[b]		
	Heterosexual	Bisexual	Homosexual	Heterosexual	Bisexual	Homosexual
Yes	27.7%	60.0%	44.8%	21.2%	55.4%	23.2%
No	72.3	40.0	55.2	78.8	44.6	76.8
	(n = 47)	(n = 70)	(n = 105)	(n = 52)	(n = 56)	(n = 56)

[a]x^2 = 12.236 df = 2 p = .002; [b]x^2 = 17.897 df = 2 p = .000. HM vs. HW: p = .006.

Table 18.7 Regret Their Sexual Preference or Practices

	Men[a]			Women[a]		
	Heterosexual	Bisexual	Homosexual	Heterosexual	Bisexual	Homosexual
Great Deal	2.5%	3.5%	1.6%	0	4.2%	0
Some-what	0	7.0	7.5	2.1%	5.2	4.3%
A Little	12.5	13.2	21.0	11.6	14.6	17.0
Not At All	85.0	76.3	69.9	86.3	76.0	78.7
	(n = 80)	(n = 114)	(n = 186)	(n = 95)	(n = 96)	(n = 94)

Collapsed table: "Not at all" vs. "a little" through "a great deal."
[a]$x^2 = 7.352$ df = 2 p = .025; [b]$x^2 = 3.428$ df = 2 p = .168.

Appendix C: Tables for the 1988 Follow-up Study

CHAPTER 19: AIDS EMERGES

Table 19.1 Demographic and Background Data, 1988 Interview Sample

Age

	1988		Combined
	Men	Women	
19-24	0	0	0
25-29	0	11.1%	5.4%
30-34	3.6%	29.6	16.3
35-39	28.5	11.1	20.1
40-44	28.5	22.2	25.6
45-49	17.9	11.1	14.4
50 +	21.5	14.8	18.0
	(n = 28)	(n = 27)	

Collapsed table: "19-34" vs. "35-44" vs. "45+." $x^2 = 11.171$ df = 2 p = .006.

Education level

	1988		Combined
	Men	Women	
High school graduate	0	7.7%	3.7%
Some college	28.6%	11.5	20.4
College graduate	28.6	46.2	37.0
Masters	17.9	23.1	20.4
Professional degree	25.0	11.5	18.5
	(n = 28)	(n = 26)	

Collapsed top two rows. $x^2 = 3.105$ df = 3 p = .386.

Income level

	1988		Combined
	Men	Women	
< $5,000	3.6%	0	1.9%
5,000-9,999	7.1	16.0%	11.3
10,000-14,999	7.1	12.0	9.4
15,000-19,999	14.3	24.0	18.9
20,000-24,999	10.7	8.0	9.4
25,000-29,999	7.1	12.0	9.4
30,000-39,999	7.1	16.0	11.3
40,000-49,999	17.9	4.0	11.3
50,000-59,999	10.7	4.0	7.5
60,000-69,999	3.6	0	1.9
70,000+	10.7	4.0	7.5
	(n = 28)	(n = 25)	

Collapsed table: "<$5,000-19,999" vs. "20,000-29,999" vs. "30,000-49,999" vs. "50,000+."
$x^2 = 3.686$ df = 3 p = .297.

Race

	1988		Combined
	Men	Women	
White	96.6%	96.3%	96.3%
Black	3.4	0	1.8
Other	0	3.7	1.8
	(n = 29)	(n = 27)	

Collapsed bottom two rows. $x^2 = .448$ df = 1 p = .508.

Currently employed

	1988		Combined
	Men	Women	
Yes	90.9%	93.3%	90.7%
No	9.1	6.7	9.3
	(n = 28)	(n = 26)	

$x^2 = .008$ df = 1 p = .933.

Religious background

	1988		Combined
	Men	Women	
Catholic	21.4%	22.2%	21.8%
Protestant	39.3	29.6	34.5
Jewish	14.3	22.2	18.2
Other	25.0	25.9	25.5
	(n = 28)	(n = 27)	

$x^2 = .860$ df = 3 p = .835.

CHAPTER 21: CHANGES IN SEXUAL PREFERENCE

Table 21.1 Kinsey Scale Score for Men

Sexual feelings

1983				1988				Row total
	0	1	2	3	4	5	6	
0								
1								
2	3.4% (1)	17.2% (5)	10.3% (3)	13.8% (4)	3.4% (1)			48.3% (14)
3		3.4 (1)	3.4 (1)	13.8 (4)		3.4% (1)		24.1 (7)
4			3.4 (1)		3.4 (1)	13.8 (4)	3.4% (1)	24.1 (7)
5						3.4 (1)		3.4 (1)
Column total	3.4 (1)	20.7 (6)	17.2 (5)	27.6 (8)	6.9 (2)	20.7 (6)	3.4 (1)	100.0 (29)

Number of respondents presented in parentheses.

Sexual behaviors

1983				1988				Row total
	0	1	2	3	4	5	6	
0	3.6% (1)	3.6% (1)						7.1% (2)
1	14.3 (4)	7.1 (2)		3.6% (1)		3.6% (1)		28.6 (8)
2		7.1 (2)			3.6% (1)			10.7 (3)
3	7.1 (2)				3.6 (1)	3.6 (1)	3.6% (1)	17.9 (5)
4		3.6 (1)		3.6 (1)	3.6 (1)	7.1 (2)		17.9 (5)
5			3.6% (1)	3.6 (1)			10.7 (3)	17.9 (5)
Column total	25.0 (7)	21.4 (6)	3.6 (1)	10.7 (3)	10.7 (3)	14.3 (4)	14.3 (4)	100.00 (28)

Number of respondents presented in parentheses.

Romantic feelings

1983	1988							Row total
	0	1	2	3	4	5	6	
0	3.4% (1)	3.4% (1)	6.9% (2)					13.8% (4)
1	13.8 (1)		10.3 (3)					24.1 (7)
2	3.4 (1)	3.4 (1)	6.9 (2)		3.4% (1)			17.2 (5)
3			3.4 (1)	13.8% (4)	3.4 (1)			20.7 (6)
4				3.4 (1)	3.4 (1)			6.9 (2)
5			3.4 (1)		6.9 (2)	6.9% (2)		17.2 (5)
Column total	20.7 (6)	10.3 (3)	27.8 (8)	17.2 (5)	17.2 (5)	6.9 (2)		100.0 (29)

Number of respondents presented in parentheses.

Table 21.2 Kinsey Scale Scores for Women

Sexual feelings

1983	1988							Row total
	0	1	2	3	4	5	6	
0								
1		3.8% (1)						3.8% (1)
2		3.8 (1)	11.5% (3)	7.7% (2)				23.1 (6)
3			3.8 (1)	19.2 (5)	11.5% (3)		3.8% (1)	38.5 (10)
4	3.8% (1)		3.8 (1)	7.7 (2)	3.8 (1)	7.7% (2)	3.8 (1)	30.8 (8)
5						3.8 (1)		3.8 (1)
Column total	3.8 (1)	7.7 (2)	19.2 (5)	34.6 (9)	15.4 (4)	11.5 (3)	7.7 (2)	100.0 (26)

Number of respondents presented in parentheses.

Sexual behaviors

1983				1988				Row total
	0	1	2	3	4	5	6	
0		8.3% (2)	4.2% (1)					12.5% (3)
1	8.3% (2)		4.2 (1)					12.5 (3)
2	8.3 (2)	8.3 (2)	8.3 (2)		4.2% (1)			29.2 (7)
3		4.2 (1)	4.2 (1)	8.3% (2)			8.3% (2)	25.0 (6)
4							12.5 (3)	12.5 (3)
5							4.2 (1)	4.2 (1)
6	4.2 (1)							4.2 (1)
Column total	20.8 (5)	20.8 (5)	20.8 (5)	8.3 (2)	4.2 (1)		25.0 (6)	100.0 (24)

Number of respondents presented in parentheses.

Romantic feelings

1983				1988				Row total
	0	1	2	3	4	5	6	
0								
1		3.8% (1)						3.8% (1)
2	7.7% (2)			3.8% (1)				11.5 (3)
3	3.8 (1)	3.8 (1)	7.7% (2)	26.9 (7)	11.5% (3)	3.8% (1)		57.7 (15)
4				3.8 (1)		3.8 (1)	7.7% (2)	15.4 (4)
5						3.8 (1)	3.8 (1)	7.7 (2)
6	3.8 (1)							3.8 (1)
Column total	15.4 (4)	7.7 (2)	7.7 (2)	34.6 (9)	11.5 (3)	11.5 (3)	11.5 (3)	100.0 (26)

Number of respondents presented in parentheses.

Table 23.1   Sexual partners

Total number of opposite-sex partners (past twelve months)

	Men[a]		Women[b]	
	1983	1988	1983	1988
0	3.6%	21.4%	7.4%	33.3%
1-4	64.3	64.3	48.1	51.9
5-14	21.4	3.6	22.2	11.1
15+	10.7	10.7	22.2	3.7
	(n = 28)	(n = 28)	(n = 27)	(n = 27)

Collapsed bottom three rows. $^a x^2 = 2.612$ df = 1 p = .109.
Collapsed top two rows and bottom two rows. $^b x^2 = 4.352$ df = 1 p = .040.

Total number of same-sex partners (past twelve months)

	Men[a]		Women[b]	
	1983	1988	1983	1988
0	14.3%	32.1%	22.2%	29.6%
1-4	32.1	32.1	63.0	59.3
5-14	32.1	25.0	7.4	7.4
15+	21.4	10.7	7.4	3.7
	(n = 28)	(n = 28)	(n = 27)	(n = 27)

Collapsed bottom three rows. $^a x^2 = 1.603$ df = 1 p = .206; $^b x^2 = 0.096$ df = 1 p = .763.

Frequency of anonymous sex (past 12 months)

	Men[a]		Women[b]	
	1983	1988	1983	1988
Never	46.4%	50.0%	74.1%	88.9%
<1 month	35.7	25.0	18.5	7.4
1-3 months	14.3	25.0	3.7	0
1 a week or more	3.6	0	3.7	3.7
	(n = 28)	(n = 28)	(n = 27)	(n = 27)

Collapsed top two rows and bottom two rows. $^a x^2 = .106$ df = 1 p = .751.
Collapsed bottom three rows. $^b x^2 = 1.105$ df = 1   p = .294.

Percentage of opposite-sex partners who were anonymous (past twelve months)

	Men[a]		Women[b]	
	1983	1988	1983	1988
None	67.9%	75.0%	63.0%	85.2%
Few	28.6	10.7	18.5	7.4
Less than half	3.6	7.1	3.7	3.7
Half	0	3.6	3.7	3.7
More than half	0	0	7.4	0
Most	0	0	0	0
All	0	3.6	3.7	0
	(n = 28)	(n = 28)	(n = 27)	(n = 27)

Collapsed top three rows and bottom four rows. $^a x^2 = .519$ df = 1 p = .481.
Collapsed bottom six rows. $^b x^2 = 2.411$ df = 1 p = .128.

Percentage of same-sex partners who were anonymous (past twelve months)

	Men[a]		Women[b]	
	1983	1988	1983	1988
None	35.7%	53.6%	80.8%	88.5%
Few	28.6	7.1	7.7	3.8
Less than half	14.3	10.7	7.7	3.8
Half	7.1	7.1	3.8	3.8
More than half	10.7	7.1	0	0
Most	3.6	14.3	0	0
All	0	0	0	0
	(n = 28)	(n = 28)	(n = 26)	(n = 26)

Collapsed bottom six rows. [a]x^2 = 1.156 df = 1 p = .285; [b]x^2 = .148 df = 1 p = .702.

Number of threesomes (past 12 months)

	Men[a]		Women[b]	
	1983	1988	1983	1988
0	53.6%	64.3%	51.9%	81.5%
1	17.9	14.3	14.8	7.4
2-3	14.3	7.1	22.2	3.7
4 or more	14.2	14.3	11.1	7.4
	(n = 28)	(n = 28)	(n = 27)	(n = 27)

Collapsed bottom three rows. [a]x^2 = 0.295 df = 1 p = .606; [b]x^2 = 4.083 df = 1 p = .045.

Number of group sex experiences (past 12 months)

	Men[a]		Women[b]	
	1983	1988	1983	1988
0	63.0%	67.9%	63.0%	88.9%
1	11.1	3.6	14.8	0
2-3	7.4	14.2	0	3.7
4-7	11.1	3.6	7.4	3.7
8+	7.4	10.7	14.8	3.7
	(n = 28)	(n = 28)	(n = 27)	(n = 27)

Collapsed bottom four rows. [a]x^2 = 0.010 df = 1 p = .925.
[b]x^2 = 3.647 df = 1 p = .059.

Table 23.2 Sexual Frequencies (Past 12 Months)

Self-masturbation

	Men[a]		Women[b]	
	1983	1988	1983	1988
<1 month	3.7%	3.7%	3.8%	15.4%
1-3 month	18.5	25.9	11.5	30.8
1-3 week	44.4	40.7	57.7	38.5
4+ week	33.3	29.6	26.9	15.4
	(n = 27)	(n = 27)	(n = 26)	(n = 26)

Collapsed bottom three rows. [a]x^2 = .520 df = 1 p = .480.
Collapsed top two rows and bottom two rows. [b]x^2 = 4.424 df = 1 p = .039.

Masturbate opposite-sex partner with goal of orgasm

	Men[a]		Women[b]	
	1983	1988	1983	1988
Never	11.1%	25.9%	14.8%	37.0%
<1 month	25.9	40.7	25.9	25.9
1-3 month	44.4	18.5	25.9	14.8
1-3 week	18.5	14.8	29.6	18.5
4+ week	0	0	3.7	3.7
	(n = 27)	(n = 27)	(n = 27)	(n = 27)

Collapsed top two rows and bottom three rows. $^a x^2$ = 3.635 df = 1 p = .059.
Collapsed bottom four rows. $^b x^2$ = 2.411 df = 1 p = .128.

Masturbate same-sex partner with goal of orgasm

	Men[a]		Women[b]	
	1983	1988	1983	1988
Never	26.9%	34.6%	16.7%	41.7%
<1 month	38.5	34.6	45.8	25.0
1-3 month	30.8	19.2	25.0	16.7
1-3 week	0	7.7	8.3	16.7
4+ week	3.8	3.8	4.2	0
	(n = 26)	(n = 26)	(n = 24)	(n = 24)

Collapsed bottom four rows. $^a x^2$ = .090 df = 1 p = .776; $^b x^2$ = 2.521 df = 1 p = .118.

Being masturbated by opposite-sex partner with goal of orgasm

	Men[a]		Women[b]	
	1983	1988	1983	1988
Never	22.2%	29.6%	11.1%	40.7%
<1 month	48.1	48.1	25.9	18.5
1-3 month	22.2	18.5	33.3	14.8
1-3 week	7.4	3.7	25.9	18.5
4+ week	0	0	3.7	7.4
	(n = 27)	(n = 27)	(n = 27)	(n = 27)

Collapsed bottom four rows. $^a x^2$ = .096 df = 1 p = .763; $^b x^2$ = 4.725 df = 1 p = .033.

Masturbated by same-sex partner with goal of orgasm

	Men[a]		Women[b]	
	1983	1988	1983	1988
Never	25.0%	32.1%	20.0%	32.0%
<1 month	42.9	42.9	44.0	32.0
1-3 month	28.6	14.3	28.0	24.0
1-3 week	0	7.1	4.0	12.0
4+ week	3.6	3.6	4.0	0
	(n = 28)	(n = 28)	(n = 25)	(n = 28)

Collapsed top three rows and bottom two rows. $^a x^2$ = .269 df = 1 p = .623.
Collapsed bottom four rows. $^b x^2$ = .416 df = 1 p = .528.

Mouth on opposite-sex partner's genitals

	Men[a]		Women[b]	
	1983	1988	1983	1988
Never	14.8%	25.9%	7.7%	33.3%
<1 month	14.8	40.7	11.5	11.1
1-3 month	29.6	22.2	23.1	33.3
1-3 week	40.7	11.1	46.2	18.5
4+ week	0	0	11.5	3.7
	(n = 27)	(n = 27)	(n = 27)	(n = 27)

Collapsed top two rows and bottom three rows. [a]x^2 = 6.008 df = 1 p = .015.
Collapsed bottom four rows. [b]x^2 = 4.110 df = 1 p = .045.

Mouth on same-sex partner's genitals

	Men[a]		Women[b]	
	1983	1988	1983	1988
Never	14.3%	46.4%	20.0%	40.0%
<1 month	39.3	39.3	32.0	28.0
1-3 month	25.0	3.6	32.0	20.0
1-3 week	21.4	10.7	16.0	12.0
4+ week	0	0	0	0
	(n = 28)	(n = 28)	(n = 25)	(n = 25)

Collapsed bottom four rows. [a]x^2 = 5.406 df = 1 p = .020; [b]x^2 = 1.524 df = 1 p = .220.

Opposite-sex partner's mouth on your genitals

	Men[a]		Women[b]	
	1983	1988	1983	1988
Never	18.5%	33.3%	14.8%	37.0%
<1 month	14.8	48.1	7.4	25.9
1-3 month	33.3	7.4	22.2	22.2
1-3 week	33.3	11.1	44.4	14.8
4+ week	0	0	11.1	0
	(n = 27)	(n = 27)	(n = 27)	(n = 27)

Collapsed top two rows and bottom three rows. [a]x^2 = 10.906 df = 1 p = .001;
[b]x^2 = 7.574 df = 1 p = .008.

Same-sex partner's mouth on your genitals

	Men[a]		Women[a]	
	1983	1988	1983	1988
Never	17.9%	42.9%	20.0%	32.0%
<1 month	32.1	35.7	36.0	36.0
1-3 month	28.6	10.7	32.0	20.0
1-3 week	14.3	10.7	12.0	12.0
4+ week	7.1	0	0	0
	(n = 28)	(n = 28)	(n = 25)	(n = 25)

Collapsed bottom four rows. [a]x^2 = 3.041 df = 1 p = .085; [b]x^2 = 0.416 df = 1 p = .528.

Vaginal intercourse

	Men[a]		Women[b]	
	1983	1988	1983	1988
Never	3.7%	29.6%	14.8%	33.3%
<1 month	18.5	18.5	11.1	11.1
1-3 month	37.0	29.6	18.5	22.2
1-3 week	33.3	18.5	48.1	14.8
4+ week	7.4	3.7	7.4	18.5
	(n = 27)	(n = 27)	(n = 27)	(n = 27)

Collapsed bottom four rows. $^a x^2 = 4.800$ df = 1 p = .032; $^b x^2 = 1.621$ df = 1 p = .203.

Perform finger-anal stimulation with opposite-sex partner

	Men[a]		Women[b]	
	1983	1988	1983	1988
Never	40.7%	63.0%	37.0%	63.0%
<1 month	25.9	22.2	29.6	22.2
1-3 month	22.2	11.1	14.8	7.4
1+ week	11.1	3.7	18.5	7.4
	(n = 27)	(n = 27)	(n = 27)	(n = 27)

Collapsed bottom three rows. $^a x^2 = 1.854$ df = 1 p = .180; $^b x^2 = 2.667$ df = 1 p = .104.

Perform finger-anal stimulation with same-sex partner

	Men[a]		Women[b]	
	1983	1988	1983	1988
Never	46.4%	64.3%	64.0%	80.0%
<1 month	17.9	25.0	28.0	12.0
1-3 month	28.6	7.1	8.0	8.0
1+ week	7.1	3.6	0	0
	(n = 28)	(n = 28)	(n = 25)	(n = 25)

Collapsed top two rows and bottom two rows. $^a x^2 = 3.606$ df = 1 p = .060.
Collapsed bottom three rows. $^b x^2 = 0.893$ df = 1 p = .358.

Receive finger-anal stimulation from opposite-sex partner

	Men[a]		Women[b]	
	1983	1988	1983	1988
Never	33.3%	59.3%	40.7%	63.0%
<1 month	33.3	37.0	25.9	25.9
1-3 month	22.2	3.7	22.2	7.4
1+ week	11.1	0	11.1	3.7
	(n = 27)	(n = 27)	(n = 27)	(n = 27)

Collapsed top two rows and bottom two rows. $^a x^2 = 6.014$ df = 1 p = .015;
$^b x^2 = 2.679$ df = 1 p = .103.

Receive finger-anal stimulation from same-sex partner

	Men[a]		Women[b]	
	1983	1988	1983	1988
Never	39.3%	71.4%	72.0%	80.0%
<1 month	25.0	25.0	20.0	16.0
1-3 month	32.1	0	8.0	4.0
1+ week	3.6	3.6	0	0
	(n = 28)	(n = 28)	(n = 25)	(n = 25)

Collapsed top two rows and bottom two rows. [a]$x^2 = 7.240$ df = 1 p = .009.
Collapsed bottom three rows. [b]$x^2 = 0.110$ df = 1 p = .747.

Perform anal intercourse with same-sex partner

	Men	
	1983	1988
Never	42.9%	67.9%
<1 month	32.1	17.9
1-3 month	14.3	10.7
1+ week	10.7	3.6
	(n = 28)	(n = 28)

Collapsed bottom three rows. $x^2 = 2.601$ df = 1 p = .110.

Receive anal intercourse from male partner

	Men[a]		Women[b]	
	1983	1988	1983	1988
Never	32.1%	82.1%	55.6%	70.4%
<1 month	42.9	14.3	25.9	22.2
1-3 month	14.3	0	11.1	7.4
1+ week	10.7	3.6	7.4	0
	(n = 28)	(n = 28)	(n = 27)	(n = 27)

Collapsed bottom three rows. [a]$x^2 = 12.323$ df = 1 p = .000; [b]$x^2 = .715$ df = 1 p = .416.

Table 23.3   Type of Partners and Frequency of Safe-Sex Behavior (1988 Only)

Opposite-sex partners

	Men	Women	Combined
Never	26.3%	23.1%	25.0%
Occasionally	5.3	0	3.1
Most of the time	36.8	23.1	31.2
Always	31.6	53.8	40.6
	(n = 19)	(n = 13)	

Collapse bottom three rows. Fisher's exact p = .587.

Same-sex partners

	Men	Women	Combined
Never	0	26.7%	12.1%
Occasionally	0	0	0
Most of the time	22.2%	26.7	24.2
Always	77.8	46.7	63.7
	(n = 18)	(n = 15)	

Collapse top three rows.   Fisher's exact p = .068.

Primary partners

	Men	Women	Combined
Never	45.0%	37.5%	41.7%
Occasionally	0	6.3	2.8
Most of the time	15.0	25.0	19.4
Always	40.0	31.3	36.1
	(n = 20)	(n = 16)	

Collapsed bottom three rows. x^2 = 1.183 df = 1 p = .280.

Secondary partners

	Men	Women	Combined
Never	0	12.5%	5.0%
Occasionally	0	12.5	5.0
Most of the time	50.0%	12.5	35.0
Always	50.0	62.5	55.0
	(n = 12)	(n = 8)	

Collapsed top two rows and bottom two rows.   Fisher's exact p = 1.000.

Casual partners

	Men	Women	Combined
Never	0	16.7%	4.8%
Occasionally	13.3%	0	9.5
Most of the time	0	0	0
Always	86.7	83.3	85.7
	(n = 15)	(n = 6)	

Collapsed bottom three rows.   Fisher's exact p = 1.000.

Anonymous partners

	Men	Women	Combined
Never	0	0	0
Occasionally	8.3%	0	7.1%
Most of the time	8.3	0	7.1
Always	83.3	100.0%	85.7
	(n = 12)	(n = 2)	

Collapsed top three rows.   Fisher's exact p = .726.

Table 23.4   Type of Partners and Extent of Care With Safe-Sex Practices

Opposite-sex partners

	Men	Women	Combined
Somewhat careful	42.9%	30.0%	37.5%
Very careful	57.1	70.0	62.5
	(n = 14)	(n = 10)	

Fisher's exact p = .418.

Same-sex partners

	Men	Women	Combined
Somewhat careful	5.6%	27.3%	13.8%
Very careful	94.4	72.7	86.2
	(n = 18)	(n = 11)	

Fisher's exact p = .268.

Primary partners

	Men	Women	Combined
Somewhat careful	27.3%	30.0%	28.6%
Very careful	72.7	70.0	71.4
	(n = 11)	(n = 10)	

Fisher's exact p = 1.000.

Secondary partners

	Men	Women	Combined
Somewhat careful	33.3%	42.9%	36.8%
Very careful	66.7	57.1	63.2
	(n = 12)	(n = 7)	

Fisher's exact p = .818.

Casual partners

	Men	Women	Combined
Somewhat careful	13.3%	0	10.0%
Very careful	86.7	100.0%	90.0
	(n = 15)	(n = 5)	

Fisher's exact p = .533.

Anonymous partners

	Men	Women	Combined
Somewhat careful	8.3%	0	7.1%
Very careful	91.7	100.0%	92.9
	(n = 12)	(n = 2)	

Fisher's exact p = 1.000.

Table 23.5   AIDS testing

Have taken test for HIV antibodies

	Men	Women	Combined
Yes	64.3%	63.0%	63.6%
No	35.7	37.0	36.4
	(n = 28)	(n = 27)	

$x^2 = 0.032$ df = 1 p = .867.

Result of test for HIV antibodies

	Men	Women	Combined
Positive (antibodies found)	16.7%	5.9%	11.4%
Negative (no antibodies found)	83.3	94.1	88.6
	(n = 17)	(n = 17)	

Fisher's exact p = .301.

CHAPTER 24: CHANGES IN RELATIONSHIPS

Table 24.1   Change Occurred in Ideal

	Men	Women	Combined
Yes	58.6%	59.3%	58.9%
No	41.4	40.7	41.1
	(n = 29)	(n = 27)	

$x^2 = . 050$ df = 1 p = .896.

Table 24.2   Primary Relationships

Currently in a primary relationship

	Men	Women	Combined
Yes	55.2%	68.0%	61.1%
No	44.8	32.0	38.9
	(n = 29)	(n = 25)	

$x^2 = 0.468$ df = 1 p = .497.

Sex of primary partner

	Men	Women	Combined
Opposite sex	75.0%	70.6%	72.7%
Same sex	25.0	29.4	27.3
	(n = 16)	(n = 17)	

Fisher's exact p = .543.

Has had a change in primary relationship(s)

	Men	Women	Combined
Yes	69.2%	80.0%	74.5%
No	30.8	20.0	25.5
	(n = 26)	(n = 25)	

$x^2 = 0.314$ df = 1 p = .594.

Nature of change in primary relationship(s) since 1983

	Men	Women	Combined
No longer in a primary relationship	55.5%	25.0%	39.5%
New primary relationship	33.3	25.0	28.9
Series of/or increase/decrease in primary relationships	11.1	50.0	31.6
	(n = 18)	(n = 20)	

$x^2 = 7.003$ df = 2   p = .033.

Table 24.3   Secondary Relationships

Currently in a secondary relationship

	Men	Women	Combined
Yes	48.1%	50.0%	49.0%
No	51.9	50.0	51.0
	(n = 27)	(n = 24)	

$x^2 = 0.022$ df = 1   p = .887.

Has had a change in secondary relationship(s)

	Men	Women	Combined
Yes	57.9%	73.7%	65.8%
No	42.1	26.3	34.2
	(n = 19)	(n = 19)	

$x^2 = 0.468$ df = 1   p = .497.

Table 24.4   Ground Rules

Ground rules in primary relationship changed

	Men	Women	Combined
Yes	50.0%	88.2%	69.7%
No	50.0	11.8	30.3
	(n = 16)	(n = 17)	

Fisher's exact p = .021.

Ground rules in secondary relationship changed

	Men	Women	Combined
Yes	30.8%	58.3%	44.0%
No	69.2	41.7	56.0
	(n = 13)	(n = 12)	

Fisher's exact p = .163.

CHAPTER 25: ADAPTING TO A NEW WORLD

Table 25.1   Changes in Attitude Toward Disclosure

	Men	Women	Combined
More wary	48.0%	30.0%	40.0%
Same	40.0	45.0	42.2
Less wary	12.0	25.0	17.8
	(n = 25)	(n = 20)	

$x^2 = 2.040$ df = 2   p = .361.

Table 25.2   Changes in Relations with Others

With homosexual men

	Men	Women	Combined
Better	37.9%	29.6%	33.9%
Same	55.2	55.6	55.4
Worse	6.9	14.8	10.7
	(n = 29)	(n = 27)	

Collapsed bottom two rows. $x^2 = .139$ df = 1   p = .713.

With homosexual women

	Men	Women	Combined
Better	37.9%	37.0%	37.5%
Same	62.1	51.9	57.1
Worse	0	11.1	5.4
	(n = 29)	(n = 27)	

Collapsed bottom two rows. $x^2 = .043$ df = 1   p = .844.

With heterosexual men

	Men	Women	Combined
Better	20.7%	14.8%	17.9%
Same	65.5	66.7	66.1
Worse	13.8	18.5	16.1
	(n = 29)	(n = 27)	

Collapsed bottom two rows. $x^2 = .050$ df = 1   p = .829.

With heterosexual women

	Men	Women	Combined
Better	24.1%	18.5%	21.4%
Same	51.7	74.1	62.5
Worse	24.1	7.4	16.1
	(n = 29)	(n = 27)	

$x^2 = 3.922$ df = 2   p = .141.

Table 25.3   Regret Being Bisexual

	Men	Women	Combined
Not at all	72.0%	90.0%	80.0%
A little	12.0	5.0	8.9
Somewhat	12.0	5.0	8.9
Very much	4.0	0	2.2
	(n = 25)	(n = 20)	

Collapsed bottom three rows. $x^2 = 1.266$ df = 1   p = .266.

Index